GLOBAL INTEGRATION AND TECHNOLOGY TRANSFER

GLOBAL INTEGRATION AND TECHNOLOGY TRANSFER

Edited by Bernard Hoekman and
Beata Smarzynska Javorcik

A copublication of Palgrave Macmillan
and the World Bank

1818 H Street NW
Washington, DC 20433
Telephone: 202-473-1000
Internet: www.worldbank.org
E-mail: feedback@worldbank.org

1 2 3 4 09 08 07 06

A copublication of The World Bank and Palgrave Macmillan.

Palgrave Macmillan
Houndmills, Basingstoke, Hampshire RG21 6XS and
175 Fifth Avenue, New York, NY 10010
Companies and representatives throughout the world

Palgrave Macmillan is the global academic imprint of the Palgrave Macmillan division of St. Martin's Press, LLC and of Palgrave Macmillan Ltd.

ISBN-10: 0-8213-6125-2 (softcover)
ISBN-10: 0-8213-6371-9 (hardcover)
ISBN-13: 978-0-8213-6125-2 (softcover)
ISBN-13: 978-0-8213-6371-3 (hardcover)
eISBN-10: 0-8213-6126-0
eISBN-13: 978-0-8213-6126-9
DOI: 10.1596/978-0-8213-6125-2

Library of Congress Cataloging-in-Publication Data

Global integration and technology transfer/edited by Bernard Hoekman, Beata Smarzynska Javorcik.
p. cm. – (Trade and development series)
Includes bibliographical references and index.
ISBN-13: 978-0-8213-6125-2
ISBN-10: 0-8213-6125-2
1. International economic integration. 2. International trade. 3. Investments, Foreign—Developing countries. 4. Technology transfer—Economic aspects—Developing countries. 5. Industrial productivity—Effect of technological innovations on—Developing countries. I. Hoekman, Bernard M., 1959- II. Javorcik, Beata K. Smarzynska. III. Series.

HF1418.5.G5715 2005
337.1—dc22

2005043451

Cover photos: Royalty Free/Fotosearch; Royalty Free/Getty.
Cover designer: Tomoko Hirata.

CONTENTS

Figures

Tables

FOREWORD

Development experience over the past 50 years suggests that trade liberalization and increased international integration are critical elements of a successful strategy to promote economic growth. The World Bank has been an advocate and supporter of a trade agenda that enables developing countries to gain as much access to foreign markets as possible, but it has also stressed the importance of developing countries opening their own markets to international trade and investment.

This volume presents a rich set of analyses exploring how trade and foreign direct investment (FDI) can help increase economic growth by allowing firms to tap into and benefit from the global pool of knowledge. The chapters demonstrate that both obtaining access to foreign markets and opening their own economies to trade and FDI are crucial to promoting economic growth in developing countries, because they stimulate international technology diffusion. The volume also identifies government policies that can facilitate technology transfer and its absorption in the developing world.

Among the conclusions emerging from the research contained in this volume, a number stand out. First, the evidence suggests that an open trade regime facilitates the diffusion of knowledge. Undistorted access to capital equipment and imported inputs that embody foreign knowledge allow firms to acquire know-how; the greater competition from imports lowers the mark-ups over costs that firms charge customers. At the same time, given that technology markets are associated with increasing returns, imperfect competition, and externalities, the argument against trade protection is not unconditional. The conclusions hinge on the scope of knowledge spillovers. International spillovers, for which there is considerable evidence, strongly tilt the balance in favor of free trade. If national spillovers are also important, there may be a potential role for intervention. Trade policy, however, is not the instrument of choice in this situation, as it does little to encourage local research and development and necessarily leads to other distortions.

Second, FDI policies in most developing countries have become more liberal in recent years. But government policies often distinguish between types of FDI, providing greater incentives to joint ventures with national firms than to fully owned subsidiaries of multinational enterprises. Such a policy stance often reflects a desire to maximize technology transfer to local firms and agents. The evidence suggests, however, that policies that discriminate between types of FDI are unlikely to promote technology transfer. By attempting to force multinational enterprises to license their technologies or engage in joint ventures, host countries may lower the quality of technologies they receive and reduce incentives for foreign firms to invest at all.

While the magnitude of international technology diffusion undertaken by multinational enterprises need not be socially optimal, evidence presented in this volume reveals that such firms are keen to transfer technology to their local suppliers. Policies that increase incentives to source locally, as opposed to regulation or legislation requiring that multinational enterprises engage in international technology diffusion to local competitors, have a greater likelihood of being successful. Examples of such a policy are supplier development programs that aim to prepare local companies to understand and meet the needs of multinational enterprises. The services provided under such programs can be effective in assisting firms, provided they are well designed, mobilize the right type of skills, and ensure that their target audience is aware of the services on offer.

Many countries actively seek to attract foreign investors through up-front subsidies, tax holidays, and other grants. A rationale for such investment incentives may be based on positive externalities generated by inflows of FDI. Local suppliers may benefit not just through expanded sales but through access to technologies provided by the investors. Such positive externalities may be enhanced by the prevalence of "follow the leader" behavior among multinational enterprises. Given the oligopolistic nature of markets within which FDI occurs, a new entrant may result in additional investments by both competitors and upstream suppliers of inputs, components, and services. An implication is that a host country may be able to unleash a sequence of investments by successfully inducing FDI from one or two major firms.

If the local economy lacks a well-developed network of potential suppliers, multinational enterprises may be hesitant to invest, and local suppliers may not develop because of lack of demand. In the presence of such interdependence, growth may be constrained by a coordination problem that can partially be resolved by initiating investments from key firms. Such coordination problems cannot be tackled solely through investment incentives. Policy efforts need to focus primarily on improving the investment climate and reducing the costs of absorbing technology, a complex task that involves building human capital and expanding national innovation systems. Thus while there may be a case for investment incentives, it is a conditional one. To be effective, the investment climate and absorptive capacity must meet certain conditions. Moreover, given competition among countries in attracting FDI through incentive packages, policymakers must carefully examine the magnitude of potential costs and benefits associated with such policies.

François Bourguignon
Chief Economist and Senior Vice-President
The World Bank

ACKNOWLEDGMENTS

The chapters in this volume are the product of a research project undertaken by the World Bank Research Department. The research was partially supported by grant PO62764 from the Bank's research support budget. The financial support provided by the U.K. Department for International Development is also gratefully acknowledged. Some of the chapters have been published in professional journals, others are published here for the first time.

The project owes much to the substantial inputs of Professor Jim Tybout (Pennsylvania State University) in the early phases. Aart Kraay helped manage the project research, working with Professor Tybout and Bernard Hoekman. The editors are grateful to both of them, as well as to Isidro Soloaga for his tireless effort to put together the data sets used in the chapters. The editors thank Stephen McGroarty, Santiago Pombo, Janice Tuten, and Nora Leah Ridolfi for their assistance in getting the manuscript through the production process in a timely manner. Thanks are also due to Barbara Karni for excellent copyediting and to Maribel Flewitt and Rebecca Martin for administrative assistance.

All views expressed in this volume are those of the contributors. They do not necessarily represent the views of the World Bank, its executive directors, or the countries they represent.

CONTRIBUTORS

Simeon Djankov, World Bank, Washington, DC, and Centre for Economic Policy Research, London

Bernard Hoekman, World Bank, Washington, DC, and Centre for Economic Policy Research, London

Beata Smarzynska Javorcik, World Bank, Washington, DC, and Centre for Economic Policy Research, London

Aart Kraay, World Bank, Washington, DC

Daniel Lederman, World Bank, Washington, DC

William F. Maloney, World Bank, Washington, DC

Giorgio Barba Navaretti, Università degli Studi di Milano and Centro Studi Luca d'Agliano, Milan

Howard Pack, Wharton School, University of Pennsylvania, Philadelphia

Kamal Saggi, Southern Methodist University, Dallas, Texas

Maurice Schiff, World Bank, Washington, DC, and Institute for the Study of Labor (IZA), Bonn

Isidro Soloaga, Universidad de las Américas, Puebla, Mexico

James R. Tybout, Pennsylvania State University, University Park, Pennsylvania and National Bureau of Economic Research, Cambridge, Massachusetts

Yanling Wang, Norman Paterson School of International Affairs, Carleton University, Ottawa, Ontario

ABBREVIATIONS AND ACRONYMS

CGE	computable general equilibrium
CMEA	Council of Mutual Economic Assistance
EBRD	European Bank for Reconstruction and Development
EU	European Union
FDI	foreign direct investment
GNP	gross national product
IRCA	innovative revealed comparative advantage
ISIC	International Standard Industrial Classification
JV	joint venture
NACE	Nomenclature générale des activités économiques dans les Communautés européennes
NAFTA	North American Free Trade Agreement
NSO	National Statistical Office (of Bulgaria)
OECD	Organisation for Economic Co-operation and Development
OEM	original equipment manufacturer
OLS	ordinary least squares
OPT	outward processing trade
PCM	price-cost margin
R&D	research and development
TFP	total factor productivity
TRIMs	Trade Related Investment Measures
UNIDO	United Nations Industrial Development Organization
USPTO	United States Patent and Trademark Office
UVI	unit values index

1

LESSONS FROM EMPIRICAL RESEARCH ON INTERNATIONAL TECHNOLOGY DIFFUSION THROUGH TRADE AND FOREIGN DIRECT INVESTMENT

Bernard Hoekman and Beata Smarzynska Javorcik

This volume examines international technology diffusion. Its chapters explore the channels through which existing knowledge is transferred across countries, the magnitude of such transfers, and their impact.

The acquisition and diffusion of knowledge or technology are of great importance for economic development, as the adoption of new techniques, machines, and production processes is a key determinant of productivity growth. Given that most research and development (R&D) and innovation is undertaken in high-income countries, most developing economies must rely largely on imported technologies as sources of new productive knowledge. This is not to say that no R&D is undertaken in developing countries; a considerable amount of follow-on innovation and adaptation does occur there, contributing to the global stock of knowledge.

The determinants of the supply of new knowledge—innovation and invention—have been the subject of a great deal of research. The chapters in this volume are not concerned with the generation of new knowledge. The focus is on the *diffusion* of knowledge through trade and direct foreign investment.

Improving our understanding of how technology diffuses across borders is important in order to identify what governments can do through policies to stimulate international technology diffusion and its absorption in their economies. Numerous policies are motivated to a greater or lesser extent by the objective of acquiring and augmenting or adapting knowledge that will enhance productivity and economic growth. Such policies range from the general to the specific. Examples include the provision of public goods such as education and infrastructure, the establishment of specific funds for the creation of technology, the granting of subsidies for capital investment, action to protect intellectual property rights, design of the structure of import protection to favor the import of equipment and machinery, the granting of tax incentives for foreign investors, export promotion schemes, the establishment of free trade or special economic zones, and specialized training programs.

The "correct" policy intervention, if any, depends critically on the channels through which technology diffuses internationally and the quantitative effects of the various diffusion processes on allocative efficiency and productivity growth. Neither is well understood. Most of the empirical evidence is based on aggregate data or cross-sectional surveys and is subject to multiple interpretations. Case studies, while informative about the case at hand, are difficult to generalize. Hence policymakers have often acted on the basis of theory, anecdotes, or instincts.

International technology diffusion can occur through numerous channels. Trade in goods and services is one. All trade bears some potential for transmitting technological information. Imported capital goods and technological inputs can directly improve productivity by being used in production processes. Alternatively, firms may learn about technologies by exporting to knowledgeable buyers, who share product designs and production techniques with them.

A second channel is foreign direct investment (FDI) or the pursuit of project-specific joint ventures. Multinational enterprises generally transfer technological information to their subsidiaries, directly affecting the productivity of these plants. In addition, some of the associated knowledge may "leak" into the host economy. In both cases—trade and FDI—technology may diffuse from firms that have acquired it internationally to other firms in the same industry or region through demonstration effects, labor turnover, mutual input suppliers, or reverse engineering.

A third channel of international technology diffusion is direct trade in knowledge through technology purchases or licensing. This may occur within firms, among joint ventures, or between unrelated firms. Licensing and FDI are often substitutes, but they may also be complements. Much of the recorded international payments and receipts for intellectual property (royalties) occur within firms, as flows between parent firms and affiliates. Which form is preferable to technology owners depends on many factors, including the strength of intellectual property rights.

In sum, new technologies can be transmitted internationally through a variety of activities. They may be embodied in goods and transferred through imports of new varieties of differentiated products or capital goods and equipment, they may be obtained through exposure to foreign buyers or foreign investors, or they may be acquired through direct trade in "disembodied knowledge"—through contracts supported by policies that protect intellectual property.

The chapters in this volume focus on the first two channels, trade and FDI.[1] They use cross-country and firm-level data sets to investigate the extent to which trade and FDI serve as channels of technology transfer. The chapters relying on cross-country data explore the importance of trade openness, absorptive and innovative capacity, and the identity of trading partners as determinants of international technology diffusion measured by total factor productivity (TFP) performance. The focus is on spillovers through trade, building on and extending the original insight of Coe and Helpman (1995). In the case of analyses based on firm-level data, the information available allows individual manufacturers to be followed over time in terms of their productivity and the types of activities they are engaged in (exporting, importing, joint ventures, and so on). Data on all potential channels of diffusion are generally not available for any of the countries studied, but together the data sets cover many of the activities of interest from the perspective of international technology diffusion.

The firm-level analyses span several Central and Eastern European transition economies, China, two Latin American countries (Colombia and Mexico), and one North African country (Morocco). They are all quite different in terms of initial conditions and endowments. The Central and Eastern European countries are relatively well endowed with human capital (skilled labor) and physical infrastructure. They also provide an unprecedented natural experiment of sudden exposure to international trade, investment, and technology. In addition, they are of interest in terms of the use that was made of regional integration agreements, with the prospect of accession to the European Union a major focal point for reform. In East Asia, China has often been cited as a country rapidly absorbing new technologies and moving up the technology ladder. China's adoption of the "open door policy" in 1979 marked the beginning of a concerted program to acquire foreign technology in order to raise growth rates and living standards. Over the course of the next 25 years, China's spectacular growth was accompanied by its emergence as the largest developing country recipient of FDI, one of the world's largest trading nations, and a major importer of capital goods. The semi-industrial countries of Latin America, where opening to trade is relatively recent and factor endowments are quite different (natural resource abundance), offer an alternative opportunity for analyzing technology diffusion patterns given the availability of high-quality firm-level data. Morocco is similar to Latin America in terms of relatively high endowments of natural resources, but it differs in terms of endowments of unskilled labor and the use of a free trade agreement with the European Union to improve export market access.

This introductory chapter sets the stage for the individual analyses. It reviews the major findings of the chapters in the volume as well as related research—with an emphasis on that undertaken by the World Bank—that focuses on contributing to the understanding of international technology diffusion. Much of the chapter focuses on the policy implications emerging from the volume's findings as well as the literature in this area more generally. The chapter ends with a brief summary of the main messages emerging from this research and suggestions for further work to address open questions. Because the first three chapters of the volume provide surveys of the literature and evidence of different dimensions of international technology diffusion—case studies and cross-country experience (chapter 2), the role of FDI (chapter 3) and the role of trade (chapter 4)—we do not review the literature in these areas in any depth here.

Technology Diffusion through Trade

Trade can contribute to international technology diffusion by providing local firms access to new technologies embodied in imported machinery and equipment and by creating opportunities for reverse engineering of products developed abroad. It can also create incentives to adopt and improve technologies through exporting opportunities.

Access to Technologies through Imports

Coe and Helpman (1995) and Coe, Helpman, and Hoffmaister (1997) find that foreign knowledge (R&D) embodied in traded goods has a statistically significant positive impact on aggregate TFP of importing countries.[2] Subsequent research finds that this impact is greater the more open a country is, the more skilled is its labor force, and, in the case of developing countries, the more trade takes place with industrial economies (Schiff and Wang forthcoming).

Knowledge accumulation means new knowledge (an increase in its quality), greater access to existing knowledge (an increase in its quantity), or both. Chapter 5, by Maurice Schiff and Yanling Wang, examines the relative contribution of these two components of knowledge to total TFP for North-North and North-South trade-related knowledge diffusion. They proxy quantity by openness (as measured by the trade to GDP ratio) and quality by the R&D content of trade (the type of products traded). The literature assumes equal contributions to TFP of openness and the R&D content of trade. Schiff and Wang show that this assumption does not correspond to reality. They find that R&D has a greater impact on TFP than openness for North-North trade and that openness has a greater impact on TFP than R&D for North-South trade. These results imply that the impact of openness on TFP in developing (developed) countries is larger (smaller) than previously obtained in this literature and that developing countries may obtain larger productivity gains from trade liberalization than previously thought.

As variations in capital goods trade can better explain cross-country differences in productivity than can overall trade (see, for example, Eaton and Kortum 1999), the distinction between quantity and quality is further examined in the context of imported capital goods by Barba Navaretti, Schiff, and Soloaga (chapter 6). Their chapter focuses on the impact of imported technologies on the productivity of countries in Central and Eastern Europe and the Southern Mediterranean. Comparing the unit values of machinery imported by each country with the unit values of machinery imported by the United States, the authors calculate the gap between the technology purchased and the technological frontier (defined by the United States). They find that the gap is persistent and in some cases even increasing. They conclude that the productivity growth in manufacturing depends on the types of machines imported (quality) and not on the share of imported equipment in total investment (quantity).

These results suggest that open trade policies are critical for developing countries in attracting technology. Openness alone, however, is not sufficient—both absorptive capacity and the ability to adapt foreign technology need to be in place, both of which are related to human capital endowments and investment in R&D-intensive industries. In developing countries, technology acquisition often amounts to adapting existing methods to local circumstances (Evenson and Westphal 1995). This process takes time. Gradual adoption of new techniques or new inputs is optimal for risk-averse producers in the face of costly adoption and uncertain returns. Producers need to learn how to apply the new technology; they often begin by applying it to a small part of their output and, if profitable, gradually increasing its application over time (Tybout 2000).

The farther the "technological distance" of a country from the global frontier (best practice), the more difficult it is to absorb information effectively into production systems (Keller 2002). As Keller (1996) argues, access to foreign technologies alone does not increase growth rates of developing countries. Countries tend to acquire international technology more readily if domestic firms have local R&D programs, there are domestic private and public research laboratories and universities, and there exists a sound basis of technical skills and human capital. In chapter 2 of this volume, Howard Pack surveys and synthesizes much of the relevant literature on this issue, concluding that the existence of such "complementary inputs" reduces the costs of imitation, adaptation, and follow-on innovation.

Access to Technologies through Exports

While it is quite plausible that repeated contacts with knowledgeable foreign buyers provide greater awareness of and access to foreign technologies, the literature survey by James Tybout (chapter 4) concludes that the jury is still out on whether firms learn from exporting. This is an important policy as well as analytical question, as many governments have export promotion programs that are at least in part premised on the existence of such learning, as well as on the presumption that such new knowledge will spill over to other firms in the local economy.

Two studies—Clerides, Lach, and Tybout (1998), using plant-level data from Columbia, Mexico, and Morocco, and Bernard and Jensen (1999), based on U.S. data—analyze the causal links between exporting and productivity. They observe that firms that are relatively efficient become exporters, while plants that cease exporting experience increasingly high costs. They find no evidence of learning from exporting and conclude that the well-documented positive association between exporting and greater efficiency is explained by the self-selection of more efficient firms into export markets.

These results contrast with the conclusions of Aart Kraay (chapter 7). Using data on 2,105 Chinese enterprises during the period 1988–92 and controlling for past performance and unobserved firm characteristics, Kraay finds that past exports lead to significant improvements in enterprise performance. Interestingly, these learning effects are most pronounced among established exporters. For new entrants to export markets, learning effects appear to be insignificant or even negative.

The reality may be more nuanced than a simple distinction between self-selection into and learning from exporting. In another World Bank study, Hallward-Driermayer, Iarossi, and Sokoloff (2002), using firm-level information from a sample of East Asian countries, conclude that firms make deliberate decisions to raise productivity in order to serve export markets. It is not simply that more productive firms self-select into exporting but that firms that explicitly target export markets consistently make different decisions regarding investment, training, technology, and the selection of inputs, thereby raising their productivity. Thus the "exporter selection" process is not necessarily driven by exogenous shocks, such as trade reforms, but reflects investments made by firms in anticipation of accessing foreign markets.

Foreign Direct Investment as a Conduit of Knowledge

Investments by multinational enterprises may provide developing country affiliates and partners with access to more efficient foreign technologies, as Kamal Saggi notes (chapter 3). Insofar as this knowledge does not remain restricted to partner firms or plants, FDI can result in technological spillovers, operating through demonstration effects (imitation) and labor turnover. Such benefits from FDI, if they arise, will be offset by more vigorous competition for local firms in the same industry, forcing down average price-cost margins.

Several chapters in Roberts and Tybout (1996) examine the potential impacts of FDI. Subsequent research, including several chapters in this volume, has continued to investigate the sign and magnitude of the effects of FDI. However, the major contribution of the more recent literature is to distinguish between horizontal spillover effects within an industry and vertical spillovers generated by linkages in the production or value chain. Research has focused on the incentives for multinational enterprises to deliberately transfer technology to upstream local firms that are (potential) suppliers of intermediates that the multinational enterprise uses in production.

Implicit to the analysis of knowledge spillovers from FDI is the assumption that foreign ownership per se conveys some intangible advantage and that proximity to foreign firms or plants can be beneficial to domestic firms. Until recently, there was relatively little robust empirical confirmation that the first part of this assumption holds, particularly in the context of developing countries. In chapter 8 Simeon Djankov and Bernard Hoekman use firm-level panel data from the Czech Republic to show that during 1992–96, FDI had a positive impact on the TFP growth of recipient firms, correcting for the fact that foreign companies may (and did) invest in firms with above-average initial productivity. More recent evidence is provided by another output of the World Bank research program in this area. Arnold and Javorcik (2005) use the Indonesian Census of Manufacturing to demonstrate that foreign acquisitions increased the productivity of the acquired Indonesian plants. To control for the possibility that foreign investors may select plants that are more productive to begin with, the study employs a matching technique and the difference in differences approach. The analysis indicates that benefits of foreign ownership are realized very quickly and that acquired plants experience large increases in productivity, investment outlays, employment, and output. Foreign ownership is also found to lead to greater integration of plants into the global economy through exports and imports.

In studying FDI spillovers, it is important to distinguish between horizontal (intraindustry) and vertical (interindustry) spillovers. In principle, vertical spillovers are more likely to occur, insofar as multinational enterprises can be expected to take actions to prevent knowledge from leaking to their competitors in the same industry. They can do so by paying efficiency wages, enforcing intellectual property rights, and imposing noncompete clauses on key staff. Vertical spillovers are more likely to occur, because foreign affiliates have an incentive to reduce sourcing costs by encouraging productivity improvements in (potential) local suppliers of inputs and services. However, the local upstream industries concerned may not always benefit, as foreign entry in a downstream sector may also give rise to greater imports of intermediates or result in entry of foreign input suppliers. Such "follow the client" FDI is quite common in sectors such as automobiles. More generally, it can be expected to be prevalent for services suppliers, for whom trade is often not technically possible.

Case studies suggest that substantial technology diffusion occurs due to FDI (Blomström and Kokko 1998). Econometric studies are more diverse. Some find that firms in sectors with a relatively high multinational enterprise presence tend to be more productive (Kokko, Tansini, and Zejan 1997), while others conclude that domestically held firms may actually do worse as the foreign presence in their industry increases (for example, Aitken and Harrison 1999). Negative spillover effects may occur in the short run if multinational enterprises siphon off domestic demand or bid away high-quality labor. As Görg and Strobl (2001) point out in a recent meta-analysis of the literature on intraindustry spillovers, studies employing panel (rather than cross-sectional) data, which can and thus do control for unobserved firm characteristics (or unobserved industry characteristics in the case of estimations at the sectoral level), are less likely to find a positive effect.

Chapter 8 falls into this category. It finds that FDI has a positive impact on TFP growth of recipient firms, controlling for selection bias, but it also concludes that joint ventures and foreign affiliates have a negative spillover effect on purely domestic firms in the same industry. Interestingly, foreign affiliates alone do not have a significant effect on the performance of domestic firms and joint ventures taken together, which the authors attribute to the better absorptive capacity of joint ventures.

The results of chapter 9, by Beata Javorcik, suggest that the findings of Djankov and Hoekman in chapter 8 could also be due to differences in technological content and thus in the spillover potential of joint ventures and foreign subsidiaries. Javorcik's analysis is based on data from 22 transition economies for the early 1990s. She shows that foreign investors with the most sophisticated technologies (relative to the industry mean) are more likely to engage in wholly owned projects than in joint ventures. These effects are present in high- and medium-technology sectors but not in low-R&D industries.

In contrast to the mixed results on intraindustry (horizontal) spillovers from FDI, the evidence on interindustry (vertical) technology transfer from multinational enterprises has been consistently positive. The literature has documented such transfers taking place through firms from industrial countries buying the output of Asian firms to sell under their own brand names. Mexico's *maquiladora* sector is an example of vertical international technology diffusion. Most *maquiladoras* began as subsidiaries of U.S. firms that shifted labor-intensive assembly operations to Mexico (Tan and Batra 1995). However, over time the *maquiladoras* adopted more sophisticated production techniques, many of which were imported from the United States (Saggi 2002). In chapter 10 Javorcik analyzes firm-level panel data from Lithuania. She produces evidence that is consistent with the hypothesis that positive productivity spillovers from FDI occur through contacts between foreign affiliates and local suppliers in upstream sectors. Her results suggest that a one standard deviation increase in FDI in the sourcing (buying) sector is associated with a 15 percent rise in output of each domestic firm in the supplying industry. Vertical spillovers are associated with projects with shared domestic and foreign ownership but not with fully owned foreign investments.

Openness to Trade and FDI: Interactions and Procompetitive Effects

The studies described looked at each channel of technology transfer in isolation. A more systematic approach requires the simultaneous consideration of the effects of various international transactions on firm productivity. This is the focus of the research undertaken by Aart Kraay, Isidro Soloaga, and James Tybout in chapter 11, which uses plant-level panel data from Colombia, Mexico, and Morocco. Their findings suggest that international activities (such as exporting, importing, and FDI) do not help much in predicting current firm performance, once past realizations on quality and marginal costs are controlled for. In the minority of cases, where significant associations emerge, international activities appear to move costs and product quality in the same direction.

The survey by Pack (chapter 2) points out that both theoretical studies and empirical evidence indicate that host-country absorptive capacity is crucial for obtaining significant benefits from FDI. This suggests that liberalization of trade and FDI policies should be complemented with appropriate measures with respect to education, R&D, and human capital accumulation. Similarly, empirical evidence supports the argument that intellectual property rights are trade related and that asymmetric intellectual property rights protection across countries distorts the pattern of world trade and alters the composition of FDI inflows (Fink and Maskus 2005; Javorcik 2004).

Lowering barriers to imports and FDI increases competition, as foreign goods and foreign producers enter the domestic market. The magnitude of the effect may be larger in developing countries, which (in part due to previous protection from international competition) lag behind industrial nations in terms of technological sophistication, quality and variety of products, and productivity. While in the short run the loss of market share to foreign goods or firms may push local firms up their average cost curve and thus lower their productivity, in the medium run weaker firms will be forced to exit and survivors will lower their cost base, upgrade their production, or both. The result is that the average productivity of indigenous firms will increase. This view is reflected in Tybout's survey (chapter 4), which concludes that mark-ups generally fall with import competition, import-competing firms cut back their production levels when foreign competition intensifies (at least in the short run), increased trade rationalizes production in the sense that markets for the most efficient plants are expanded but large import-competing firms tend to simultaneously contract, and exposure to foreign competition often improves intraplant efficiency.

Policies to liberalize trade and FDI are, of course, not the only policy instruments for increasing competition. Complementary policies are important as well, including in particular hard budget constraints to ensure that financial resources are allocated to activities with the highest expected return and that firms failing to perform are forced to exit. Djankov and Hoekman explore the importance of such complementary sources of market discipline on TFP in chapter 12. They investigate the effects of demonopolization and restructuring of conglomerates, the imposition of harder budget constraints, and trade liberalization on the performance of Bulgarian firms between 1991 and 1995. They find a positive relationship between the increased market discipline associated with these reforms and subsequent productivity growth, a result that is consistent with the general finding in the literature. But they also conclude that different types of market discipline have different impacts on exporting and nonexporting (import-competing) enterprises. Trade policy reform has less of an impact on the exporters in the sample—indeed, trade reforms are found to have a significant effect only once firms that export most of their output are removed from the sample. Moreover, access to subsidies (soft budget constraints, as measured by the magnitude of arrears) and more concentrated markets reduce the impact of trade liberalization. Not considering such factors can lead to either over- or underestimation of the effects of trade reforms on TFP.

The impact of increased competition as a result of FDI inflows is documented in several chapters in this volume, as well as in many of the investment climate–related surveys undertaken in developing countries and transition economies since the late 1990s. In a 2003 survey of 391 domestic enterprises in the Czech Republic, almost half of respondents reported that foreign entry had increased the level of competition in their sector. Twenty-nine percent of respondents claimed to have lost market share as a result of foreign entry. Similar responses were obtained from a survey of 396 enterprises in Latvia (Javorcik and Spatareanu 2005). Many studies' finding of a negative correlation between the presence of foreign firms and the productivity of domestic enterprises in the same sector may be attributed to the fact that the increased competition resulting from foreign entry outweighs the effect of knowledge transfer (Aitken and Harrison 1999). The percentage of Czech firms reporting multinationals siphoning off the demand for their products exceeded the share of Czech firms reporting learning from foreign firms through the demonstration effect (figure 1.1). These findings help explain the negative spillovers obtained by Djankov and Hoekman in chapter 8.

FIGURE 1.1 Perceived Effects of FDI in the Czech Republic and Latvia

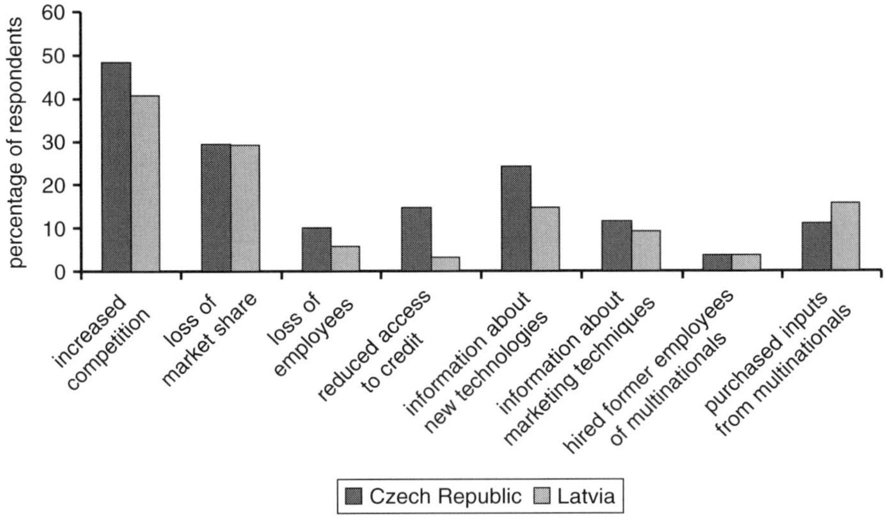

Source: Javorcik and Spatareanu (2005).

Policy Implications of Research Findings

Market transactions in technology are affected (constrained) by three major problems: asymmetric information, market power, and externalities (Hoekman, Maskus, and Saggi 2005). These market failures imply that there is a *potential* for policies to increase welfare by encouraging international technology diffusion insofar as these failures result in a (globally) suboptimal amount of international technology diffusion. Such policies may be constraining (for example, a prohibition on trade) or encouraging (for example, subsidies). To be effective, a policy must alter the incentives of private agents that possess innovative technologies in the "right" way. In practice, the potential for welfare-improving policy may not be realized due to mistakes or rent-seeking activity. In this area of policy, perhaps more than in others, the potential for government failure is high.

Asymmetric Information

Technology transfer involves the exchange of information between those who have it and those who do not. Firms that have information cannot fully reveal their knowledge without destroying the basis for trade, creating a problem of asymmetric information: buyers cannot fully determine the value of the information before buying it. This can lead to large transactions costs, which stifle market-based technology transfer. In the international context, information problems are more severe, are and the enforcement of contracts more difficult to achieve. The theory of the multinational firm posits that one motivation for such firms to establish foreign subsidiaries is the desire to overcome the difficulty of using markets to profit from their proprietary technologies.

Market Power

Owners of (new) technologies typically have substantial market power, resulting from lead time, fixed (sunk) costs, and learning by doing or from the granting of intellectual property rights. This necessarily implies that the price of technology will exceed the socially optimal level (that is, the marginal cost). While this divergence between price and cost allows innovators to profit from their innovation, it also implies a reduction in the (static) national welfare of those importing technologies.

Externalities

Externalities may arise if the costs and benefits of technology exchange are not fully internalized by those involved. A major share of benefits to recipient countries of international technology diffusion is likely to arise from uncompensated spillovers. Positive spillovers exist whenever technological information is diffused into the wider economy and the technology provider cannot extract the economic value of that diffusion. Spillovers can arise from imitation, trade, licensing, FDI, and the movement of people.

Import-competing industries register productivity gains following trade and FDI reforms (with much of the gains due primarily to factor use reallocation effects), and price-cost margins fall as a result of greater competition. But the policy implications of these phenomena are not straightforward, as Erdem and Tybout (2004) note, because it is not clear what drives improved performance. Is it due to better management incentives? To greater returns to innovation? To incentives to shed redundant labor as a result of greater competitive pressures? Understanding the drivers is important from a policy perspective, as policies need to target the specific incentives that matter. There is significant uncertainty here. One conclusion that seems robust is that domestic distortions (market failures) and general economywide policies motivated by public good rationales (education, infrastructure) may be more important than trade and investment policies in determining long-run outcomes. However, trade and investment policies are a key determinant of international technology diffusion. The analyses in this volume of the effects of trade and FDI suggest that the impacts of trade and investment policies differ depending on circumstances in the importing or host country, such as the ability to supply intermediate inputs, the country's absorptive capacity, and the extent of any bias toward or against types of imports and types of FDI.

Technology-Related Trade Policies and International Technology Diffusion

Many countries have historically engaged in infant industry protection, but the evidence suggests that diffusion of knowledge is facilitated by an open trade regime (Hoekman, Maskus, and Saggi 2005). Undistorted access to capital equipment and imported inputs that embody foreign knowledge allows firms to acquire know-how. The case for open markets extends to other products as well, as greater competition will reduce price-cost markups. At the same time, given that technology markets are associated with increasing returns, imperfect competition, and externalities, the argument against trade protection is not unconditional. The conclusions hinge on the scope of knowledge spillovers. International spillovers, for which there is considerable evidence (see chapter 5 and the references cited there), strongly tilt the balance in favor of free trade. If national spillovers are also important, however, there may be a potential role for intervention. For example, if productivity improvements depend only on a country's own R&D, a case can be made for policies that ensure that industries in which such improvements occur at a rapid rate are not all located elsewhere. The relative magnitudes of international as opposed to national spillovers are not known, and much will depend on the specific circumstances of countries. What matters is that research indicates there is a geographic component to knowledge flows, suggesting that agglomeration-type effects can be important (Jafee, Trajtenberg, and Henderson 1993; Sjoholm 1996; Audretsch and Feldman 1996). From a policy perspective, restricting trade is neither efficient nor effective. Instead, general policies are needed to increase the incentives of agents to undertake activities that generate social benefits that exceed private returns, without simultaneously creating additional distortions. Such policies include measures to improve the investment climate and reduce the "costs of doing business," as well as investments in education. Trade policy does little to encourage local R&D and necessarily leads to other distortions.[3]

One set of trade policies that is often motivated on the basis of international technology diffusion objectives is trade related investment measures (TRIMs). Examples are local content and technology transfer requirements for foreign investors. TRIMs discriminate against imports by creating incentives (other than import tariffs) to source inputs from domestic producers. In the international technology diffusion context, a motivation for TRIMs is often that foreign firms might be expected to transfer knowledge to ensure that local inputs satisfy their specifications. TRIMs act as an implicit tariff on intermediate goods imports, because manufacturers are forced to use higher-cost local inputs. They are inferior to tariffs in welfare terms, as no tariff revenue is generated. More important from the perspective of international technology diffusion is the fact that TRIMs provide little or no incentive for the protected producers of intermediate goods to acquire more knowledge.

FDI-Related Policies

Historically, restrictive trade policies were often complemented by restrictions on FDI, in part to prevent the use of investment as a way of getting around trade protection (so-called tariff-jumping investment). Japan, the Republic of Korea, and Taiwan (China) have all imposed restrictions on FDI. Policies were often more welcoming toward other modes of international technology diffusion, including liberal trade policies affecting imports of machinery and equipment (through zero duty ratings or drawback mechanisms) and the licensing of foreign technology. In recent years national FDI policies have generally become more liberal in developing countries, but governments often differentiate between joint ventures and fully owned subsidiaries of multinational enterprises. China, among others, has been more encouraging of joint ventures than inward FDI, although it has become the leading destination for FDI. Such a policy stance may reflect a desire to maximize technology transfer to local agents.

The literature suggests that such discriminatory policies are unlikely to achieve the desired effects with respect to technology transfer. Survey evidence indicates that technologies transferred to wholly owned subsidiaries are of a newer vintage than licensed technologies or those transferred to joint ventures (Mansfield and Romeo 1980). Foreign investors also tend to devote more resources to technology transfer to their wholly owned subsidiaries than to partially owned affiliates (Ramachandaram 1993). Moreover, the econometric analysis conducted by Javorcik (chapter 10) shows that multinational enterprises with advanced technologies tend to enter a host country through wholly owned subsidiaries rather than joint ventures. By forcing multinational enterprises to license their technologies or engage in joint ventures, host countries may be lowering the quality of technologies they receive as well as reducing the incentives to invest in the country at all.

While the magnitude of international technology diffusion undertaken by multinational enterprises need not be socially optimal, considerable evidence suggests that such firms are keen to transfer technology to their local suppliers.[4] Policies that facilitate this process, as opposed to regulation or legislation requiring that multinational enterprises engage in international technology diffusion to local competitors, have a greater likelihood of being successful. A concrete example of such a policy would be a supplier development program that aims to prepare local companies to understand and meet the needs of multinational enterprises. Such programs have been successfully implemented in Ireland and more recently the Czech Republic. Participation is offered to promising small- and medium-size enterprises that undergo an evaluation process followed by intense work with outside consultants to make improvement to areas in which they are lacking. Participants are given opportunities to start new business relationships by meeting with multinationals looking for local suppliers.

If they are well designed, the services provided in this way can be effective in assisting firms, especially smaller enterprises, mobilize the right type of skills (consultants) and ensure that their target audience is aware of the services on offer. In practice these conditions are frequently not met. As a result, such programs often have limited impact at best, because the services provided do not generate much value-added or because the "wrong" firms—those with limited potential for improvement or those that had access to market-based consulting services (in the process creating a potential negative spillover on the private consulting industry)—get help.[5]

Many countries actively seek to attract foreign investors through up-front subsidies, tax holidays, and other grants. For there to be a rationale for such investment incentives, host countries must enjoy positive externalities from inward FDI. The prevalence of "follow the leader" behavior among multinational enterprises provides another potential case for FDI incentives. Given the oligopolistic nature of markets within which FDI occurs, a new entrant may attract investments by both competitors and upstream suppliers. If it does, competition at multiple stages of production may increase, raising efficiency, overall output, and employment. An implication is that a host country may be able to unleash a sequence of investments by successfully inducing FDI from one or two major firms.

If the local economy lacks a well-developed network of potential suppliers, multinational enterprises may be hesitant to invest, and local suppliers may not develop because of lack of demand. In the presence of such interdependence, growth may be constrained by a coordination problem that can be partially resolved by initiating investments from key firms. Of course, such coordination problems cannot be tackled solely through investment incentives. Policy efforts need to focus primarily on improving the investment climate and reducing the costs of absorbing technology, a complex task that involves building human capital and expanding national innovation systems.

Thus while there may be a case for incentives, it is a conditional one. To be effective, the investment climate and absorptive capacity must meet certain conditions.[6] Moreover, given competition among countries in providing FDI incentive packages, one must carefully examine the magnitude of potential costs and benefits associated with such policies. Before considering FDI incentives, it probably makes sense to invest in marketing a country as a profitable location for investment and ensuring that potential investors can easily access all relevant information (Rodriguez-Clare 2004).

General versus Specific Domestic Technology-Related Subsidy Policies

A substantial share of R&D benefits may be local in nature, leading to a concentration of innovative activity, often around academic research centers (Audretsch and Feldman 1996). Such agglomeration externalities may provide a case for specific R&D–related subsidies as well as support of basic research and training to expand absorptive capacity in a country, which many studies indicate is crucial for obtaining significant spillover benefits from trade or FDI.[7] Without adequate human capital or investments in R&D, spillovers may simply be infeasible. The implication is that liberalization of trade and open FDI policies need to be complemented by policies with respect to education, R&D, and human capital accumulation if countries are to take full advantage of international technology diffusion.

Daniel Lederman and William Maloney (chapter 13) examine the evolution of Mexican technological progress in the past few decades, devoting special attention to the role of trade, FDI, and the national innovation system. Their main message is that trade liberalization (North American Free Trade Agreement) was helpful but insufficient to help Mexico catch up to the levels of innovation and the pace of technological progress in its North American partners. The evidence suggests that, given its level of development, Mexico suffers from low levels of R&D expenditure and low levels of patenting activity. It severely underperforms when compared with successful developing economies, such as the Republic of Korea. The authors argue that Mexico's national innovation system—how the private sector, universities, and public policies interact to produce economically meaningful innovation—is inefficient. Without addressing this weakness, Mexico is unlikely to be able to converge on the pace of innovation in North America. Improving the quality of research institutions; providing incentives for researchers to get involved with the productive sector, in particular incentives for appropriating innovations emanating from technical research; negotiating the cofinancing of research exchange programs with NAFTA partners; and developing domestic credit markets are among the policy responses suggested.

Of great relevance to international technology diffusion is the role that subsidies can play to facilitate learning, technology acquisition, and dynamic comparative advantage where returns to such activities cannot be appropriated by private agents and hence will not be undertaken by any individual firm. Amsden (1989) and others argue that policy interventions, including implicit or explicit subsidies, lay behind the economic "miracles" in the Republic of Korea and Taiwan (China). They claim that carefully targeted subsidies allowed these governments to stimulate key sectors that became efficient in their own right and provided positive spillovers for the economy as a whole.

In considering this infant industry argument for government support, it is important to differentiate between sector-specific subsidies and general policies facilitating learning and the development of private enterprise. In a recent comprehensive retrospective on the East Asian development experience, Noland and Pack (2003) argue that sector-specific policies did not result in high rates of TFP growth for manufacturing. In the Republic of Korea and Taiwan (China), TFP growth was not much higher than in the Organisation for Economic Co-operation and Development (OECD). In India selective interventionist policies were associated with declining TFP growth rates, while the opening of the economy led to an increase in TFP growth rates (Krishna and Mitra 1998). Specific interventions will often get it wrong, in part as a result of rent-seeking, and in part as a result of general equilibrium effects (a subsidy for one activity usually implies a tax on all others).

The case for general policy support for certain types of activity, including innovation, education, transport infrastructure, and similar public goods, is uncontroversial. The same is true for policies aimed to promote socially beneficial activities. Markets do fail, and there may be good rationales for governments to provide incentives for firms and agents to undertake activities that would otherwise be undersupplied. An important example that has a direct bearing on the subject at hand is the learning externality analyzed by Hausmann and Rodrik (2003). The market undersupplies investment by firms in new (nontraditional) activities because of appropriability problems: as soon as an entrepreneur is successful in identifying a profitable new production opportunity, entry by imitators prevents him or her from recouping costs. In this case, a subsidy or similar incentive can help expand innovation and risk taking.[8] As Rodriguez-Clare (2004) notes, this argument provides a rationale for supporting innovating projects leading to new knowledge about the country's comparative advantage, but it is not a justification for a general policy of supporting small and medium-size enterprises. Nor is it a case for trade policy, which is too blunt an instrument (Baldwin 1969). Any support program should be aimed at young firms, as firms that have remained small for many years most likely exhibit low productivity and are therefore not promising targets for intervention (see, for example, Srinivasan 2004).

The efficient use of support policies requires that governments be effective at identifying cases that justify intervention, designing effective instruments, and implementing these instruments appropriately. In practice, governments may fail in doing all three; the policy problem is to assess the relative sizes of government failure and the failures of the market for knowledge. Potential problems include the fact that subsidies can serve to support inefficiency; firms may behave strategically (by underinvesting, for example) in order to win subsidies; and subsidies can result in corruption, bad corporate governance, and rent-seeking behavior. The greatest challenge of implementing subsidies is that they are difficult to control. Government needs to establish an effective and credible exit strategy that weeds out unsuccessful efforts from successful ones. The capabilities and autonomy of the state play a fundamental role in implementing subsidy policies effectively (Rodrik 1993), suggesting that this is a policy path that countries with weak governance and institutions pursue at their peril. Greater policy consistency and credibility may be obtained through international cooperation (for example, trade agreements), which can help government be more credible in ensuring that a policy to assist a particular activity is temporary (Hoekman 2005).

Intellectual Property Rights and Technology Transfer

Intellectual property rights can support markets in technology, including international technology diffusion, and they are likely to interact with both trade and FDI flows (Arora, Fosfuri, and Gambardella 2001; Javorcik 2004; Maskus 2000). Although the chapters in this volume do not address intellectual property rights issues, we discuss briefly some of the stylized facts and findings from the literature from the perspective of international technology diffusion.[9]

Absent intellectual property rights, firms would be less willing to engage in technology transactions. Patents and trade secrets provide a legal basis for revealing the proprietary characteristics of technologies to subsidiaries and licensees, supporting the formation of licensing contracts.[10] Patent protection both increases flows of international technology diffusion to countries with technological capacity and shifts incentives for investors between FDI and licensing.

The empirical literature yields several results. First, patent applications from foreign nations are strongly associated with productivity growth in recipient countries (Eaton and Kortum 1996). Except in the United States, more than half of productivity growth in OECD countries comes from importing technologies (patents). This proportion is higher for small economies. Thus trade in ideas is a major factor in world economic growth.

Second, patent citations reflect knowledge flows across borders. While there is a limited amount of diffusion overall (because of distance, borders, and differences across regions in technological specialization [Peri 2003]), the most significant patents and knowledge in highly technological sectors are widely diffused. There is a strongly positive impact of knowledge flows on international innovation.

Third, international trade flows, especially in patent-sensitive industries, respond positively to increases in patent rights among middle-income and large developing countries. An important reason is that these countries represent a competitive imitation threat with weak intellectual property rights, and stronger patents expand the market for foreign exporters (Smith 2001). However, trade flows to poor countries are not responsive to patent rights.

Fourth, the evidence on patent protection and inward FDI is mixed, but recent studies uniformly find positive impacts among middle-income and large developing countries (Fink 2005). However, in poor countries, patents do not expand FDI (Smith 2001; Blyde and Acea 2002). Stronger patent protection also shifts the focus of FDI from distribution of imported products to local production (Javorcik 2004).

Fifth, strengthening intellectual property rights shifts international technology diffusion from exports and FDI toward licensing and positively affects knowledge inflows, measured as R&D expenditures undertaken on behalf of affiliates. These findings apply only to recipient countries with strong imitative abilities; the impact is zero in other countries (Smith 2001).

Sixth, a common finding of the literature is that the poorest countries are unlikely to benefit from strong intellectual property rights (McCalman 2001). Stronger patent rights may be expected to raise considerably the rents earned by international firms as intellectual property rights become more valuable, obliging developing countries to pay more for the average inward flow of protected technology. In some countries international technology diffusion-related spillovers are likely to be small at best, given limited absorptive capacity. The implications are that in poor countries, policy should aim to lower the costs of imports of intellectual property rights–intensive goods and technology and raise the capacity to absorb and adapt technologies through programs aimed at bolstering human capital.

Conclusion: Main Messages and Outstanding Research Questions

The relationship between trade and investment policies and international knowledge flows associated with trade in goods and FDI is complex. What are the main messages that emerge from the research summarized above and the chapters in this volume? Without implying any rank ordering in terms of importance, we would stress the following:

- Trade and FDI are major indirect channels for diffusing technology. Following increases in competition due to liberalization of trade and FDI, price-cost mark-ups and the output of import-competing firms fall. In the medium term, more efficient plants expand, while others contract or exit. These procompetitive effects generate the incentives to invest in new equipment and apply new methods of production, to innovate, to improve quality and differentiate products, and so forth. A liberal policy stance toward such flows is, therefore, an important instrument with which to encourage international technology diffusion. This conclusion is bolstered by the finding that restrictive policies will not generate incentives for greater investments in the acquisition and absorption of knowledge.

- The procompetitive effects of trade and FDI may be attenuated or not realized in the absence of complementary policies that ensure that domestic firms confront hard budget constraints and markets are contestable. If domestic banks are encouraged by the government (whether explicitly or implicitly) to build up nonperforming assets and finance inefficient loss-making firms, greater competition from imports may not induce firms to apply new techniques and improve productivity. It is also important that pro-FDI policies be accompanied by open trade policies, in order to confront foreign investors with competition. The negative association between concentration and TFP performance in Bulgaria found by Djankov and Hoekman in chapter 12 is indicative in this regard.
- Trade is also an important direct conduit for technology diffusion, in that much knowledge and R&D is embodied in goods that are imported, especially capital goods and machinery. From the perspective of international technology diffusion, this helps explain the structure of protection often observed in developing countries, where the effective rate of protection on investment or intermediate goods is much lower than on consumer or final goods. Such a structure of protection may encourage international technology diffusion, but it is unlikely to be optimal, because it may distort both production and consumption decisions—leading, for example, to the use of excessively capital-intensive techniques and creating incentives for domestic investment in the more protected industries. The finding that vertical spillovers are more likely (and larger) if there is a local supplying industry points in the same direction: a policy that explicitly favors imports of inputs over other imports may act as a disincentive for the development of such activities, reducing the benefits of FDI.
- The positive effect of trade on developing countries depends not just on trade policies (openness) but also on the prevailing conditions in an economy. As several chapters emphasize, the impact of openness on productivity growth is a function of absorptive capacity, broadly defined. The absorptive capacity combined with the structure of trade policy has implications for the composition of trade taking place and its effects. The findings suggest that a neutral trade policy is called for in order not to distort the investment choices of domestic firms and that complementary policies enhancing technological capacity are important.
- Exporters tend to be more efficient than nonexporters, suggesting that they have acquired and apply better technologies, both hard and soft. However, the

lack of strong evidence of learning by exporting in the literature, in conjunction with the robust finding that imports are an important source of knowledge and market discipline again suggests that governments should pursue a neutral trade policy. The finding that firms that target export markets are (or become) more productive because they take explicit actions regarding investment, training, technology, and so forth in anticipation of accessing foreign markets suggests a likely additional payoff in terms of international technology diffusion from engaging in market access negotiations with trade partners.

- FDI is an important direct source of technology and knowledge. Almost all studies show that firms with foreign capital perform better than indigenous firms, even after controlling for the fact that foreign investors generally select plants that are more productive in the first place. This suggests that FDI results in a more effective use of resources in host countries.
- The evidence on externalities associated with FDI is less clear cut. Intraindustry spillovers are less likely to take place than vertical spillovers. Vertical spillovers are a function of the capacity of the domestic industry to satisfy the demands of multinational enterprises at lower cost than imports, which in turn will depend in part on the nationality of investors and the location of host countries relative to the home countries of the multinational enterprises concerned.
- While open trade and investment policies help developing countries in attracting and accessing technology, openness alone is not sufficient. Absorptive capacity matters, as does the initial level of technological capacity of domestic firms. This finding points to the importance of education, infrastructure, access to finance for small firms and start-ups, and low regulatory costs of entry (World Bank 2004). To a large extent, the policy agenda here overlaps that of improving the investment climate in general and reducing the costs of doing business. However, there may also be a need for specific supply-related policies, that is, horizontal measures to encourage investment in innovative new activities, such as supplying world markets and local multinational enterprises, and activities oriented toward final demand. Determining what types of complementary policies are needed to enhance technological capabilities requires country-specific analysis (Sutton 2002). Effective monitoring and evaluation of programs that aim to improve the supply of knowledge is critical to ensure that they are beneficial.

This summary of the main messages emerging from the research has already indicated that several areas remain in which the state of knowledge cannot provide reliable policy guidelines, either because a particular question has not been addressed or because the analyses suffer from shortcomings. We conclude with four areas in which more policy-relevant research is needed on the trade- and investment-related dimensions of international technology diffusion.[11]

First, more research is needed on the question of selection versus learning in the context of domestic firms that are successful in exporting. The jury is still out on whether exporters learn from selling abroad. Whether or not they do is important for policy, as it will determine whether there is a case for subsidizing such activities. A related policy-relevant question concerns the role of the prospect of export market access on investments by local firms and, in particular, what it implies for the design and content of trade agreements, increasingly used as the instrument through which to obtain and lock in access to export markets.

Second, not enough is known about the mechanism behind vertical spillovers from FDI. While anecdotal evidence suggests that such spillovers may be due to deliberate knowledge transfer from multinational enterprises, it is plausible that the prospect of receiving a lucrative deal from a foreign company creates a powerful incentive for domestic firms to restructure and increase their efficiency. If local suppliers do learn from interactions with foreign affiliates, using policy instruments to attract FDI or establishing supplier development programs may be justified. If, however, what matters is the prospect of obtaining more profitable contracts than those available from local firms and customers, then a similar outcome could be achieved by, for instance, facilitating access to foreign markets through multilateral or preferential trade agreements or facilitating the flow of information about foreign markets and business opportunities available there. Many governments use both types of policy instruments; a better understanding of the mechanisms that drive vertical spillovers could help identify whether some of these programs are redundant.

Third, the literature has largely ignored the impact of what is increasingly an important input category when estimating firm productivity: services. These range from management consulting to finance, design, marketing, and transportation. Assessing the importance of services-related policies on firm performance, the impact of access to domestic versus foreign services providers, and the effect of international integration on the number of service varieties available in the market are examples of important questions in their own right that also have strong policy implications for international technology diffusion. There is likely to be a close connection between the functioning of services industries and the realization of the agglomeration externalities that have been stressed in the recent literature on industrial development as a precondition for sustained growth of specific sectors or activities (Burgess and Venables 2004).

Fourth, as trade and investment policies and the transfer of technology are on the agenda of trade negotiations at both the regional and multilateral level, enhancing the understanding of the impacts of policy reform suggestions made in such negotiations is another subject for further research. The chapters in this volume indicate that the origin of investment and trade may matter in terms of the likely magnitude of international technology diffusion.

Notes

1. The role of intellectual property rights is discussed in another volume in this series (Fink and Maskus 2005). We summarize some of the major findings of the empirical literature on the links between intellectual property rights and trade/FDI later in this chapter.

2. See Keller (1996) for a criticism of earlier work by Coe and Helpman (1995). The fact that trade embodies technical information is evident in numerous studies, including Eaton and Kortum (1996) and Keller (2002).

3. Deardorff and Stern (1987) note that the use of trade policy in such situations is analogous to doing acupuncture with a fork—one prong may "hit the spot," but the others can only do harm.

4. Moran (1998) discusses several case studies that document this process.

5. See, for example, ADE, IBM, and EPU-NTUA (2003) for an evaluation of EU assistance programs to Mediterranean countries in this area, as well as Batra and Mahmood (2003). Hoekman and Javorcik (2004) discuss some of the literature in this area.

6. Care must also be taken in terms of the sectoral allocation of incentives. In particular, the case for seeking to attract "high-tech" investments is weak. Investment in simple activities (such as transportation and other fundamental services) may yield larger returns, even if it does not result in large technology spillovers. Industrialization can be subject to large coordination failures, which investment in local infrastructure can help resolve, improving the investment climate in the process.

7. The first avenue raises concerns regarding the ability of government to identify the right activities and prevent capture of policies.

8. In many cases, intellectual property rights will be ineffective, as the technologies that will be used are likely to be known in the rest of the world.

9. What follows draws on Hoekman, Maskus, and Saggi (2005). For a more detailed discussion of these issues, see Fink and Maskus (2005).

10. See Correa (2005) for the counterargument that strong intellectual property rights may stifle international technology diffusion as firms exploit market power.

11. Research is also needed on other dimensions of international technology diffusion, especially the supply side. This includes innovation policies and systems, including the impact of protection of intellectual property rights and the preconditions for benefiting from strengthening them (see Fink and Maskus 2005).

References

ADE (Aide à la Dećision Économique), IBM Business Consulting Services, and EPU-National Technical University of Athens. 2003. "Evaluation of Economic Cooperation between the European Commission and the Mediterranean Countries." Final Report, Vol. 1 (November). Belgium.

Aitken, Brian J., and Ann E. Harrison. 1999. "Do Domestic Firms Benefit from Direct Foreign Investment? Evidence from Venezuela." *American Economic Review* 89 (3): 605–18.

Amsden, A. 1989. *Asia's Next Giant: South Korea and Late Industrialization*. New York: Oxford University Press.

Arnold, Jens Matthias, and Beata Smarzynska Javorcik. 2005. "Gifted Kids or Pushy Parents? Foreign Acquisitions and Plant Productivity in Indonesia." Policy Research Working Paper 3597, World Bank, Washington, DC.

Arora, Ashish, Andrea Fosfuri, and Alfonso Gambardella. 2001. *Markets for Technology: The Economics of Innovation and Corporate Strategy*. Cambridge, MA: MIT Press.

Audretsch, David, and Maryann Feldman. 1996. "R&D Spillovers and the Geography of Innovation and Production." *American Economic Review* 86 (3): 630–40.

Baldwin, Robert. 1969. "The Case against Infant Industry Protection." *Journal of Political Economy* 77 (3): 295–305.

Batra, Geeta, and Syed Mahmood. 2003. "Direct Support to Private Firms: Evidence on Effectiveness." Policy Research Working Paper 3170, World Bank, Washington, DC.

Bernard, Andrew B., and J. Bradford Jensen. 1999. "Exceptional Exporter Performance: Cause, Effect, or Both?" *Journal of International Economics* 47 (1): 1–25.

Blomström, Magnus, and Ari Kokko. 1998. "Multinational Corporations and Spillovers." *Journal of Economic Surveys* 12 (2): 247–78.

Blyde, J.S., and C. Acea. 2002. The Effects of Intellectual Property Rights on Trade and FDI in Latin America." Inter-American Development Bank, Washington, DC.

Burgess, Robin, and Anthony Venables. 2004. "Toward a Microeconomics of Growth." In *Accelerating Development: Annual World Bank Conference on Development Economics*, ed. François Bourguignon and Boris Pleskovic, 105–39. Washington, DC: World Bank.

Clerides, Sofronis K., Saul Lach, and James R. Tybout. 1998. "Is Learning by Exporting Important? Micro-Dynamic Evidence from Colombia, Mexico, and Morocco." *Quarterly Journal of Economics* 113 (3): 903–47.

Coe, David T., and Elhanan Helpman. 1995. "International R&D Spillovers." *European Economic Review* 39 (5): 859–87.

Coe, David T., Elhanan Helpman, and Alexander W. Hoffmaister. 1997. "North-South R&D Spillovers." *Economic Journal* 107 (440): 134–149.

Correa, Carlos M. 2005. "Can the TRIPS Agreement Foster Technology Transfer to Developing Countries?" In *International Public Goods and Transfer of Technology under a Globalized Intellectual Property Regime*, ed. Keith E. Maskus and J.H. Reichman, 227–56. Cambridge: Cambridge University Press.

Deardorff, Alan, and Robert Stern. 1987. "Current Issues in Trade Policy: An Overview." In *U.S. Trade Polices in a Changing World Economy*, ed. Robert Stern, 15–77. Cambridge, MA: MIT Press.

Eaton, Jonathan, and Samuel Kortum. 1996. "Trade in Ideas: Patenting and Productivity in the OECD." *Journal of International Economics* 40 (3–4): 251–78.

———. 1999. "International Technology Diffusion: Theory and Measurement." *International Economic Review* 40 (3): 537–70.

Erdem, Erkan, and James Tybout. 2004. "Trade Policy and Industrial Sector Responses in the Developing World: Interpreting the Evidence." In *Brookings Trade Forum 2003*, ed. Susan Collins and Dani Rodrik, 1–43. Washington, DC: Brookings Institution Press.

Evenson, Robert, and Larry Westphal. 1995. "Technological Change and Technology Strategy." In *Handbook of Development Economics*, Vol. 3, ed. T.N. Srinivasan and Jere Behrman, 2209–99. Amsterdam: North-Holland.

Fink, Carsten. 2005. "Intellectual Property Rights and U.S. and German International Transactions in Manufacturing Industries." In *Intellectual Property and Development: Lessons from Recent Economic Research*, ed. Carsten Fink and Keith Maskus, 75–110. Washington, DC: World Bank.

Fink, Carsten, and Keith Maskus, eds. 2005. *Intellectual Property and Development: Lessons from Recent Economic Research*. Washington, DC: World Bank.

Görg, Holger, and Eric Strobl. 2001. "Multinational Companies and Productivity Spillovers: A Meta-Analysis." *Economic Journal* 111 (475): 723–39.

Hallward-Driermayer, M., G. Iarossi, and K. Sokoloff. 2002. "Exports and Manufacturing Productivity in East Asia: A Comparative Analysis with Firm-Level Data." NBER Working Paper 8894, National Bureau of Economic Research, Cambridge, MA.

Hausmann, Ricardo, and Dani Rodrik. 2003. "Economic Development as Self-Discovery." *Journal of Development Economics* 72 (2): 603–33.

Hoekman, Bernard. 2005. "Operationalizing the Concept of Policy Space in the WTO: Beyond Special and Differential Treatment." *Journal of International Economic Law* 8 (2): 405–24.

Hoekman, Bernard, and Beata Smarzynska Javorcik. 2004. "Policies to Encourage Firm Adjustment to Globalization." *Oxford Review of Economic Policy* 20 (3): 457–73.

Hoekman, Bernard, Keith Maskus, and Kamal Saggi. 2005. "Transfer of Technology to Developing Countries: Unilateral and Multilateral Policy Options." *World Development*, 33(10): 1587–1602.

Jaffe, Adam, Manuel Trajtenberg, and Rebecca Henderson. 1993. "Geographic Localization of Knowledge Spillovers as Evidenced by Patent Citations." *Quarterly Journal of Economics* 108 (3): 577–98.

Javorcik, Beata Smarzynska. 2004. "The Composition of Foreign Direct Investment and Protection of Intellectual Property Rights: Evidence from Transition Economies." *European Economic Review* 48 (1): 39–62.

Javorcik, Beata Smarzynska, and Mariana Spatareanu. 2005. "Disentangling FDI Spillover Effects: What Do Firm Perceptions Tell Us?" In *Does Foreign Direct Investment Promote Development?* ed. Theodore Moran, Edward Graham, and Magnus Blomström, 45–71. Washington, DC: Institute for International Economics.

Keller, Wolfgang.1996. "Are International R&D Spillovers Trade-Related? Analyzing Spillovers among Randomly Matched Trade Partners." *European Economic Review* 42 (8): 1469–81.

———. 2002. "Geographic Localization of International Technology Diffusion." *American Economic Review* 92 (1): 120–42.

Kokko, Ari, Ruben Tansini, and Mario Zejan. 1997. "Trade Regimes and Spillover Effects of FDI: Evidence from Uruguay." Stockholm School of Economics.

Krishna, Pravin, and Devashish Mitra. 1998. "Trade Liberalization, Market Discipline, and Productivity Growth: India." *Journal of Development Economics* 56 (2): 447–62.

Mansfield, Edwin, and Anthony Romeo. 1980. "Technology Transfer to Overseas Subsidiaries by U.S.-Based Firms." *Quarterly Journal of Economics* 95: 737–49.

Maskus, Keith E. 2000. *Intellectual Property Rights in the Global Economy*. Washington, DC: Institute for International Economics.

McCalman, Phillip. 2001. "Reaping What You Sow: An Empirical Analysis of International Patent Harmonization." *Journal of International Economics* 55 (1): 161–86.

Moran, Theodore. 1998. *Foreign Direct Investment and Development*. Washington, DC: Institute for International Economics.

Noland, Marcus, and Howard Pack. 2003. *Industrial Policy in an Era of Globalization: Lessons from Asia*. Washington, DC: Institute of International Economics.

Peri, G. 2003. "Knowledge Flows, R&D Spillovers, and Innovation. Department of Economics, University of California, Davis.

Ramachandaram, Vijaya. 1993. "Technology Transfer, Firm Ownership, and Investment in Human Capital." *Review of Economics and Statistics* 75 (4): 664–70.

Rhee, Yung Whee. 1990. "The Catalyst Model of Development: Lessons from Bangladesh's Success with Garment Exports." *World Development* 18 (2): 333–46.

Roberts, Mark, and James Tybout, eds. 1996. *Industrial Evolution in Developing Countries*. New York: Oxford University Press for the World Bank.

Rodriguez-Clare, Andres. 1996. "Multinationals, Linkages, and Economic Development." *American Economic Review* 86 (4): 852–73.

———. 2004. "Microeconomic Interventions after the Washington Consensus." Inter-American Development Bank, Research Department, Washington, DC.

Rodrik, Dani. 1993. "Taking Trade Policy Seriously: Export Subsidization as a Case Study in Policy Effectiveness." NBER Working Paper 4567, National Bureau of Economic Research, Cambridge, MA.

Saggi, Kamal. 1996. "Entry into a Foreign Market: Foreign Direct Investment versus Licensing." *Review of International Economics* 4 (1): 99–104.

———. 2002. "Trade, Foreign Direct Investment, and International Technology Transfer: A Survey." *World Bank Research Observer* 17 (2): 191–235.

Schiff, Maurice, and Yangling Wang. Forthcoming. "North-South and South-South Trade-Related Technology Diffusion." *Journal of Development Studies*.

Sjoholm, Fredrik 1996. "International Transfer of Knowledge: The Role of International Trade and Geographic Proximity." *Weltwirtschaftliches Archiv* 132 (1): 97–115.

Smith, P. J. 2001. "How Do Foreign Patent Rights Affect U.S. Exports, Affiliate Sales, and Licenses?" *Journal of International Economics* 55 (2): 411–40.

Srinivasan, T.N. 2004. "Entrepreneurship, Innovation, and Growth." In *Accelerating Development: Annual World Bank Conference on Development Economics*, ed. F. Bourguignon and B. Pleskovic, 73–104. Washington, DC: World Bank.

Sutton, John. 2002. "Rich Trades, Scarce Capabilities: Industrial Development Revisited." *Economic and Social Review* 33 (1): 1–22.

Tan, Hong, and Geeta Batra. 1995. "Technology Spillovers from FDI: Evidence from Mexico." Working Paper, Private Sector Development Department, World Bank, Washington, DC.

Tybout, James. 2000. "Manufacturing Firms in Developing Countries: How Well Do They Do and Why?" *Journal of Economic Literature* 38 (1): 11–44.

World Bank. 2004. *Doing Business*. Washington, DC: World Bank.

PART I

LITERATURE SURVEYS

2

ECONOMETRIC VERSUS CASE STUDY APPROACHES TO TECHNOLOGY TRANSFER

Howard Pack

There has been increasing interest in recent years in econometric estimates of the productivity performance of firms in developing countries (Aw, Chen, and Roberts 2001; Pack and Paxson 2001; Tybout 2000; Kraay [chapter 7 of this volume]; and Kraay, Soloaga, and Tybout [chapter 11 of this volume]). There is also a large and rich case study literature, insights from which could help frame relevant questions and point out the limitations of econometric studies as a result of the absence of information in censuses of manufacturing that has been shown to be relevant in extensive case study interviews.

The major determinant of productivity is the effort of firms to absorb technology. The core of the industrial development process consists of obtaining and assimilating new technology. New technologies consist of hardware—machinery and new buildings—and the knowledge or software with which to run a plant. Firms can obtain new technology by purchasing new or used equipment (foreign or domestic), by engaging in technology licensing agreements, and by hiring consultants. Firms improve purchased technology and occasionally develop their own through formal and informal research and development (R&D) programs. They may learn about the software component from newly hired workers with experience at other firms or recent university studies, from informal interaction with more advanced firms (domestic and foreign), and from purchasers of their output or purveyors of their inputs.

Most econometric studies that attempt to study these processes rely on observations obtained from periodic manufacturing censuses, 5 or 10 years apart, or from annual industrial surveys that collect measures of output (gross receipts, value-added) and inputs (number of workers, capital value at historic acquisition costs, current price intermediates) but do not permit fine-grained analysis of the determinants of a firm's evolution. Analyses in which researchers collect the data themselves have generated richer data, which allow the exploration of more complex issues (Bigsten and others 2002).

In contrast, case studies provide a rich source of evidence on the details of the transfer and absorption process and offer important clues to the type of microeconomic detail that would contribute to deeper understanding of the process. Econometric and case studies are complementary. It is impossible for econometric studies to include many of the subtle insights that have been obtained from exhaustive case studies of firms. But case studies are often concerned largely with the process of technology acquisition and neglect to measure the ultimate productivity effect on the firms studied or the impact on other firms and sectors, phenomena that are better addressed by analyzing large industrial surveys and censuses of manufacturing. Moreover, given the substantial time spent in each plant, it is difficult to generate a sample that is sufficiently large that the results can be viewed as robust.

Modes of Technology Transfer for Individual Firms

Importation of New Equipment or Purchase of Locally Produced Equipment

The purchase of equipment is a potent potential source of new technology if the new vintage is more productive than that currently employed by the firm. Gains can be obtained from both new and used equipment, but I concentrate on new equipment purchases. The gain in efficiency embodied in equipment is reflected in designs that (potentially) allow greater output per machine hour, reduced labor and material inputs per unit of output, or both. Such physical productivity improvements are realized in the country of origin by plants that purchase the machine and by firms in other countries with similar per capita income. Whether these potential gains lead to actual reductions in unit cost depends on the price of the equipment and the ability of firms to realize the gains in efficiency embodied in the design of the equipment, referred to as technology assimilation or absorption.

Given the very limited capacity of domestic machinery producers, in all but a handful of developing countries, firms in the modern manufacturing sector buy most of their equipment abroad. Any equipment (including secondhand equipment) that reduces unit costs should be viewed as providing a productive infusion of technology. Despite the potential improvement in efficiency bestowed by new machinery, as measured by the reduction in unit labor inputs, there is abundant evidence in many developing countries that the productivity improvements offered by superior equipment are frequently not realized in local factories.[1] In many cases, local firms lack the technological knowledge or do not face competitive pressure to achieve the technical efficiency that characterizes firms in industrial countries that have adopted identical machinery. Knowledge and incentives are important.

Knowledge Obtained in Market Transactions

A considerable literature documents the efforts of firms to increase productivity with a given complement of plant and equipment (Dahlman and Westphal 1981; Hobday 1995; Katz 1987; Kim 1997; Lall 1987; Matthews and Cho 2000; Pack 1987). They hire consultants, engage in technology licensing agreements, and undertake R&D in order to realize the potential of new equipment.

Firms in most successful newly industrialized countries have undertaken one or more but not necessarily all of these actions. A voluminous case study literature describes such efforts, but many studies provide no numerical evidence of their effectiveness in raising productivity or reducing costs. The emphasis in many studies is on the process of knowledge acquisition rather than its results. The evidence shows that firms in the Republic of Korea and Taiwan (China) perform much better than firms in Argentina, India, Kenya, and the Philippines, where the policy environment often discouraged productive technology transfer, since acceptable levels of profits were obtainable in the narrow, protected domestic market.[2] In contrast, in countries in which large export growth has been realized, the cost of innovative efforts to assimilate technology could be allocated over the very large quantities sold in the world market.

Foreign direct investment may be another source of new knowledge for the manufacturing sector. Foreign firms that establish new production are typically closer than local firms to world best practice. They may improve the performance of the sector in which they operate or in vertically related sectors. Improvement may occur voluntarily, if they allow other firms to observe their operations, or involuntarily, as firms respond to increased competition. If managers and workers leave for locally owned firms or provide knowledge in informal settings, the diffusion of knowledge will affect productivity of firms in the same and related industries.[3] Joint ventures, though somewhat different in ownership structure from FDI, may yield similar benefits.

Knowledge Obtained as a Byproduct of Market Transactions

Firms may obtain technology at low or zero cost from transactions in both their output and input markets. A considerable case study literature (Hobday 1995, Kim 1997) documents that knowledge has been acquired by firms that engage in exporting. The knowledge obtained consists of process improvements, quality control, and information about markets that allows firms to achieve a more profitable product mix.[4] Foreign firms may provide this proprietary knowledge in anticipation of its beneficial effects on their own profits (Pack and Saggi 2000).

It is also possible that firms obtain knowledge from newly hired workers as a result of their experience with previous employers (Marshall 1890; David and Rosenbloom 1990; Henderson 1999; Quigley 1998). These are pecuniary external economies to the firm and the industry.[5] Resurrecting Marshallian economies, some studies argue that firms obtain knowledge from other firms in the same and related industries in an informal manner, through social interaction and direct observation of other plants, a process that is much more difficult to document. The most convincing case studies are based on interviews in Taiwan (China) (Saxenian and Hsu 2001) and in Silicon Valley in the United States (Saxenian 1994).

Determinants of the Demand for and Productivity of Technology Transfers

Despite the potential for firms in all countries to participate in the international and intranational transfer of technology, such transfer has occurred mainly in countries that have enjoyed sustained growth in per capita income and attempted to penetrate international markets. This impression may or may not reflect the reality of transfer. The Republic of Korea and Taiwan (China) have been subjected to much more intensive analysis than Colombia, Nepal, or Tanzania, but one suspects that casual empiricism is correct. Economies that have had rapid growth are likely to have experienced greater technology inflows. Sustained growth has usually been characterized by limited fluctuations in GDP and by low inflation. The productivity benefits from the search for and adoption of new technology are more likely to translate into improved profitability with such stability than in an environment characterized by sustained decline, as in Tanzania or by substantial cycles and inflation, as in many Latin American economies.[6] In a country such as Argentina, with persistent high inflation and unpredictable changes in relative prices of purchased inputs and output, increases in productivity may be offset by unanticipated changes in relative prices. The rewards to astute financial transactions and negotiated lower input prices are likely to far outweigh the reductions in cost obtainable from improved total factor productivity (TFP) in the plant. In periods of significant downturns in GNP, a not infrequent occurrence, the existence of extensive excess capacity also militates against the systematic search for new technology. This implies that even if some of a firm's efforts to acquire and utilize technology cannot be precisely measured, inclusion of the rate of growth and the standard deviation of output of the sector to which it belongs may provide some indication of unmeasurable technology transfers, with larger transfers occurring in less volatile environments.

The international trade regime has several effects on incentives to obtain new technology. High rates of effective protection, especially if they are viewed as likely to continue, may reduce incentives to reduce cost. Although in principle, cost-reducing technology transfer will always improve profits, firms may be satisfied with existing rates of return and not undertake the required effort.[7] Conversely, in economies in which exporting offers profitability similar to sales in domestic markets, firms may attempt to acquire new technology that allows them to produce at lower cost, higher quality for the international market, or both. Thus differences in the trade regime may account for some of the observed variation in technology transfer levels across less developing countries. In addition, firms in countries that have more favorable incentives for exporting may more easily obtain the foreign exchange necessary to import equipment and pay for knowledge, in the form of technology licenses, consultants' fees, or other expenses.

A third set of policy variables that may affect the ability of firms to acquire technology is the intellectual property rights regime of the country. Difficulty in enforcing contracts dealing with technology licenses or the ability to obtain legal redress if the local partner in a joint venture illegally transfers proprietary knowledge to other firms may limit the supply of technology (or raise its price), even in the presence of good macroeconomic and trade policies.

The Case Study Literature

In discussing the case studies, I consider their treatment of three different aspects of the development of technological abilities. The first is technology transfer per se (T), including the source of equipment, the hiring of consultants, the signing of technology licensing agreements. The second is efforts to absorb the technology transferred (A). Typically, this measure is descriptive, recording the number of personnel engaged in the transfer effort, the resources expended, and the involvement, if any, of the foreign suppliers. The third measure, R, is the result of T and A. It captures increases in TFP or declines in production cost.

Few of the case studies document T in great detail, most describe A extensively, and few attempt to assess R. The descriptive studies that omit R are nevertheless useful for the insight they provide, though their ultimate impact cannot be reliably assessed by objective measures but must rely on the reader's evaluation of the evidence.

Case Studies of Technology Transfer in Countries with Trade Restrictions

Most of the early case studies were conducted in India and Latin America. This reflects the fact that these studies were initiated in the 1970s, when the greatest concentration of industrial skills in the developing world existed in these regions, following two decades or more of import-substituting industrialization. As production abilities were accumulated in the fast-growing Asian countries, a new round of research in the 1990s focused on them.

The most extensive set of case studies in Latin America is that published in a volume edited by Katz (1987). The studies in his volume fit mainly in the A category. The major question they address is how plants were able to improve the productivity of equipment already in place. There is very little detailed discussion of the nature of the technology transfer process, such as the source of equipment or whether suppliers provided technical help, the emphasis being on the adaptive efforts of local firms. Except in one case, there is no measure of the impact of these efforts. Neither the magnitude of TFP growth nor unit cost reduction is calculated.

The firms studied in Katz's volume did not use the foreign equipment purchased according to the guidelines equipment manufacturers provided. Local engineers altered the production process to increase the output obtained from a given complement of machines by adjusting the production process to allow the use of local raw materials, given their lower cost or the difficulty of obtaining foreign exchange to pay for imported inputs. A few new products were developed. As in Organisation for Economic Co-operation and Development (OECD) countries in their early stage of industrialization, most of the technological breakthroughs were "minor" rather than major (Rosenberg 1976), involving small changes in machine settings, the addition of devices that improved performance, and the use of equipment at speeds that differed from manufacturers' recommendations. Such innovations had been observed earlier in the Republic of Korea and Taiwan (China) in response to a low user cost of capital, which provided an inducement to search for capital-saving innovations (Ranis 1973).[8]

Cumulatively, the impact of these technical changes altered the imported processes significantly (in an engineering sense) and occasionally yielded new products. The case studies in Katz's volume provide detailed histories about the process of assimilation in two or three firms, but they do not provide systematic insights into either the initial process of technical acquisition or the economic impact of the efforts they describe. There is no attempt to determine the impact of these adaptations on other firms in the same local industry or the effect on downstream purchasers when the products are intermediates. Where the output is consumer goods or final investment goods, no welfare assessment is made of the impact of price changes or the greater or more appropriate variety of goods. Nor is there evidence on the interactions of the firms with upstream suppliers. The detailed descriptions of the process of technical adaptation are thus bereft of a systematic analysis of the economic impact of the process. One is informed of the process of how firms move along or shift to another isoquant, but there is no confrontation of such moves with an isocost curve that would indicate whether allocative efficiency or TFP improved.

An analysis of the petrochemical industry by Cortes and Bocock (1986) concentrates on the process of transfer, T, but does not consider A or R. Cortes and Bocock are concerned mainly with the terms of transfer, reflecting a large literature that argues that the charges to less developing countries for technology licensing agreements were excessive. The agreements analyzed were executed in the 1960s and 1970s. The final contracts ranged from simple licensing (with an initial lump-sum payment plus a royalty equal to a percentage of annual revenue) to majority ownership by the local firm in a joint venture. In some agreements, the foreign firm provided all of the equipment and software; in others local companies assembled many of the elements from different suppliers. The evidence on the quality of technology transferred and its price was too ambiguous to permit any clear-cut conclusions. There is no evaluation of the effectiveness of the transfer as opposed to the details of the negotiations and price. The productivity of the initial transfer is not investigated.

Lall (1987) conducted a comprehensive set of studies in India. The emphasis is on A, rather than on the technology transferred to India or further diffusion within India. There is little evaluation of the impact on costs or productivity, though there is a discussion of exports as a potential indicator of the ability to learn sufficiently to compete on world markets.[9] The study's exhaustive taxonomy is helpful in underlining the difficulties that may occur in econometric estimates of the impact of technology diffusion. Table 2.1, taken from Lall, provides a description of the technological activities of a firm, divided into six categories: project preparation, project execution, process engineering, product engineering, industrial engineering, and technology transfer. Any of the activities listed in the table can reduce costs or increase productivity.

These are much finer categories than can be investigated in a large-scale survey involving many firms. Case studies have been based on spending days or weeks at individual firms to obtain this type of qualitative information. Few of these activities will be recorded in the questionnaires typically employed in industrial surveys and censuses. Thus it is unlikely that any census or survey in less developing countries captures the details of production developments that constitute the basis of technology transfers.

Does the absence of such elements present special difficulties for econometric studies based on broader categories of technology transfer? For example, improved knowledge of quality control by a U.S. firm may be part of the knowledge package transferred in a licensing agreement, or it may be embodied in new equipment. The results of successful efforts are packaged in the technology, embodied or disembodied, and the price will reflect the productivity of the components, assuming a relatively competitive market for knowledge. Significant positive coefficients on a variable such as payments for technology cannot distinguish

TABLE 2.1 Technological Capabilities in Manufacturing Industry

Project preparation	Project execution	Process engineering	Product engineering	Industrial engineering	Technology transfer
Identification of suitable project	Basic process design	Debugging of new plant	Assimilation of product design	Improved workflow and scheduling	Development of local suppliers and subcontractors
Feasibility studies	Basic and detailed engineering	Initial adaptation to local raw materials, working conditions, layout	Initial adaptation of product to local market needs	Better working practices, time and motion studies	Transfer of know-how to buyers
Specification of product range, input testing requirements	Equipment search, procure-ment, testing	Balancing of facilities to remove bottlenecks, reach full capacity	Improvement in product design	Monitoring of capacity utilization, coordination of different stages	Setting up of turnkey projects for customers (capital goods firms)
Specification of scale of production	Civil construction	Achievement of adequate quality control	Adaptation to export market needs	Evaluation of in-house and bought-out components	Setting up of projects for similar plants
Identification of technology	Ancillary services	Equipment maintenance and repair; some in-house spares and manufacture	New product introduction by licensing		Taking out of patents; development of blueprints

Source: Lall (1987).

among the sources of the productivity boost that could affect any of the activities, but such detailed knowledge may not be relevant for many issues, such as whether governments should loosen their restrictions on technology licensing payments.

Lall's studies and those contained in Katz's volume show that in some, but not all, categories of technology development, there has been progress that should have led to TFP growth for firms. However, growth accounting research at the industry level in India and several Latin America countries suggests that TFP growth in individual industrial sectors between 1960 and 1985, when the case studies were conducted, was either very low or negative (Ahluwalia 1985; Bonelli 1992). Thus even if there were cost-reducing innovations in individual firms, something never explicitly investigated, their significance was not a general phenomenon leading to industrywide TFP growth.[10] This may partly have reflected the limited demand for technological knowledge by most firms in the 1960s and 1970s in the highly protected domestic environment of countries engaged in import substituting industrialization.

A few case studies have been conducted in Sub-Saharan Africa. Pack (1987) analyzes the determinants of TFP in manufacturing firms in Kenya. In four multinational textile plants, equipment of various vintages was transferred to local subsidiaries. All of the plants employed a large contingent of managers from the home firm. The plants achieved surprisingly high levels of TFP relative to British textile mills using identical equipment. The precise level of TFP relative to British plants depends on the assumed elasticity of substitution, σ. With an assumed value of 0.5, the Kenyan firms achieve relative TFP of roughly 70 percent, most of the shortfall being due to the protectionist policy regime, which encouraged short production runs. This implies that the combined transfer of hardware and knowledge was quite successful in the case of fully owned subsidiaries. TFP in multinational firms was almost certainly greater than it would have been in locally owned companies. A comparison with local firms producing similar products but relying on technology licensing would provide an important contrast, as it might be argued that more knowledge is transferred where the potential for unauthorized knowledge seepage is lower (see chapter 3 by Saggi for analysis and references).

The most complete study examines a sample of firms in Ghana's manufacturing sector (Lall and others 1994). It measures "technological competence," proxied by the number of educated owners and employees, but it does not discuss the sources of hardware, and software, or the productivity with which they are deployed.

Case Studies in Export-Oriented Countries

Case studies in protectionist countries explore mainly heavy industries characterized by relatively large-scale, capital-intensive production. The firms were often enclaves in which forward and backward linkages to the rest of the industrial sector were limited. Very often there were only a few firms in the sector, and many of the skills generated in production had limited uses in other industrial branches.

In contrast, the literature on the export economies emphasizes technology transfer in newer sectors, such as electronics (Bloom 1991; Hobday 1995; Kim 1997; Matthews and Cho 2000). There was also extensive study in the 1970s and early 1980s of sectors, such as clothing and footwear, in which low-wage countries developed their early industrial skills based on an initial comparative advantage. In these studies, considerable attention is given to the source of equipment, the hiring of consultants, the signing of technology licensing agreements, and the transfer to the firms of knowledge from foreign purchasers. Much of this research also considers the success of absorption of technology by individual firms.

Relatively few studies evaluate either TFP growth or cost reduction. Given that much of the output of the firms was directed to export markets, there is a presumption that the T and A activities were economically efficient, although problems of intrafirm cross-subsidization leave this an open question.

In almost all of the Asian countries analyzed, the development trajectory of industry was rapid. Equipment (often secondhand) was imported in the early years, relevant knowledge was obtained, and both were improved upon by firms (Enos and Park 1988). In the period of labor-intensive development, the 1960s and early 1970s, most of the machinery was readily available on the world market, and the knowledge necessary to produce textiles, clothing, wigs, and simple sporting equipment was easily obtained from freely available industry knowledge, consultants, and trade publications. Rhee, Ross-Larson, and Pursell (1984) document these modes of technology transfer to the Republic of Korea. Their study was the first to suggest another mode of transfer, as exporters obtained information, design, and production engineering free of charge. Hou and Gee (1993) obtained similar results in Taiwan (China).

The authors did not examine whether free transfers of knowledge from OECD importers to Asian exporters improved TFP. Moreover, there was no systematic investigation of whether these knowledge transfers were diffused to other firms (nonexporters) through worker mobility, communication among owners, or other modes.[11] Whatever its quantitative importance, knowledge transfer occurred over a relatively short period, roughly the decade after the (then) small firms had begun operations. The firms were largely in labor-intensive sectors in which the "software" of production had a relatively simple mechanical or chemical engineering base. There was little in the way of complicated science that needed to be transferred. Much of the relevant knowledge was tacit but widely diffused within OECD countries. As the sectoral structure changed in the 1970s to more capital- and technology-intensive branches, much of the critical knowledge had a more complex scientific base, was more codified, and was subject to stricter proprietary rules. Informal transfer of such knowledge in transactions between buyers and sellers became less likely, as they were replaced by more formal licensing agreements. The free knowledge obtained from exporting was thus largely limited to a short period for new firms in labor-intensive sectors.

In econometric analyses attempting to establish a link between exports and productivity growth, the relatively short period during which learning is supported by OECD importers implies that the dates of the census or survey data are critical. It would be surprising if Korean or Taiwanese exporters from, say, 1985 onward acquired quantitatively significant free knowledge from their purchasers. But economies currently at lower stages of industrialization might still derive such benefits. The feedback from exporting to productivity growth needs to be investigated with suitably chosen dates, but this is usually impossible given the necessarily arbitrary dates of officially collected data. Moreover, unless such knowledge transfers are augmented by internal efforts at absorbing these transfers, their impact will be low. Interaction effects are important, but the variables available in manufacturing censuses usually cannot capture these complex interactions.

Descriptions of Asian technology acquisition in the succeeding period in newer, more technology-intensive sectors demonstrate that international transfers have been supplemented by intensive internal absorption efforts (Enos and Park 1988; Hobday 1995; Kim 1997). Complex equipment and seemingly complete technology license agreements rarely produce a fully saleable end product or productivity improvement unless resources are devoted to mastering the imported technology. Such efforts may take several years, even for well-specified products with a codified knowledge base. A major lesson to be drawn for econometric studies is the need for precise measures of absorptive effort and the difficulty of specifying the correct lag structure. The promised productivity improvement (or product) may come to fruition several years after the technology transfer, the timing being a function of the absorptive effort after the transfer.

A more detailed summary of some of the case studies underlines a number of the implications for econometric estimates of the case studies.

Case Studies in Export-Oriented Economies Kim's 1997 study of the Republic of Korea is the most detailed analysis of the transfer and absorption of technology in Asia. The categories he uses in describing the transfer process are similar to those in table 2.1. For individual firms in automobiles, microwave ovens, and semiconductor production, he identifies the foreign firms that supplied the equipment and discusses many aspects of the transfer process, including the product blueprints acquired, the technical manuals providing instructions about operating individual machines, the expenses incurred to train local engineers abroad, and the extensive visits by foreign engineers to Korean firms.

How will such activities enter a standard econometric analysis? If firm accounts are precise, many of the payments for productivity-enhancing transfers will be reported. If each of these activities is priced at its marginal productivity for the firm, total expenses incurred will be an economically meaningful measure. Two caveats apply. First, some expenditures, such as the living expenses of Korean auto engineers sent abroad for training, are unlikely to be fully or correctly recorded. In many instances, firms will not have recorded actual expenditures but simply acknowledge that workers were sent abroad; the variable used will have to be dummies indicating whether such activities occurred, regardless of the time period and presumably the total learning that occurred. In a similar vein, several of the firms report that they employed engineers who flew to Seoul from Japan on weekends. It is not clear that such outlays were correctly recorded, yet they are viewed as having had high productivity. If all firms in a large sample had similar omissions, there would not be a problem. However, a reasonable conjecture is that firms differ in the accuracy of their accounting.

Second, insofar as the price of technology licenses is the outcome of a bilateral bargaining process in an imperfect market, as suggested by the empirical work of Cortes and Bocock (1986), recorded expenditures across firms may have a tenuous relation to the productivity of the technology transferred. Prices may not reflect marginal productivity. While it is impossible to retrospectively deal with such omissions, interpretation of the econometric results must necessarily take these into account. In either case it is possible that precise estimates of the impact of technology transfer may not be obtainable given measurement errors and critical interactions among the relevant variables.

The preceding paragraph assumes that accounting standards in the Asian countries are high. However, as the Asian financial crisis unfolded, it became clear that accurate measures of such standard ratios as the rate of return on equity capital were grossly deficient even in the largest firms with listed equity shares. With lax general accounting standards, it requires a leap of faith to assume that reported technology transfer costs and internal R&D measures are accurate, yet these measures are often used in econometric studies. Moreover, many Asian countries are characterized by a conglomerate corporate structure in which the allocation of expenses and income among their component firms is arbitrary. Some have centralized research and may charge only part of the expenses for technology acquisition to component firms.[12] Relatively arbitrary transfer pricing among the affiliates may diminish the accuracy of value-added measures in any given product line or plant. If the dependent variable in production function estimation is value-added by product line or plant, problems arise, as the absolute measure is not accurate and changes in transfer pricing among related firms may lead to changes in value-added that are unrelated to any changes in productivity.

Even if expenditures are recorded correctly, the measured impact of technology expenditures may be misleading. One or two foreign managers may make nonmarginal contributions. One of the Korean auto companies reports that an indispensable manager was hired from British Leyland. Without this individual, the entire production process might not have been established; the production function would not have existed without him. In contrast, estimates of the impact of technology transfers assume the existence of a production function to which a given technology purchase makes a contribution at the margin. This suggests that the impact of some transfers will be underestimated.

All of these are standard problems encountered in attempting to obtain generalizable econometric results. If the estimates contradict those of intensive case studies, it is necessary to consider whether problems of measurement and specification have led to misleading results. Consider, for example, the problem of determining which technology variables are the sources of productivity growth. Say that firms that purchase equipment from abroad also send workers for training to the plant of the equipment manufacturers but that this training, which is the key to increased productivity, is not recorded. Growth in firm productivity may then be mistakenly attributed to the imported machinery rather than the additional training of the workers. The entire process should be thought of in Bayesian terms, in which the case studies provide priors while the sample information leads to revisions that form the base for the posterior estimates. While this cannot be implemented formally, viewing the case studies and econometric evidence in such a framework is likely to lead to more nuanced insights than sole reliance on either. In the instance just cited, evidence from case studies would suggest that any econometric findings that most of the growth of productivity was attributable to new equipment should lead to a re-examination of the survey results to determine whether the recorded responses on training or expenditures on consultants are consistent (and believable) across firms. The purchase of large amounts of new equipment, for example, will usually be accompanied by significant training. If the data do not indicate such training, caution in interpreting the results is warranted.

Occasionally, case studies are too concerned with establishing their own views; the hypotheses they suggest must be carefully evaluated to ensure that the insight derived from them has validity given the information they report. For example, Kim (1997) contrasts the experience of Hyundai, which utilized many sources of external technology in establishing auto production, with Daewoo, which relied entirely on General Motors. His plausible view is that Hyundai was forced to learn to integrate different types of knowledge and became more self-reliant, a process that showed up in higher labor productivity. My own calculation with the data he presents leads to the conclusion that the two firms had almost identical TFP. Thus even if one could obtain measures on whether the technology was packaged, as in the case of Daewoo, or unpackaged (from many sources), as in the case of Hyundai, it may well be the case that the interpretation of the author is open to question. Estimates from a large number of firms are likely to reduce such problems.

Many of the case studies suggest plausible views that an early weaning from reliance on technology suppliers is superior to minor adaptation of packaged technology and will lead to improved productivity in the long run. Lall (1987) makes this argument, and it is a recurrent theme in the studies in Katz (1987). But there is no quantitative confirmation of this view in any of the studies, and attempting to test such hypotheses would require a degree of detail that is unlikely to be present in officially collected information. If any measures were available, even a measure as simple as the number of foreign technology suppliers, they would be worth including.

Kim (1997) describes other aspects of the process by which technology transfer translates into increased TFP. First, there are long lags between transfer and successful production. Second, even having obtained a seemingly foolproof combination of hardware and software from foreign suppliers, productive use of these requires intensive internal efforts by the firm. The interaction of transfer and internal effort is thus a very important component of the payoff from technology transfer, yet there is no obvious way of modeling this, since a multiplicative variable constructed from multiplying contemporaneous measures of technology transfer and absorptive effort may not be adequate. Equipment or software obtained in one year may become productive only after application of internal absorption efforts in succeeding years, with highly variable lag structures. Interviews in case studies can focus on such interactions; the data in censuses and surveys are less likely to be amenable to such fine-grained testing. All the researcher can do is be aware of the issues raised by the case studies and interpret the quantitative findings judiciously. Kim describes the problems:

> Despite the training and consulting services of Ricardo [a foreign consultant] and the three experts, Hyundai engineers repeated trials and errors for fourteen months before creating the first prototype. But the engine block broke into pieces at its first test. New prototype engines appeared almost every week, only to break in testing. No one on the team could figure out why the prototypes kept breaking down, casting serious doubts, even among Hyundai management, on its capability to develop a competitive engine. The team had to scrap eleven more broken prototypes before one survived the test. There were 288 engine design changes, 156 in 1986 alone. Ninety-seven test engines were made before Hyundai refined its natural aspiration and turbo-charger engines; 53 more engines were produced for durability improvement, 88 more for developing a car, 26 more for developing its transmission, and 6 more for other tests, totaling 324 test engines. In addition more than 200 transmissions and 150 test vehicles were created before Hyundai perfected them in 1992. (1997, p. 122)

These efforts, over a long period of time, partly reflect Hyundai's inability to obtain all of the relevant knowledge from its Japanese licensor. The time lags and the interaction between imported knowledge and internal effort that are recorded in such a case study are difficult to replicate in econometric ones, even those based on careful survey instruments.

Exports and Productivity Even with all of the information obtained from a licensor and equipment obtained from firms in industrial nations, Samsung, another major Korean firm, had considerable difficulty producing the central mechanism of a microwave oven.

> It took the team a year of eighty-hour weeks to complete the first prototype (in 1976) but the plastic in the cavity melted in a test. . . . Finally in June 1978, after two years, the

> team developed a model that survived the test; but it was too crude to compete in the world market. Samsung incrementally improved the product and developed a makeshift production line, producing one over a day, then two, which it placed in local bakeries for feedback from users. (Kim 1997, p. 137)

Samsung's effort in microwaves was in response to a large order from J.C. Penney. In this case, the export order catalyzed the increase in productivity, with the growth in TFP (or product quality) preceding the actual export but with the causality going from an export order to productivity. The order provided the inducement mechanism that stimulated the internal effort just described that was complementary to imported technology. In principle, a domestic order of comparable size could have warranted the investment in product development. However, the size of the initial foreign order and its prospective magnitude if there was an initial success, dwarfed any forecast of domestic sales, as Korean per capita income was then still far from levels at which microwaves were purchased. Although a production subsidy could have been offered by the government and achieved a similar result, officials would have had to be implausibly well informed to have chosen both the right product and the appropriate firm. This anecdote suggests that causality may be very difficult to identify in standard econometric models. No data will be available on the date of orders, productivity-augmenting efforts to make the sales profitable, and the eventual producer surplus generated by the sales.

Kim reports another phenomenon that suggests that externalities from exports manifest themselves in ways that standard production function estimates do not capture. The success of Samsung in producing microwaves and its ability to fulfill export orders benefited another Korean firm, Lucky Goldstar. Lucky Goldstar had attempted to reach a technology licensing agreement with Hitachi, a Japanese firm, but Hitachi had declined to provide knowledge to a potential competitor. After Samsung's demonstrated success at technological development and exporting, Hitachi agreed to a licensing agreement with Lucky Goldstar, concluding that it might as well obtain some royalties rather than watch as Lucky Goldstar duplicated Samsung's success or hire workers from Samsung (Kim 1997). While it is difficult to classify this externality, it is clear from this and other case studies that potential technology licensors are more willing to enter agreements once local technical ability has been demonstrated, particularly if it meets the market test of sales in export markets. Lucky Goldstar decreased its costs, and presumably increased its profits, as a result of Samsung's innovation cum exporting. In this case, technology development by firm x confers an external economy on y as it obtains a license, probably at a lower cost.[13] Such interactions among a small number of key firms cannot be investigated in large samples of firms, yet they may play a role. Econometric estimates can thus be fruitfully supplemented by the findings, judiciously interpreted, of case studies.

Examining the development of the electronics industry in several newly industrial countries in Asia, Hobday (1995) also underlines the importance of the contracts local firms obtained as original equipment manufacturers to companies such as Radio Shack. The firms not only received technical aid from their customers but, given the large size of the orders, were more willing to undertake their own R&D. Thus the original finding of Rhee, Ross-Larson, and Pursell (1984) about "externalities" from exporting is supplemented in newer industries by the role that large foreign sales play in encouraging R&D. Such effects are likely to be difficult to sort out in large multiproduct firms, but nevertheless the case studies are quite uniform in suggesting this link between exports and productivity growth, mediated by R&D.

With perfect data sets, exports would be divided into those occurring under original equipment manufacturer (OEM) agreements and those on the spot market, with OEM agreements more likely to generate technology transfers than the latter; absent such information, any results on exporting have to be carefully interpreted. But it is clear from these examples that testing the relation between productivity and exports can be difficult. Studies such as Clerides, Lach, and Tybout (1998), while suggestive of the directions of causation between productivity and exports, have to be viewed as tentative in light of case studies that delineate a very complex relationship between export orders and efforts to enhance productivity. The inability to differentiate among such micro phenomena may account for the widely disparate results found on the productivity-export link.

More generally, differences across countries in management's ability to perceive and exploit opportunities underscores the need for extremely careful formulation of hypotheses. Lucky Goldstar management noted and acted on the potential to go back to Hitachi. Many of these managers had experience in high-technology firms in the United States and were immersed in an entrepreneurial culture. In many countries, such opportunities will not be seized, and researchers could mistakenly conclude there were no externalities. Differences in the entrepreneurial abilities of firms may lead to very different outcomes given the same options.[14]

Tests of such phenomena can be suggested, for example, making the time profile of licensing in firm A in industry i a function of the exports of firm B in industry i. But given the multiple determinants of licensing decisions, this is likely to be too simple. Such case studies highlight the need for caution in reaching conclusions about the presence or absence of benefits from exporting in addition to the standard gains in national income.

Foreign Direct Investment Consider the impact of FDI on firm productivity in the recipient country. Some authors (Rodríguez-Clare 1996; Rivera-Batiz and Rivera-Batiz 1990) suggest that many of the (pecuniary) externalities will manifest themselves in upstream and downstream firms rather than in firms in the same industry. FDI can lead to an increase in productivity in the host economy through diffusion to firms in the same sector as well as backward and forward linkages. The starting point in understanding the role of linkages is the market imperfection that exists in the production of new goods or the use of new technologies in an economy. Given that some of the benefits of FDI will be captured by consumers as consumer surplus (or by downstream producers as producer surplus), there is a positive externality that generally leads to insufficient innovation or entry. Goods with positive net social surplus will not be introduced if the profits from doing so do not compensate for the costs of introducing the good (Romer 1994). The local market may be too small to establish a profitable firm.

If these inputs are nontradable, domestic firms will not have access to them, making them less productive. All firms would be better off if these inputs were produced locally, but given the market imperfection noted above, this does not happen. When inputs are tradable, local firms often contend that imported inputs raise their costs, because of transportation costs, the higher costs of altering the design of the input if the supplier is far away, the need to maintain higher inventories because of the fixed costs of transportation, uncertainty about timely delivery, excess costs imposed by the absence of local agents able to set up machinery or advise firms on the most suitable intermediates, and so forth.[15] Say that a multinational corporation establishes production. Under these conditions, it generates a positive externality enjoyed by the local firms that now have access to this input (see Rodríguez-Clare 1996).[16] This holds for both backward and forward linkages. That is, both users other than the foreign firm and supplying firms now have access to the locally supplied input, increasing their producer surplus. Firms could not charge for all of these impacts, and the impact of the technology transfer will be underestimated using econometric evidence alone.

A number of case studies consider one or more of these issues. Ranis and Schive (1985) document the impact in Taiwan (China) of the establishment of a branch of the Singer Sewing Machine Company. With Singer's help, new firms were established and existing ones improved. While the qualitative evidence indicates that there were benefits from Singer's entry, no estimate of their size is provided.[17]

Meyanathan (1994) provides interesting anecdotal evidence, including a brief discussion of the impact of Intel's plant in Singapore. Lim and Fong (1991) cite a large number of instances of foreign firms having beneficial effects in Malaysia, Singapore, Taiwan (China), and Thailand. None of these studies considers the full set of benefits just noted, however, or estimates the magnitude of the benefits. While the qualitative properties have the right signs, they provide no guidelines to policymakers about the magnitude of benefits and thus offer little guidance about the desirable size of appropriate subsidies.

Labor Mobility Saxenian (1994) and others have documented that firms acquire considerable amounts of knowledge from the workers they hire. Such processes are indeed one of the central tenets of discussions of agglomeration economies (David and Rosenbloom 1990; Quigley 1998; Henderson 1999). Quigley (1998) and Henderson (1999) rely on a variety of indirect productivity tests and do not try to sort out the various sources of agglomeration economies. Some of the case studies, particularly Kim (1997) and Hobday (1995) identify and describe the role of key personnel, largely nationals educated and employed abroad, who have returned and provided critical knowledge. Kim reports that a U.S.–educated Korean who had achieved a high managerial position at Motorola played a critical role in the development of Samsung's semiconductor operations. He also describes the impact of four Korean graduates of major U.S. universities with considerable work experience who were recruited to return to the Republic of Korea. Such critical personnel are difficult to model within a statistical study, as their numbers and total salaries will be tiny within large firms, but their impact may be disproportionately large.[18] Further complicating the picture is the possibility that a key role of FDI stems from the movement of foreign workers to locally owned firms. Schive (1990) provides evidence of the considerable movement of engineers and scientists from foreign-owned firms in Taiwan (China) to locally owned firms.

Summary and Conclusions

Given the many questions that cannot be answered by case studies, more systematic evidence is needed to confirm or disprove some of the conjectures based on them. But caution is needed in drawing strong inferences from econometric studies about the impact of technology diffusion on productivity.

The case studies raise several modeling issues:

1. It would be useful to estimate the rate of growth and fluctuations in output for the sector to which individual firms belong, as they provide an important conditioning variable through their effect on the decisions of firms on the probable benefit of productivity improvements relative to financial asset management.
2. Knowledge provided by OECD importers is likely to be a function of whether the product is old (and proprietary knowledge less important) or new (and most of the relevant knowledge is proprietary). Some indication of each firm's product mix is important.
3. Measures of internal absorption efforts are critical, as the payoff to technology transfers is contingent on them.

4. Given the likelihood of a variable lag structure between efforts to improve productivity and their realization, measures of absorptive effort need to be cumulative rather than for a "typical" year.
5. The numbers of foreign technology suppliers, if available, would be useful to test views that technological independence yields greater TFP growth than technological dependence.
6. The feedback from exporting to productivity growth needs to be investigated with suitably chosen dates rather than simply abiding by the necessarily arbitrary dates of officially collected data. It may be more important early in the industrialization process and for a relatively short period of time.
7. Where possible, it would be helpful to divide exports into those occurring under OEM agreements and those on the spot market; absent such information, any conclusions on the relation between exporting and productivity have to be carefully hedged. The timing of export orders and increased R&D in the product area is also important to consider, though the collection of suitable data is clearly a major problem. It is necessary to have precise measures of absorptive effort and to specify a plausible lag structure, as the promised productivity improvement (or product) may come to fruition several years after the technology transfer, the timing being a function of the absorptive effort after the transfer.
8. The benefits of FDI in sector j may appear in firms in all sectors that purchase the same inputs from sector k, established as a result of new demand by the foreign firm. Establishing a causal link would require information not only on input-output relations but also on whether firms in sectors $a \ldots j, l - z$ were begun as an indirect consequence of FDI in j.
9. If TFP growth is partly due to technology licensing agreements, measured by royalty payments, the availability of licenses may itself be endogenous, a response to the perception of the growing technological ability of developing countries. An examination of the time profile of licensing in firms in industry i as a function of the exports of other firms in the same industry would be of interest in testing the possibility that greater competence of local firms induces foreign firms to be more forthcoming in their licensing behavior.

Notes

1. See Pack (1987) for examples from Kenya, the Philippines, and Tanzania.

2. Protection raises the profitability of domestic sales, generating an income effect that may discourage cost-reducing efforts. However, the greater opportunity cost of not engaging in such effort may spur innovation, a substitution effect. Arguments that import-substituting industrialization discourages the search for productivity growth assume that the income effect outweighs the substitution effect.

3. Saggi (chapter 3 in this volume) contains an excellent review of the extensive literature.

4. Westphal (2002) contains a very thorough review of the issues and evidence.

5. If workers pay for these gains by accepting reduced wages, there is no welfare gain. Workers are more likely to have paid for such acquisition of knowledge if it consists of general rather than firm-specific knowledge.

6. One of the explanations for the convergence between per capita income in the United States and the other OECD countries toward U.S. per capita income in the years 1950–75 is the better management of domestic demand and the greater willingness of firms to invest in equipment and software, often acquired from the United States.

7. As noted earlier, there are income and substitution effects that work in opposite directions.

8. In principle, an increase in the user cost of capital should lead to a search for any cost-saving innovation, but it often seems to lead to a focus on using equipment more intensively (employing second or third shifts, for example).

9. Exports do not prove definitive evidence of such ability, as they may be cross-subsidized by profits in the domestic market earned by other products within a multiproduct firm or encouraged by government subsidies.

10. The failure of industrywide estimates of TFP growth to confirm individual firm studies probably reflects a selection bias. The choice of firms in individual studies may reflect the authors' prior perceptions that these firms were engaged in intensive technological effort—it would not be very interesting to spend six months in a firm that undertook no such activities.

11. Pack and Saggi (2000) analyze the determinants of such diffusion.

12. See the excellent discussion of some of the accounting practices among the Korean *chaebol* in Chang (2003).

13. Such instances of changed willingness of foreign technology suppliers to offer local firms proprietary technologies are not found in case studies in India or Latin America, presumably reflecting the inability of local firms there to offer a credible threat to international licensors. Most of the innovation efforts were directed to redressing various handicaps imposed by the import-substituting industrialization regime itself.

14. Nelson and Pack (1999) emphasize the role of entrepreneurship as part of the explanation of the rapid productivity growth in some Asian countries.

15. For estimates of the costs of some of these, see Pack (1987).

16. Following the lead of Scitovsky and Chenery, Pack and Westphal (1986) argue that such a situation may justify investment coordination.

17. Permission for multinational corporations to operate within the local economy may be conditioned on their achieving a given share of local value-added, motivating foreign firms to build up local suppliers. Such a condition runs the danger of encouraging inefficient local suppliers and ensuring demand by multinational corporations for protection in the domestic market or for export subsidies to offset the excess cost of domestic sourcing of inputs. Multinational corporations may voluntarily attempt to promote local suppliers in order to reduce their own costs.

18. See the *Wall Street Journal* (1990) for one description of the role of returning Taiwanese nationals.

References

Ahluwalia, Isher Judge. 1985. *Industrial Growth in India.* Delhi: Oxford University.

Aw, Bee Yan, Hiaomin Chen, and Mark J. Roberts. 2001. "Firm-Level Evidence on Productivity Differentials and Turnover in Taiwanese Manufacturing." *Journal of Development Economics* 66 (1): 51–86.

Bigsten, A., P. Collier, S. Dercon, M. Fafchamps, B. Gauthier, J.W. Gunning, A. Isaksson, A. Oduro, R. Oostendorp, C. Pattillo, M. Soderbom, F. Teal, and A. Zeufack. 2002. "Do African Manufacturing Firms Learn from Exporting?" Centre for the Study of African Economies Working Paper Series 2002–09 (September): 1–33. Oxford University, Oxford.

Bloom, Martin. 1991. *Technological Change in the Korean Electronics Industry.* Paris: Organisation for Economic Co-operation and Development.

Bonelli, R. 1992. "Growth and Productivity in Brazilian Industries: Impacts of Trade Orientation." *Journal of Development Economics* 39 (1): 85–111.

Chang, Sea-Jin. 2003. *Financial Crisis and Transformation of Korean Business Groups: The Rise and Fall of Chaebols.* Cambridge: Cambridge University Press.

Clerides, Sofronis, Saul Lach, and James R. Tybout. 1998. "Is Learning by Exporting Important? Micro-Dynamic Evidence from Columbia, Mexico, and Morocco." *Quarterly Journal of Economics* 113 (3): 903–47.

Cortes, Mariluz, and Peter Bocock. 1986. *Transfer of Petrochemical Technology to Latin American.* Baltimore: Johns Hopkins University Press.

Dahlman, Carl J., and Larry F. Westphal. 1981. "Technological Effort in Industrial Development: An Interpretative Survey of Recent Research." World Bank, Washington, DC.

David, Paul, and Joshua L. Rosenbloom. 1990. "Marshallian Factor Market Externalities and the Dynamics of Industrial Location." *Journal of Urban Economics* 28: 349–70.

Enos, John, and W.H. Park. 1988. *The Adoption and Diffusion of Imported Technology: The Case of Korea.* London: Croom Helm.

Henderson, Vernon. 1999. "Urbanization in Developing Countries." Department of Economics, Brown University, Providence, RI.

Hobday, Mike. 1995. *Innovation in East Asia: The Challenge to Japan.* London: Edward Elgar.

Hou, Chi-Ming, and San Gee. 1993. "National Systems Supporting Technical Advance in Industry: The Case of Taiwan." In *National Innovation Systems: A Comparative Analysis*, ed. Richard R. Nelson. New York: Oxford University Press.

Katz, Jorge, ed. 1987. *Technology Generation in Latin American Manufacturing Industries.* London: Macmillan.

Kim, Linsu. 1997. *From Imitation to Innovation: Dynamics of Korea's Technological Learning.* Boston: Harvard Business School Press.

Lall, Sanjaya. 1987. *Learning to Industrialize.* London: Macmillan.

Lall, Sanjaya, Giorgio Naveretti, Simon Teitel, and Wignaraja Ganeshan. 1994. *Technology and Enterprise Development: Ghana under Structural Adjustment.* New York: St. Martin's Press.

Lim, Linda Y.C., and Pang Eng Fong. 1991. *Foreign Direct Investment and Industrialization in Malaysia, Singapore, Taiwan, and Thailand.* Paris: Organisation for Economic Co-operation and Development.

Marshall, Alfred. 1890. *Principles of Economics.* London: MacMillan.

Matthews, John A., and Dong-Sung Cho. 2000. *Tiger Technology.* Cambridge: Cambridge University Press.

Meyanathan, Saha Dhevan. 1994. *Industrial Structures and the Development of Small and Medium Enterprise Linkages.* Washington, DC: World Bank.

Nelson, Richard R., and Howard Pack. 1999. "The Asian Growth Miracle and Modern Growth Theory." *Economic Journal* 109 (457): 416–36.

Pack, Howard. 1987. *Productivity, Technology, and Industrial Development.* New York: Oxford University Press.

Pack, Howard, and Christina Paxson. 2001. "Is African Manufacturing Skill Constrained?" In *The Industrial Experience of Africa*, ed. A. Szirmai. London: Macmillan.

Pack, Howard, and Kamal Saggi. 2000. "Vertical Technology Transfer via International Outsourcing." *Journal of Development Economics* 65 (2): 389–415.

Pack, Howard, and Larry E. Westphal. 1986. "Industrial Strategy and Technological Change: Theory vs. Reality." *Journal of Development Economics* 22 (1): 87–128.

Quigley, John M. 1998. "Urban Diversity and Economic Growth." *Journal of Economic Perspectives* 12 (2): 127–38.

Ranis, Gustav. 1973. "Industrial Sector Labor Absorption." *Economic Development and Cultural Change* 21 (3): 387–408.

Ranis, Gustav, and Chi Schive. 1985. "Direct Foreign Investment in Taiwan's Development." In *Foreign Trade and Investment: Economic Development in the Newly Industrializing Asian Countries*, ed. Walter Galenson. Madison: University of Wisconsin Press.

Rhee, Yung W., B. Ross-Larson, and Garry Pursell. 1984. *Korea's Competitive Edge: Managing Entry into World Markets.* Baltimore: Johns Hopkins University Press.

Rivera-Batiz, F., and L. Rivera-Batiz. 1990. "The Effects of Direct Foreign Investment in the Presence of Increasing Returns Due to Specialization." *Journal of Development Economics* 34 (1–2): 287–307.

Rodríguez-Clare, Andrés. 1996. "Multinationals, Linkages, and Economic Development." *American Economic Review* 86 (4): 852–73.

Romer, Paul. 1994. "New Goods, Old Theory, and the Welfare Costs of Trade Restrictions." *Journal of Development Economics* 43 (1): 5–38.

Rosenberg, Nathan. 1976. *Perspectives on Technology.* Cambridge: Cambridge University Press.

Saxenian, A. 1994. *Regional Advantage: Culture and Competition in Silicon Valley and Route 128.* Cambridge, MA: Harvard University Press.

Saxenian, A., and Jinh-Yuh Hsu. 2001. "The Silicon Valley–Hsinchu Connection: Technical Communities and Industrial Upgrading." *Industrial and Corporate Change* 10 (4): 893–920.

Schive, Chi. 1990. *The Foreign Factor: The Multinational Corporations' Contribution to the Economic Modernization of the Republic of China.* Stanford, CA: Hoover Institution Press.

Tybout, James R. 2000. "Manufacturing Firms in Developing Countries: How Well Do They Do, and Why?" *Journal of Economic Literature* 38 (1): 11–44.(March)

Wall Street Journal. 1990. "Taiwan, Long Noted for Cheap Imitations, Becomes an Innovator." June 1.

Westphal, Larry E. 2002. "Technology Strategies for Economic Development in a Fast Changing Global Economy." *Economics of Innovation and New Technology* 11 (4–5): 275–320.

3

FOREIGN DIRECT INVESTMENT, LINKAGES, AND TECHNOLOGY SPILLOVERS

Kamal Saggi

Intrafirm trade (trade between subsidiaries and headquarters of multinational firms) may account for one-third of total world trade today; sales of subsidiaries of multinational firms now exceed worldwide exports of goods and services (UNCTAD 2003). Thus foreign direct investment (FDI) is the dominant channel through which firms serve customers in foreign markets. While much FDI occurs between industrial countries, developing countries are becoming increasingly important host countries for FDI. In fact, about 33 percent of the global stock of FDI today is in developing countries (UNCTAD 2003).

The effects of FDI on technology transfer, backward and forward linkages, local employment, and overall industrial development of host countries have always been an important policy concern in developing countries. Many Asian and Latin American countries whose model for economic development was based on a strategy of import substitution actively intervened with respect to international trade and FDI.

This chapter evaluates the analytical arguments and related empirical evidence regarding the effects of FDI on local industrial development (particularly with respect to linkages and technology transfer) in order to derive implications for policy. It argues that while substantial evidence supports the view that multinationals facilitate international technology transfer, evidence regarding the diffusion of technologies introduced by them to local firms is more mixed. The most recent empirical studies of FDI have shown convincingly that suppliers and buyers that interact with multinationals experience substantial improvements in technology and productivity. These studies have established a strong connection between technology transfer and linkage effects of multinational firms, an insight that has begun to gain currency only in recent years.

Effects of FDI on Linkages and Technology Transfer

A basic tenet of the theory of the multinational firm is that such firms rely heavily on intangible assets, such as superior technology and well-established brand names, to successfully compete with local firms that are better acquainted with the host country environment. Arnold and Javorcik (2005) provide direct evidence of the impact of FDI. Using data from Indonesia's manufacturing sector for 1983–96, they find that foreign ownership leads to significant improvements in acquired plants: after three years total factor productivity (TFP) in the acquired plants outperformed TFP in the control plants by 34 percent. Arnold and Javorcik control for the self-selection problem: while it is true that multinationals typically acquire firms that are relatively more productive, they contribute to the future productivity of acquired firms. Thus multinational firms surely transfer technology to their subsidiary and can therefore potentially play a crucial role in the international diffusion of technology to other firms. The question is, do they?

Casual evidence supports the view that multinationals are intimately involved in international technology transfer. In 1995 more than 80 percent of global royalty payments for international transfers of technology were made from subsidiaries to their parent firms (UNCTAD 1997). And the intrafirm share of technology flows has increased over time. Of course, royalty payments record only the explicit sale of technology and do not capture the full magnitude of technology transfer through FDI caused by technology transfer through imitation, trade in goods, and other channels.

By encouraging inward FDI, developing countries hope not only to import more efficient foreign technologies but also to improve the productivity of local firms through technological spillovers to them. Not surprisingly, a large literature tries to determine whether host countries enjoy spillovers (that is, positive externalities) from FDI.

Before turning to this evidence, it is useful to be clear about the potential channels through which such spillovers may arise. Any discussion of spillovers from FDI needs to tackle the difficult question of whether it is reasonable to expect such spillovers to occur: multinationals have much to gain from preventing the diffusion of their technologies to local firms, and one would expect them to take actions that help preserve their technological superiority. Of course, this argument does not apply when their technologies diffuse vertically to potential suppliers of inputs or buyers of goods and services sold by multinationals (as discussed later).

In general, however, a skeptical a priori position on spillovers from multinationals to their local competitors seems appropriate.

At the micro level, the literature suggests three potential channels of spillovers: demonstration effects, labor turnover, and vertical linkages.

Demonstration Effects

The demonstration effect argument states that exposure to the superior technology of multinational firms may lead local firms to update their own production methods. The implicit assumption behind this argument is that it may be too costly for local firms to acquire the information required to adopt a new technology if it is not first introduced in the local economy by a multinational (and hence demonstrated to succeed in the local environment). As the payoff from the adoption of a foreign technology can be highly uncertain, the successful application of such a technology in the local environment by a multinational can help reduce uncertainty, thereby generating informational spillovers for local firms. Hausmann and Rodrik (2003) argue that such externalities are pervasive in the process of economic development and can provide a rationale for an activist industrial policy.

Suppose FDI does lower the cost of technology adoption and leads to faster adoption of new technologies by local firms. Does that imply that, relative to trade (that is, a scenario in which foreign firms export to the domestic or world market), inward FDI necessarily improves productivity in the local economy? Technology diffusion may strengthen the competitors of foreign firms. Foreseeing the consequences of such diffusion, foreign firms may alter the terms of their original technology transfer. For example, a foreign firm may choose to transfer technologies of lower quality when there is a risk of leakage to local firms. Thus while demonstration effects may exist in principle, strategies undertaken by multinational firms may limit their effect.

Labor Turnover

Researchers have extensively studied direct imitation and reverse engineering as channels of interfirm technology diffusion but neglected the role of labor turnover. Labor turnover differs from the other channels, because knowledge embodied in the labor force moves across firms only through the physical movement of workers. The relative importance of labor turnover is difficult to establish, because it would require tracking individuals who once worked for multinationals and then determining the impact of their job choices on the productivity of new employers. Few empirical studies attempt to measure the magnitude of labor turnover from multinationals to local firms.

The evidence on labor turnover is mixed. Gershenberg's (1987) study of Kenyan industries finds limited evidence of labor turnover from multinationals to local Kenyan firms, but several other studies document substantial labor turnover from multinationals to local firms. Consider the case of the garment industry in Bangladesh (see Rhee 1990 for details). Korea's Daewoo supplied Desh (the first Bangladeshi firm to manufacture and export garments) with technology and credit. Thus Desh was not a multinational firm in the strict sense but a domestic firm that benefited substantially from its connection with Daewoo. Eventually, 115 of the 130 initial workers left Desh to set up their own firms or to join other newly established garment companies. The remarkable speed with which the former Desh workers transmitted their know-how to other factories clearly demonstrates the role labor turnover can play in technology diffusion. Rodrik (2004) argues that imitative entry through labor turnover drove industry growth not only in the case of garments in Bangladesh but also in information technology in India and salmon farming in Chile (where the first successful investment was actually made by the government).

Pack (1997) discusses evidence documenting the role of labor turnover in disseminating the technologies of multinationals to local firms. In the mid-1980s, almost 50 percent of all engineers and about 63 percent of all skilled workers who left multinationals in Taiwan (China) joined local Taiwanese firms. Gershenberg's (1987) study of Kenyan industry reports smaller figures: of the 91 job shifts studied, only 16 percent involved turnover from multinationals to local firms.

In order to synthesize these empirical findings, the cross-country variation in labor turnover rates itself requires explanation. One possible generalization is that in economies such as the Republic of Korea and Taiwan (China), local competitors are less disadvantaged relative to their counterparts in many African countries, thereby making labor turnover possible. Thus the ability of local firms to absorb technologies introduced by multinationals may be a key determinant of whether labor turnover occurs as a means of technology diffusion in equilibrium (see Glass and Saggi 2002 for a formal model). Furthermore, the local investment climate may be such that workers looking to leave multinationals in search of new opportunities (or other local entrepreneurs) find it unprofitable to start their own companies, implying that the only alternative opportunity is to join existing local firms. The presence of weak local competitors probably goes hand in hand with the lack of entrepreneurial efforts, because both may result from the underlying structure of the economic environment.

Labor turnover rates may vary at the industry level as well. Casual observation suggests that industries with a fast pace of technological change (such as the computer industry in Silicon Valley) are characterized by very high turnover rates relative to more mature industries. Therefore, cross-country variation in labor turnover from multinationals could stem simply from the global composition of FDI: developing countries are unlikely to host FDI in sectors subject to rapid technological change.

Using firm-level data from the manufacturing sector in Ghana, Görg and Strobl (2002) provide recent evidence on labor turnover and its effect on the productivity of local firms. They show that firms run by owners who worked for multinationals in the same industry before establishing their own firms have higher productivity growth than other domestic firms. This finding implies that entrepreneurs bring with them some of the knowledge accumulated in the multinational. Görg and Strobl do not find any positive effects on firm-level productivity if the owner had experience in multinationals in other industries or received training by multinationals, suggesting that the knowledge imparted to workers may sometimes be of a very specific type, with limited applicability to other industries.

Linkages and Vertical Technology Transfer

A voluminous informal and empirical literature exists on backward linkages. The 2001 *World Investment Report* was devoted entirely to the effects of FDI on backward linkages in host countries. Analytical models exploring the relationship between multinationals and backward linkages in the host country are hard to come by, however. Two prominent examples of such models are Markusen and Venables (1999) and Rodriguez-Clare (1996). Both provide important insights regarding the two-way relationship between multinationals and linkages. The intermediate goods sector is modeled as monopolistically competitive, so that the effects of foreign investment occur by altering the incentives for entry into such markets.

Barrios, Görg, and Strobl (2004) construct a model in which the competition effect generated by a multinational is eventually dominated by the positive externalities it generates. Using plant-level panel data from the manufacturing sector in Ireland (a country whose economic development has been greatly influenced by multinational firms), they show that such a model indeed describes the Irish experience with FDI.

Alfaro and Rodriguez-Clare (2004) use plant-level data from several Latin American countries to evaluate the linkage effects of multinationals. Their work, motivated by a modified version of the model presented in Rodriguez-Clare (1996), makes the important point that many empirical studies lack a tight link to existing theoretical models and often use inappropriate measures to evaluate the linkage effects of multinationals. More specifically, empirical studies often use the share of inputs purchased locally by a multinational to measure the impact on linkages. Alfaro and Rodriguez-Clare argue that the proper measure (as implied by theory) is the ratio of the value of inputs bought domestically to the total number of workers hired by a multinational. The distinction between the two types of measures is important, because multinationals typically source a smaller percentage of their inputs locally than their local competitors. This does not imply that their linkage effects are necessarily negative, since their production techniques may require more inputs in relation to the workers they hire. Alfaro and Rodriguez-Clare find that the linkage coefficient of multinationals is actually higher than that of local firms in Brazil, Chile, and República Bolivariana de Venezuela, where it is no different (statistically) than in Mexico.

Lin and Saggi (2005a, 2005b) consider how the entry of multinationals might affect the supply side of the intermediate good sector. They raise the following questions: What is the relationship between vertical technology transfer from a multinational to its local suppliers and the equilibrium degree of backward linkages? How does the nature of contractual relationships between multinationals and their local suppliers affect the degree of backward linkages in the local industry? To address these questions, they develop a two-tier model in which the production of a final good requires an intermediate good, and the market structure at both stages of production is oligopolistic. Upon entry the multinational sources the intermediate good locally and engages in vertical technology transfer to its suppliers, guided by a contractual agreement. Two types of contractual relationships are considered, one in which its suppliers must abide by an exclusivity condition that precludes them from serving its local rivals and another in which they face no such restriction. The major point of their paper is that the linkage effects that result from the multinational's entry depend crucially on the nature of the contractual agreement that emerges in equilibrium: under exclusivity, the multinational's entry can even lower linkages (and welfare) relative to autarky.

As Rodriguez-Clare (1996) notes, multinationals improve welfare only if they generate linkages over and beyond those generated by the local firms they displace. The question of relevance here is whether the generation of linkages is expected to result in productivity improvements, technology diffusion, or both. Vertical technology transfer has been documented to occur when firms from industrial countries buy the output of firms in Asian economies in order to sell it under their own name (Hobday 1995). Companies such as Radio Shack and Texas Instruments have commissioned firms in developing countries to produce components or entire products, which are then sold under the retailer's name. Pack and Saggi (2001) summarize evidence that shows that in the late 1970s many Korean firms directly benefited from the technical information foreign buyers provided. The knowledge transfers involved were multifaceted and included not only manufacturing knowledge but also information on sizes, colors, labels, packing materials, and instructions to users. Over the same period, many importing firms from industrial countries maintained very large staffs in the Republic of

Korea and Taiwan (China), who spent considerable time helping their local manufacturers meet required quality specifications.

Based on this evidence, Pack and Saggi (2001) develop a model that explores the interdependence between production of manufactures in developing countries and their marketing into industrial country markets. In their model, a buyer from an industrial country can transfer technology to producers in a developing country in order to outsource production. Since firms in developing countries often lack the ability to successfully market their products internationally, technology leakage in the developing country market actually benefits the industrial country firm, since it increases competition among the developing country suppliers. An interesting implication of their analysis is that fully integrated multinational firms may be more averse to technology diffusion than firms involved in international arm's length arrangements.

More recent evidence regarding vertical technology transfer is provided by Mexico's experience with the *maquiladora* sector and its automobile industry. Mexico started the *maquiladora* sector as part of its Border Industrialization Programme, designed to attract foreign manufacturing facilities along the U.S.–Mexico border. Most *maquiladoras* began as subsidiaries of U.S. firms that shifted labor-intensive assembly operations to Mexico because of its low wages relative to the United States. The industry evolved over time, and the *maquiladoras* now employ sophisticated production techniques, many of which have been imported from the United States.

The development of the automobile industry in Mexico provides an excellent example of the pitfalls and benefits of policies that seek to maximize local benefits from FDI. Historically, Mexico's motivations for intervening in the automobile industry were similar to those of a host of other developing countries that intended to use trade policy to encourage industrial development. During the 1960s and 1970s, Mexico required multinationals in the automobile industry to have a domestic content of 60 percent. Mexico did not allow foreign firms to vertically integrate with their local suppliers. The goal of this nationalistic stance may have been to ensure that domestic firms captured the benefits generated by the backward linkages of FDI.

An interesting aspect of Mexico's export performance requirement in the automobile industry (as noted by Moran 1998) was that, unlike countries such as Malaysia, Mexico did not require that foreign firms export a particular product (such as a finished car) but only that the value of exports bear some relation to the value of imports. Clearly, such a policy let the car companies decide what to export, leaving them free to make their own calculations based on comparative advantage considerations.

The export performance of the industry has improved so much that Mexico has become the world's largest developing country exporter in the automobile sector (Moran 1998). Mexico's experience in the automobile industry is also illustrative of how FDI can contribute to industrial development in the host country. Initial investments by U.S. car manufacturers in Mexico were followed by investments by Japanese and European car manufacturers and by firms that made automobile parts and components. As a result, competition in the automobile industry increased at multiple stages of production, thereby improving efficiency. This pattern of FDI behavior, in which investment by one firm is followed by investment by others, probably reflects strategic considerations. Most multinational firms compete in highly concentrated markets and are highly responsive to one another's decisions. An important implication of this interdependence of competing multinationals is that a host country may be able to unleash a sequence of investments by successfully inducing FDI from one or two major firms.

Extensive backward linkages resulted from FDI in the Mexican automobile industry: within five years of investments by major auto manufacturers, there were 300 domestic producers of parts and accessories, 110 of which had annual sales of more than $1 million. Multinationals in the Mexican automotive sector conducted production audits, held weekly coordination meetings, and provided technical training to their suppliers (Perez-Nunez 1990). Foreign producers also transferred technology to these domestic suppliers: industry best practices, zero-defect procedures, production audits, and others practices were introduced to domestic suppliers, improving their productivity and the quality of their products. As a result of increased competition and efficiency, Mexican exports in the automobile industry boomed.

A comprehensive case study of the effects of Intel's investment in Costa Rica (Larrain, Lopez-Calva, and Rodriguez-Clare 2000) finds evidence that local suppliers benefited substantially from Intel's investment. Similar evidence exists for other sectors and countries, as discussed in detail in Moran (1998, 2001). In the electronics sector, Moran (2001) notes that foreign investors in Malaysia helped their local subcontractors keep pace with modern technologies by assigning technicians to the suppliers' plants to help set up and supervise large-volume automated production and testing procedures. In chapter 10 in this volume Javorcik examines backward linkages and technology spillovers using data from Lithuanian manufacturing sector during the period 1996–2000. She finds that firm productivity is positively affected by a sector's intensity of contacts with multinational customers but not by the presence of multinationals in the same industry. Thus her results support vertical spillovers from FDI but not horizontal ones. Blalock (2001) uses a panel data set from Indonesian manufacturing establishments to check for the same effects. He finds strong evidence of a positive impact of FDI on the productivity growth of local suppliers, indicating that technology transfer from multinationals indeed takes place. He plausibly suggests that since multinationals tend to source inputs that require relatively simple technologies relative to the final products they produce, local firms that produce such intermediates may be in a better position to learn from multinationals than those that compete with them.

The evidence on vertical technology transfer from multinationals to their suppliers is positive and robust (as long as local policies do not restrict multinationals from taking advantage of comparative advantage considerations in making their sourcing decisions). What do we know about the effects of multinationals on their local competitors? One should not expect the picture to be too rosy from the perspective of host countries. A challenge facing the positive view regarding technology spillovers from FDI is to explain how such spillovers can ever be in the interest of multinational firms. Clearly, under most circumstances multinationals would rather limit diffusion in the local economy. In fact, the heart of the theory that seeks to explain the emergence of multinationals is that such firms can successfully compete with local firms precisely because they possess superior technologies, management, and marketing. Why, then, would multinationals not take actions to ensure that such advantages do not diffuse to local competitors? With that in mind, let us now turn to the empirical evidence.

Do Horizontal Spillovers from FDI Exist?

Early efforts in search of horizontal spillovers from FDI proceeded by relating the interindustry variation in productivity to the extent of FDI (Caves 1974; Globerman 1979; Blomström and Persson 1983; Blomström 1986). By and large, these studies find that sectors with a higher level of foreign involvement (as measured by the share of the labor force in the industry employed by foreign firms or the extent of foreign ownership) tend to have higher productivity, higher productivity growth, or both. The fact that these studies involve data from different countries (Australia for the Caves study, Canada for Globerman, Mexico for Blomström) lends a strong degree of robustness to this positive correlation between the level of foreign involvement and local productivity at the sector level.

Of course, correlation is not causation and, as Aitken and Harrison (1999) note, this literature may overstate the positive impact of FDI on local productivity. Investment may have been attracted to the more productive sectors of the economy instead of being the cause of the high productivity in such sectors. In other words, the studies ignore an important self-selection problem. Both trade and FDI help ensure an efficient allocation of global resources by encouraging investment in those sectors in which an economy enjoys comparative advantage. In this sense, Aitken and Harrison's point is almost necessarily implied by traditional trade theory. However, if trade protection encourages investment in sectors in which a host economy does not enjoy comparative advantage, trade protection may be welfare reducing. This possibility was relevant for countries that sought to industrialize by following a strategy of import substitution.

Only plant-level studies can control for the self-selection problem that may plague industry-level studies. Taking the argument a step further, the self-selection problem may also arise in plant-level studies: more productive plants may be the ones that attract foreign investment (as Arnold and Javorcik 2005 find).

What do plant-level studies find with respect to spillovers from FDI? Haddad and Harrison's (1993) study is the first to employ a comprehensive data set at the level of the individual firm over several years. The data come from an annual survey of all manufacturing firms in Morocco. An important result of the study is that foreign firms exhibit higher levels of TFP but their rate of TFP growth is lower than that of domestic firms. As Haddad and Harrison note, at first glance such a finding suggests that perhaps there is some sort of convergence between domestic and foreign firms. In fact, this is not the case. Although there is a level effect of foreign investment on the TFP of domestic firms, such an effect is missing for the growth rate of the TFP of domestic firms. In addition, when sectors are divided into high and low tech, the effect of FDI at the sector level is more positive in low-tech sectors. The authors interpret this result as indicative of the lack of absorptive capacity on the part of local firms in the high-tech sector, where they may lag behind multinationals and be unable to absorb foreign technology.

Aitken, Harrison, and Lipsey (1996) undertake a somewhat different approach to measuring spillovers from FDI. The idea behind their study is that technology spillovers should increase the marginal product of labor, which should show up as higher wages. The study employs data from manufacturing firms in Mexico, the United States, and República Bolivariana de Venezuela. For both Mexico and República Bolivariana de Venezuela, a higher share of foreign employment is associated with higher overall wages for both skilled and unskilled workers, and royalty payments to foreign firms from local firms are highly correlated with wages. The study finds no positive impact of FDI on the wages of workers employed by domestic firms. In fact, the authors report a small negative effect for domestic firms, whereas the overall effect for the entire industry is positive. These findings differ from those for the United States, where a larger share of foreign firms in employment is associated with both a higher average wage as well as higher wages in domestic establishments. Putting Aitken, Harrison, and Lipsey's (1996) findings into the context of previous work, it is clear that wage spillovers (from foreign to domestic firms) are associated with higher productivity in domestic plants. Conversely, the absence of wage spillovers appears to accompany the existence of productivity differentials between domestic and foreign firms.

Using annual census data on more than 4,000 Venezuelan firms, Aitken and Harrison (1999) provide another test of the spillover hypothesis. Since each plant was observed over a period of time, their study avoids the self-selection problem of earlier sector-level studies. The authors find a positive relationship between foreign equity participation and plant performance, implying that foreign participation does indeed benefit domestic plants. However, this own-plant effect is robust only for plants that employ fewer than 50 employees. For larger plants, foreign participation results in no significant improvement in productivity relative to domestic plants. More interestingly, Aitken and Harrison find that productivity in domestic plants declines with an increase in foreign investment, that is, they find evidence of negative spillovers from FDI, which they suggest could result from a market stealing effect. That is, foreign competition may have forced domestic firms to lower output, thereby forgoing economies of scale. Nevertheless, on balance Aitken and Harrison find that the effect of FDI on the productivity of the entire industry is weakly positive. They also note that similar results are obtained for Indonesia, except that the positive effect on own plants is stronger and the negative effect on domestic plants weaker, suggesting a stronger overall positive effect.

Haskel, Pereira, and Slaughter (2002) use plant-level panel data for all U.K. manufacturing from 1973 through 1992 to re-examine the issue of spillovers from FDI. As they note, there can be little doubt that local firms in the United Kingdom possess sufficient absorptive capacity to benefit from the introduction of newer technologies by multinationals, so if spillovers do not materialize, they cannot be attributed to the limitations of domestic firms. Across a wide range of specifications, the authors find that a 10 percent increase in foreign presence in a U.K. industry raises the TFP of that industry's domestic plants by about 0.5 percent. However, they note that the large tax breaks and incentive packages given to multinationals seem out of proportion relative to the magnitude of spillovers they generate.

To recapitulate, while some studies cast doubt on the view that FDI generates positive spillovers for local firms, others reach more positive conclusions. Regardless of one's view regarding these findings, domestic firms should be expected to suffer from the increase in competition that often results from FDI; part of the benefit of inward FDI is that it can help weed out relatively inefficient domestic firms. Resources released in this process will be put to better use by foreign firms with superior technologies, efficient new entrants (both domestic and foreign), or some other sectors of the economy. However, such reallocation of resources cannot take place instantaneously. Studies of spillovers may not cover a long enough period to be able to accurately determine how FDI affects entry and exit. Furthermore, their design limits such horizontal studies, because they cannot clarify linkages and spillovers that may result from FDI in industries other than the one in which FDI occurs.

Moran (2005) argues that there is a substantial difference between the operating characteristics of subsidiaries that are integrated into the international sourcing networks of the parent multinationals and subsidiaries that serve protected domestic markets and are prevented by policy barriers (such as mandatory joint venture and domestic content requirements) from being so integrated. These different characteristics include plant size, the proximity of technology and quality control procedures to industry best practices, the speed with which production processes are brought to the frontier, the efficiency of operations, and the cost of output. Subsidiaries that are well integrated into the global network of their parent firms have a greater positive impact on the host country, often accompanied by vertical backward linkages and externalities. Subsidiaries that cater primarily to protected local markets have a much less positive—and sometimes demonstrably negative—impact on the local economy.

Drawing on a wealth of case studies and econometric evidence, Moran argues that this contrast in performance holds across different industries, countries, and time periods. He notes that the inability of earlier studies to isolate the influences of FDI on host country welfare stems from their failure to differentiate between export-oriented FDI and import-substitution FDI, between foreign investors free to source from wherever they wish and foreign investors operating with domestic content requirements, or between foreign investors obliged to operate as minority shareholders and those with whole or majority ownership accounts.

The entry of multinationals may benefit host countries even if it fails to result in spillovers for local firms, for several reasons. First, spillovers to local firms that directly compete with the multinationals would be the most elusive of benefits that host countries may expect to enjoy from FDI. Second, local agents other than domestic competitors of multinationals (for example, local workers and local suppliers) may enjoy positive externalities from FDI. If they do, the total effect of FDI on local welfare may be positive despite the lack of technology spillovers. Third, spillovers may be of an entirely different nature: local firms may enjoy positive externalities from foreign firms that make it easier for them to export. Such externalities may come about because better infrastructure (transportation, storage facilities, and ports) emerges in regions with a high concentration of foreign exporters.

Aitken, Hanson, and Harrison (1997) provide direct evidence on this issue. In their detailed study of 2,104 manufacturing plants in Mexico, 28 percent of firms had foreign ownership and 46 percent of foreign plants exported. They find that the probability of a Mexican-owned plant exporting is positively correlated with its proximity to foreign-owned exporting plants. Such spillovers may result from informational externalities and may lower the fixed costs of accessing foreign markets rather than the marginal costs of exporting.

Concluding Remarks

There is no simple way of describing the policy environment that multinationals face in developing countries. A roughly accurate statement is that while FDI in services markets faces a multitude of restrictions, FDI into the manufacturing sector is confronted with both restrictions and incentives, often in the same country. In regions that historically emphasized import-substituting industrialization—most of Africa, Latin America, and Southeast Asia—countries either prohibited FDI or forced multinational firms to operate under severe restrictions. Even in economies in which technology acquisition was a major policy objective, multinationals were rarely permitted to operate fully owned subsidiaries; Japan, the Republic of Korea, and Taiwan (China) all imposed restrictions on FDI at various points in time. In other words, "outward-oriented" economies were not particularly keen on allowing multinational firms into their markets.

Despite the prevalence of various types of restrictions on FDI, multinationals do not necessarily face an entirely hostile environment in developing countries. In fact, many countries try to lure large multinational firms with investment incentives. Interestingly, it is not unusual for investment incentives to be offered in conjunction with performance requirements and other restrictions on FDI, perhaps to partially offset the negative impact of the restrictions on the likelihood of investment by multinationals. Is there a case for the use of such incentives? In principle, a case for the use of tax incentives for FDI could be made on the basis of positive spillovers of FDI to local firms, but the argument is difficult to make, because the evidence is not sufficiently conclusive. The complicated restrictions and requirements often imposed by developing countries on multinational firms also seem counterproductive. While multinationals can possess significant market power, restrictive trade polices are likely to increase, not reduce, such market power (once they are in the local market). A relatively open policy regime toward trade and FDI, with adequate safety nets for those likely to be hurt by foreign competition, seems the most reasonable way forward. A restrictive trade regime coupled with an open FDI regime creates incentives for tariff-jumping FDI, the welfare effects of which are far from obvious.

References

Aitken, Brian, and Ann E. Harrison. 1999. "Do Domestic Firms Benefit from Direct Foreign Investment?" *American Economic Review* 89 (3): 605–8.

Aitken, Brian, Gordon H. Hanson, and Ann E. Harrison. 1997. "Spillovers, Foreign Investment, and Export Behavior." *Journal of International Economics* 43 (1–2): 103–32.

Aitken, Brian, Ann E. Harrison, and Robert E. Lipsey. 1996. "Wages and Foreign Ownership: A Comparative Study of Mexico, Venezuela, and the United States." *Journal of International Economics* 40 (3–4): 345–71.

Alfaro, Laura, and Andrés Rodriguez-Clare. 2004. "Multinationals, Linkages, and Economic Development." *Economica*: 113–69.

Arnold, Jens M., and Beata S. Javorcik. 2005. "Foreign Acquisitions and Plant Performance in Indonesia." Policy Research Working Paper 3597, World Bank, Washington, DC.

Barrios, Salvador, Holger Görg, and Eric Strobl. 2004. "FDI, Competition, and Industrial Development in the Host Country." CORE Discussion Paper 2004/11, Louvain-la-Neuve, Belgium.

Blalock, Garrick. 2001. "Technology from Foreign Direct Investment: Strategic Transfer through Supply Chains." Department of Applied Economics and Management, Cornell University, Ithaca, NY.

Blomström, Magnus. 1986. "Foreign Investment and Productive Efficiency: The Case of Mexico." *Journal of Industrial Economics* 35 (1): 97–110.

Blomström, Magnus, and Hakan Persson. 1983. "Foreign Investment and Spillover Efficiency in an Underdeveloped Economy: Evidence from the Mexican Manufacturing Industry." *World Development* 11 (6): 493–501.

Caves, Richard E. 1974. "Multinational Firms, Competition, and Productivity in Host-Country Industries." *Economica* 41: 176–93.

Gershenberg, Irving. 1987. "The Training and Spread of Managerial Know-How: A Comparative Analysis of Multinational and Other Firms in Kenya." *World Development* 15 (7): 931–39.

Glass, Amy J., and Kamal Saggi. 2002. "Multinational Firms and Technology Transfer." *Scandinavian Journal of Economics* 104 (4): 495–513.

Globerman, Steve. 1979. "Foreign Direct Investment and 'Spillover' Efficiency Benefits in Canadian Manufacturing Industries." *Canadian Journal of Economics* 12 (1): 42–56.

Görg, Holger, and Eric Strobl. 2002. "Spillovers from Foreign Firms through Worker Mobility: An Empirical Investigation." IZA Discussion Paper 591, Institute for the Study of Labor, Bonn.

Haddad, Mona, and Ann Harrison. 1993. "Are There Positive Spillovers from Direct Foreign Investment? Evidence from Panel Data for Morocco." *Journal of Development Economics* 42 (1): 51–74.

Haskel, Jonathan E., Sonia Pereira, and Matthew J. Slaughter. 2002. "Does Inward Foreign Direct Investment Boost the Productivity of Domestic Firms?" Working Paper 8724, National Bureau of Economic Research, Cambridge, MA.

Hausmann, Ricardo, and Dani Rodrik. 2003. "Economic Development as Self-Discovery." *Journal of Development Economics* 72 (2): 603–33.

Hobday, Michael. 1995. *Innovation in East Asia: The Challenge to Japan*. Cheltenham: Edward Elgar.

Larrain, B., Felipe, Luis F. Lopez-Calva, and Andrés Rodriguez-Clare. 2000. "Intel: A Case Study of FDI in Central America." Working Paper 58, Center for International Development, Harvard University, Cambridge, MA.

Lin, Ping, and Kamal Saggi. 2005a. "Multinational Firms and Backward Linkages: A Critical Survey and a Simple Model." In *Does Foreign Direct Investment Promote Development?* ed. Theodore Moran, Edward Graham, and Magnus Blomström. 159–74. Washington, DC: Institute for International Economics.

———. 2005b. "Multinational Firms, Exclusivity, and the Degree of Backward Linkages." Department of Economics, Southern Methodist University, Dallas, TX.

Markusen, James R., and Anthony Venables. 1999. "Foreign Direct Investment as a Catalyst for Industrial Development." *European Economic Review* 43 (2): 335–56.

Moran, Theodore. 1998. *Foreign Direct Investment and Development*. Washington, DC: Institute for International Economics.

———. 2001. *Parental Supervision: The New Paradigm for Foreign Direct Investment and Development*. Washington, DC: Institute for International Economics.

———. 2005. "How Does Foreign Direct Investment Affect Host-Country Development: Do We Already Know the Answer? Using Industry Case Studies to Make Reliable Generalizations." In *Does Foreign Direct Investment Promote Development?* ed. Theodore Moran, Edward Graham, and Magnus Blomström. 281–314. Washington, DC: Institute for International Economics.

Pack, Howard. 1997. "The Role of Exports in Asian Development." In *Pathways to Growth: Comparing East Asia and Latin America*, ed. Nancy Birdsall and Frederick Jaspersen. Washington, DC: Inter-American Development Bank.

Pack, Howard, and Kamal Saggi. 2001. "Vertical Technology Transfer via International Outsourcing." *Journal of Development Economics* 65 (2): 389–415.

Perez-Nunez, Wilson. 1990. *Foreign Direct Investment and Industrial Development in Mexico.* Paris: OECD.

Rhee, Yung Whee. 1990. "The Catalyst Model of Development: Lessons from Bangladesh's Success with Garment Exports." *World Development* 18 (2): 333–46.

Rodriguez-Clare, Andrés. 1996. "Multinationals, Linkages, and Economic Development." *American Economic Review* 86 (4): 852–74.

Rodrik, Dani. 2004. "Industrial Policy for the Twenty-First Century." Department of Economics, Harvard University, Cambridge, MA.

UNCTAD (United Nations Commission on Trade and Development) 1997. *World Investment Report: Transnational Corporations, Market Structure, and Competition Policy.* New York: United Nations.

—— 2001. *World Investment Report. Promoting Linkages.* NewYork United Nations.

—— 2003. *World Investment Report: FDI Policies for Development: National and International Perspectives.* New York: United Nations.

4

PLANT- AND FIRM-LEVEL EVIDENCE ON "NEW" TRADE THEORIES

James R. Tybout

Two decades ago, in an effort to become more relevant, trade economists began developing models with imperfectly competitive product markets. The result was a richer body of theory that describes how commercial policy might affect price-cost mark-ups, firm size, productivity, exports, and profitability of domestic producers. The literature also yielded formal representations of the channels through which commercial policy might influence growth. This chapter selectively surveys and interprets the firm- and plant-level evidence that has emerged on the basis of these theories.

The first section focuses on three static predictions of the "new" trade theory that have attracted attention from empiricists. First, protection can change firms' pricing behavior, thereby affecting the allocative efficiency of the economy and the distribution of real income. Second, when trade policies affect prices, they generally also change the set of active producers, their output levels, or both. These adjustments induce productivity changes through scale effects and market share reallocations. Third, changes in the intensity of foreign competition, in firms' opportunities to export, or both can affect their technical efficiency.[1]

The second section continues to discuss firm-level responses to policy reforms in terms of pricing decisions, output levels, exports, and productivity. However, rather

This chapter previously appeared in the 2003 *Handbook of International Economics*, edited by James Harrigan and E. Kwan Choi. The author is grateful to James Harrigan and to participants in workshops at the National Bureau of Economic Research, the University of Toronto, and the University of Texas-Austin for many useful comments.

than focus on comparative statics, the models and evidence in this section are explicitly dynamic. They allow for sunk entry costs, firm heterogeneity, and uncertainty. They thus highlight the relation between responses, expectations, and initial conditions.

The last section of the chapter briefly recaps what is known and what we would like to know. It also indicates some directions for future research.

Static Results: Mark-Ups, Scale, and Productivity

Pricing

Theory Except when collusive equilibria are considered, trade models with imperfect competition treat firms' pricing decisions as determined by static profit maximization. Accordingly, the ratio of output prices (p) to marginal costs (c) is typically a decreasing function of the elasticity of demand (η) that firms face:

$$\frac{p}{c} = \left(\frac{\eta}{\eta - 1}\right). \tag{4.1}$$

It follows that when trade liberalization increases η, mark-ups should fall.

This kind of elasticity effect has been generated by a variety of modeling devices. For example, under the "Armington assumption" that foreign and domestic goods are imperfect substitutes, the demand elasticity for domestic goods rises as the relative price of foreign goods falls (see, for example, Devarajan and Rodrik 1991). When protection takes the form of nontariff barriers, the removal of a quota can create heightened competitive pressures (Bhagwati 1978). When liberalization makes more product varieties available (Krugman 1979), reduces the market share of domestic firms (Helpman and Krugman 1985), or both, producers may perceive their demand elasticities to rise.

When collusive equilibria are modeled, trade liberalization can change the payoff to defecting, change firms' ability to punish defectors, or make it more difficult to detect them (Prusa 1992; Staiger and Wolak 1989).[2] It is possible that cooperative behavior will become unsustainable and mark-ups will fall. Some have argued that collusive firms are likely to use the (exogenous) tariff-distorted price of imports as a reference price.[3] By construction, models that begin from this pricing rule predict that trade liberalization will depress the price of import-competing goods.

Evidence Several simple methodologies have been used to link mark-ups to import competition. Prices and marginal costs are rarely observable, so each technique infers mark-ups indirectly. The most common approach is to use the price-cost margin—sales net of expenditures on labor and materials over sales. If one assumes that unit labor and material costs are flat with respect to output, and we interpret c as short-run marginal costs, this statistic is a monotonic transformation of the mark-up in equation (4.1):

$$PCM_{it} = \frac{p_{it}q_{it} - c_{it}q_{tt}}{p_{it}q_{it}} = \frac{p_{it} - c_{it}}{p_{it}},$$

where q_{it} is the physical output of the ith firm in period t. The PCM is also current economic profits (π_{it}) over sales plus the competitive return on capital over revenues:

$$PCM_{it} = \frac{\pi_{it}}{p_{it}q_{it}} + \frac{(r_t + \delta)k_{it}}{p_{it}q_{it}},$$

where k_{it} is the capital stock, r is the market return on capital, and δ is the depreciation rate.[4] By this logic, after controlling for the ratio of capital stocks to sales, variables that measure the intensity of foreign competition should contribute nothing to the explanation of price-cost margins in industries in which free entry drives profits to zero. If, however, economic profits are present ($\pi_{it} > 0$), these variables should correlate negatively with *PCM* whenever trade liberalization increases demand elasticities or destroys collusive equilibria.

Most analyses of mark-ups based on the PCM begin from a simple regression, such as

$$PCM_{it} = \beta_0 + \beta_1(k_{it}/p_{it}q_{it}) + \beta_2 I_{it} + \cdots + \varepsilon_i \tag{4.2}$$

where i may index either firms or industries, and I_{it} is a proxy for the intensity of import competition (the import penetration rate, the effective protection rate, or a license coverage ratio). (Import competition can be observed only at the industry level, so when firm-level data are used, I_{it} takes the same value for all firms within each industry.) When industry-level cross-sectional data are used, the typical finding is that "the ratio of imports to domestic consumption tends to be negatively correlated with the profitability of domestic sellers, especially when domestic concentration is high" (Schmalensee 1989, p. 976).[5]

A handful of studies has implemented equation (4.2) with plant-level panel data, controlling for permanent cross-industry differences in technology with industry dummies and controlling for efficiency-related variation in mark-ups by including plant-level market shares.[6] Results for Chile (1979–86), Colombia (1977–85), Mexico (1985–90), and Morocco (1984–89) all reveal the same basic pattern: "In *every* country studied, relatively high industrywide exposure to foreign competition is associated with lower [price-cost] margins, and the effect is concentrated in larger plants" (Roberts and Tybout 1996, p. 196, their italics). This pattern seems robust with respect to measures of import competition. In the case of Mexico, where it was possible to explore alternative measures of protection, the pattern appears whether one uses import penetration rates, effective protection rates, or license coverage ratios (Grether 1996).

The standard interpretation for these price-cost margin findings is that large firms or concentrated industries enjoy the most market power, hence their prices are the most responsive to heightened foreign competition. But other explanations are also plausible. For example, it may be that "relatively efficient industries are more profitable, and thus better able to compete against potential imports (low import penetration)" (Roberts and Tybout 1996, p. 195). Or concentration may reflect large sunk entry costs instead of market power (see, for example, Hopenhayn 1992). In this case, rather than squeezing monopoly profits, unanticipated foreign competition cuts into the revenues that firms had expected would cover their entry costs, making them sorry, ex post, that they entered (see, for example, Albuquerque and Rebelo 2000). If this interpretation is correct, one should observe output contractions in industries with high sunk costs and exit in the others when trade is liberalized. There is some evidence that this happens, as I show below.

An alternative methodology for linking foreign competition and pricing begins from the standard Tornqvist growth decomposition. Suppose the ith firm produces output according to $q_{it} = A_{it}h(v_{it})$, where $v_{it} = (v_{it}^1, v_{it}^2 \ldots v_{it}^J)$ is the vector of J factor inputs it uses and A_{it} measures its productivity level at time t. Then, suppressing time subscripts, output growth can be decomposed into a weighted-average of growth rates in the factor inputs and a residual productivity growth term:

$$d\ln(q_i) = \sum_{j=1}^{J} \frac{\partial \ln(h)}{\partial \ln(v_i^j)} d\ln(v_i^j) + d\ln(A_i). \tag{4.3}$$

Hall (1988) notes that when product markets are imperfect, this expression can be combined with equation (4.1) and the cost-minimization conditions $c_i = w_j/(\partial q_i/\partial v_i^j)$, $\forall j$, to link output growth, input growth, productivity growth, and mark-ups:

$$d\ln(q_i) = \left(\frac{\eta}{\eta - 1}\right) \sum_{j=1}^{J} \left(\frac{v_i^j w_j}{p_i q_i}\right) d\ln(v_{ij}) + d\ln(A_i). \tag{4.4}$$

He argues that a regression of output growth on the share-weighted rate of input growth, treating $d\ln(A_i)$ as the mean productivity growth rate plus noise, should reveal the price-cost mark-up as the slope coefficient.

By allowing η to vary through time with trade reforms, one can test whether import competition affects mark-ups. Similarly, one can look for trade-related shifts in the mean rate of productivity growth. Several analysts have performed these exercises by fitting generalized versions of equation (4.4) to plant-level panel data.[7]

Studying Turkey and Côte d'Ivoire, respectively, Levinsohn (1993) and Harrison (1994) conclude that certain protected sectors had significant mark-ups during the sample period and that these mark-ups fell with trade liberalization or exchange rate appreciation. Repeating the exercise using a panel of Indian firms, Krishna and Mitra (1998) report "strong evidence of an increase in competition (as reflected in price-marginal cost mark-ups)" after the 1991 trade liberalization (p. 447). Thus studies based on Hall's approach are consistent with studies based on the price-cost margin. Both methodologies suggest that heightened foreign competition forces down mark-ups by domestic firms.

Hall's approach is subject to several criticisms. First, profit-maximizing firms should adjust their factor demands in response to productivity shocks. Hence consistent estimators of the slope coefficient in equation (4.4) require instruments that are correlated with factor stock growth but not with transitory productivity growth. It is difficult to argue that any available instruments satisfy this criterion, so mark-up estimates are probably biased upward and may exhibit spurious correlation with the trade regime (Abbott, Griliches, and Hausman 1989).

Second, the framework presumes that firms face no adjustment costs. If some or all of the factors are subject to such costs, they will be paid less than their marginal revenue product during upswings (when factor inputs are growing rapidly) and more during downswings (when factor inputs are growing slowly or shrinking). Since measurement error is countercyclical and productivity growth tends to be procyclical, the estimated mark-up may be understated. Furthermore, if import competition depresses demand for domestic goods, it may appear to eliminate monopoly power when it merely creates underutilization of capacity.

Third, inputs and outputs are typically poorly measured, and year-to-year fluctuations in these variables are particularly noisy. For example, due to gestation lags and changes in capacity utilization, growth in capital stocks is quite different from growth in capital services.[8] Perhaps more important, growth rates in physical output are not really observed; what we observe is growth in nominal revenue deflated by a broad price index. If firms that expand rapidly also tend to drive their output prices down relatively rapidly, as one would expect in a differentiated product market, then true output growth is understated when input growth is rapid, and the mark-up estimate should be biased downward.[9] I discuss these measurement problems below.

Firm Size Distribution and Its Effects on Productivity

Theory The output changes that accompany price adjustments depend on whether markets are segmented and whether entry or exit barriers inhibit adjustments in the number of producers. Head and Ries (1999) provide a useful synopsis of some alternative theories. In the absence of collusive behavior, unilateral trade liberalization either reduces firm size (when there are entry or exit barriers or markets are segmented) or leaves it unchanged (when entry and exit are free).[10] Alternatively, when firms collude to slightly undercut the tariff-inclusive price of imports, trade liberalization cum free entry and scale economies force import-competing firms that remain in the market to operate on a larger scale.

As Head and Ries (1999) acknowledge, the invariance of firm size under free entry and no collusion is an artifact of the Dixit-Stiglitz demand system used in the models they consider. More generally, free entry is consistent with firm-size adjustments whenever trade liberalization induces changes in the demand elasticities (η) that domestic firms perceive. In particular, when demand elasticities rise with liberalization, price-cost mark-ups are squeezed according to equation (4.1), which should induce exit until the remaining firms can make up on volume what they lost on margin.

Business and labor groups care about policy-induced output adjustments, because they are generally accompanied by job creation or destruction and by capital gains or losses. But trade economists have focused mainly on the ways the changes in the size distribution affect productivity. To summarize these effects, I adopt Tybout and Westbrook's (1995) decomposition of industrywide productivity growth. As before, let output at the *i*th firm in year *t* be given by $q_{it} = A_{it}h(v_{it})$, but now write $h(v_{it}) = \gamma(g(v_{it}))$, where $g(v_{it})$ is a constant-returns homothetic function of the input vector, v_{it}, and $\gamma(\cdot)$ captures any scale economies. Let $S_{it} = g(v_{it}) / \Sigma_{i=1}^{n_t} g(v_{it})$ be this firm's market share in terms of its input use, and let $B_{it} = q_{it} / g(v_{it})$ be its productivity level. Then the rate of growth in industrywide average productivity, $B_t = \Sigma_{i=1}^{n_t} B_{it} S_{it}$, can be decomposed as

$$\frac{dB_t}{B_t} = \sum_{i=1}^{n_t} \left(\frac{dg_{it}}{g_{it}} \right) (\mu_{it} - 1) \left(\frac{q_{it}}{q_t} \right) + \sum_{i=1}^{n_t} dS_{it} \left(\frac{B_{it}}{B_t} \right) + \sum_{i=1}^{n_t} \left(\frac{dA_{it}}{A_{it}} \right) \left(\frac{q_{it}}{q_t} \right), \quad (4.5)$$

where $\mu_{it} = d\ln(q_{it}) / d\ln(g_{it})$ measures returns to scale at the *i*th plant in year *t*. The first right-hand-side term quantifies efficiency gains due to scale economies at the margin, the second term quantifies gains due to market share reallocations toward relatively efficient producers, and the last term picks up residual intrafirm average efficiency changes that are unrelated to internal scale economies. I hereafter refer to these three quantities as *scale effects, market share effects*, and *technical efficiency effects.*

In most trade models, all firms within an industry are characterized by a common technology and face identical demand conditions, so they expand or contract together in response to liberalization. Productivity gains or losses, when they are present, thus come exclusively from scale effects.[11]

Several models deal explicitly with intraindustry heterogeneity and show how size adjustments (including entry or exit) might affect productivity through the market share effects. For example, Bond (1986) shows how heterogeneous workers might endogenously allocate themselves between entrepreneurial positions and salaried employment. In his "normal" case, protection of the industrial sector increases firm heterogeneity and lowers average productivity by drawing low-quality entrepreneurs into managerial roles.

Melitz (2003) obtains a related set of results in a forward-looking model of steady state trade with firm heterogeneity and imperfect competition. Movement toward freer trade increases a country's imports and erodes each domestic firm's domestic sales and profits.[12] Firms at the lowest end of the productivity distribution contract or exit, while firms at the high end of the productivity distribution expand their exports more than they contract their domestic sales. Accordingly, aggregate productivity improves.

Still another version of the same basic idea can be found in Bernard and others (2000), who use a static model to study the effects of liberalization on the size and productivity mix of producers. They show that when firms use Bertrand pricing rules to compete, trade liberalization expands the market shares of the most efficient firms by providing them with larger export markets, and it forces firms at the low end of the productive efficiency spectrum to shut down as they face competition from abroad.

The Evidence, Part 1: Size Distributions and Trade What do we know empirically about size distributions and trade? Many analysts have fit cross-sectional regressions that relate firm-size measures to the intensity of import competition, controlling for a few other factors, such as domestic market size.[13] Whether the competition proxy is the import penetration rate or a measure of the industry-wide rate of protection, this literature finds that import competition reduces the average plant size, if it has an effect at all. Studies that include export shares in the explanatory variable set find that average plant sizes are relatively large in the export-oriented industries.

One limitation of this literature is that domestic output appears in the denominator of import penetration rates, so there may be a spurious negative correlation between output per firm and this foreign competition proxy. A second problem is that causality may run from size to protection. Concentrated industries that are dominated by a few large producers may have an easier time coordinating lobbying efforts, because they face less of a free-rider problem. A third problem is that most of these studies presume that firms in all industries will adjust to foreign competition in the same way. This runs contrary to theory, which tells us that industries with low entry barriers, such as apparel, are likely to show relatively less size adjustment and more adjustment in the number of active firms.

Several studies handle the first two criticisms by measuring exposure to foreign competition with policy variables such as tariff rates and license coverage ratios and by focusing on intraindustry changes in average firm size rather than cross-industry differences. Comparing industrial census data before and after Chile's trade liberalization, Tybout, de Melo, and Corbo (1991) find that plants in "sectors with relatively large declines in protection have shown a greater tendency toward employment reductions" (p. 236).[14] Tybout and Westbrook (1995) find that during Mexico's unilateral trade liberalization of 1984–89, firms in the sectors that underwent relatively large reductions in license coverage ratios tended to grow relatively slowly, while firms grew quickly in sectors with rapid export growth.[15]

A subset of studies that deals with the first two criticisms also deals with the third by allowing intraindustry changes in firm size to vary with entry costs (proxied by industry-specific plant turnover rates). Perhaps the best is Head and Ries's (1999), which uses the Canada–U.S. Free Trade Agreement as a natural experiment. Their regressions suggest that "Canadian tariff reductions lowered scale [in Canada], while U.S. tariff reductions increased scale" (p. 309). They confirm that entry barriers affect the way firms respond: industries with high turnover (low entry costs) show relative mild reductions in scale in the face of heightened import competition. Roberts and Tybout (1991) obtain similar findings by contrasting industry-specific size distributions in Chile and Colombia and relating them to cross-country, industry-specific differences in effective protection.

The Evidence, Part 2: Trade-Induced Size Adjustments and Scale Efficiency

The finding that foreign competition is associated with smaller firms in import-competing industries seems robust. There is also some evidence that foreign liberalization increases the size of exporting firms. How dramatically have these trade-induced adjustments affected scale efficiency?

Most of the studies that address this question are based on computable general equilibrium (CGE) models. They suggest that the scale-based efficiency gains when trade is liberalized range from 1 to 5 percent of GDP.[16] These findings are suspect, for two reasons. First, while CGE models often predict firm-size expansion in all traded goods industries, the econometric evidence clearly suggests that firms in import-competing sectors contract when import competition intensifies, at least in the short run. Second, even if exporter expansion were the dominant effect of liberalization, it is unlikely that the gains in scale efficiency would amount to much. Although CGE studies often presume returns to scale of 1.10–1.25 at the margin, this is probably a gross overstatement of the extent of unexploited scale economies.

Exporting plants tend to already be the largest in their industry (Bernard and Jensen 1995, 1997; Bernard and Wagner 1997; Aw, Chen, and Roberts 1997; Das, Roberts, and Tybout 2001). They are thus not likely to exhibit much potential to further exploit scale economies. Similarly, since most of the production in any industry comes from large plants, scale efficiency losses due to contraction in import-competing sectors are also typically minor (Tybout and Westbrook 1996).

As an alternative to CGE analysis, Tybout and Westbrook (1995) used panel data on Mexican firms to estimate returns to scale (μ_{it}) as a function of size. They then combine these estimates with the firm-specific growth rates observed during Mexico's unilateral trade liberalization of 1984–89 to implement equation (4.5). Although the cumulative weighted-average growth rate in output is 53 percent for the manufacturing sector, they find that the associated productivity growth rate due to scale efficiency effects is only 0.5 percentage points. This reflects the fact that large plants were operating in the flat portions of their average cost schedule and these plants accounted for the bulk of the output adjustments.

The Evidence, Part 3: Market Shares and Productivity Effects Of course, scale effects are not necessary to link size adjustments and productivity growth. Trade-induced market share reallocations can affect industrywide performance as long as firms are heterogeneous in terms of A_{it}. What do we know empirically about these effects?

A simple way to address this question is to view firms' sizes as reflecting their productivity.[17] If liberalization causes large firms to expand while small firms contract or exit, the associated market share reallocations should improve efficiency. From this perspective, the very robust finding that larger firms are more likely to export suggests that access to foreign markets allows the most efficient firms to become larger, thus pulling up industrywide productivity levels.

Studies that associate changes in trade protection with changes in the intraindustry size distribution deliver mixed evidence. Head and Ries (1999) find that large Canadian firms grew the most dramatically with U.S. tariff reductions and shrank the most dramatically in response to Canadian tariff reductions. Roberts and Tybout (1991) find that shrinkage in response to import competition—proxied by import penetration rates or effective protection rates—was relatively dramatic among large firms in Chile (1979–85) and Colombia (1977–87). But Dutz (1996) finds that as Morocco dismantled nontariff barriers during the 1980s, small plants shrank more than larger plants and their exit probabilities increased relative to others'. Tybout, de Melo, and Corbo (1991) find that in Chile, reductions in effective protection between 1967 and 1979 were associated with balanced percentage reductions in employment across the entire size distribution.

These mixed findings could mean that the selection effects emphasized by Melitz (2003) are not robust, that size is a poor proxy for productivity, or both. To determine whether size is a poor proxy for productivity, several studies measure share effects directly by constructing firm- or plant-specific B_{it} trajectories. Tybout (1991) uses revenue per worker as his productivity measure and measures share-based gains for Chile (1979–85), Colombia (1977–1987) and Morocco (1984–87).[18] He finds that market share reallocations contribute to productivity growth among tradable goods, but his data span periods of major macro shocks rather than major trade liberalization episodes, so it is difficult to argue that the gains are trade induced. Using the same Chilean data set, Pavcnik (2002) measures TFP much more carefully. She finds that the shifting of market shares toward more efficient plants was an important source of efficiency gain during the sample period. However, she does not investigate the link between market share reallocations and foreign competition. Tybout (1991), Liu (1993), Liu and Tybout (1996), and Pavcnik (2002) all find that exiting plants were substantially less productive than surviving plants in Chile (and elsewhere), but none of these studies links this gap to import competition or exporting opportunities.

Tybout and Westbrook (1995) have a better basis for inference in the unilateral Mexican liberalization of 1984–89. Using equation (4.5), as well as a similar decomposition based on cost functions, they find that this liberalization was associated with efficiency gains and that some of these gains were due to market share reallocations. However, they do not find strong evidence that rationalization effects were concentrated in the tradable goods industries. Trefler (2001), studying the Free Trade Agreement between Canada and the United States, finds little evidence that turnover-based productivity gains were concentrated in the industries subjected to the largest tariff reductions.[19]

In sum, market share reallocations (including entry and exit) do matter, but it is difficult to find empirical studies that convincingly link these processes to the trade regime.[20] This is not surprising, given that the effects of import competition on industrial evolution are inherently dynamic and poorly captured by contemporaneous, reduced-form correlations. I return briefly to this issue when I discuss transition dynamics in the last section.

Other Intrafirm Productivity Gains

Leaving aside productivity effects due to adjustments in the firm size distribution, there are many other linkages between commercial policy and efficiency gains. These are bundled together in the third right-hand-side (technical efficiency) term in equation (4.5). Some have to do with changes in the incentives to innovate or eliminate waste. For example, foreign competition or access to foreign markets may change the effort that a firm's managers put forth or the rate at which they improve their products and processes. However, a diverse body of theory suggests that the direction of change in efficiency hinges critically upon model specifics (Corden 1974; Goh 2000; Hart 1983; Miyagawa and Ohno 1995; Rodrik 1992; Scharfstein 1988; Voustden and Campbell 1994).

Other effects on intrafirm productivity are more robust. As Ethier (1982) notes, intrafirm productivity gains may accompany trade liberalization if it expands the menu of intermediate inputs available to domestic firms. This allows each producer to match its input mix more precisely to the desired technology or product characteristics. Similar comments apply concerning access to capital goods, as de Long and Summers (1991) stress.

Trade may also act as a conduit for disembodied technology diffusion if firms learn about products by observing imported varieties or exporting to knowledgeable buyers who provide them with blueprints and give them technical assistance (see, for example, Grossman and Helpman 1991). Similar knowledge transfers may occur when domestic firms enter into joint ventures or sell equity to foreign multinationals, although these activities are less directly related to commercial policy.

Domestic knowledge spillovers further confound the picture. If learning externalities are generated by experience producing a good, then changes in a country's product mix induced by commercial policy can change the rate at which domestic efficiency grows (see, for example, Krugman 1987; Young 1991). Whether trade liberalization helps or hurts in this respect depends on which productive processes generate the most positive externalities and whether they expand or contract as protection is dismantled.

The Evidence, Part 1: Product Variety and Productivity Very little firm-level empirical work has been done on the popular notion that increases in the menu of available inputs improve productivity. This lack of micro evidence reflects the practical difficulties in identifying a firm's desired input mix, observing the actual input mix, and relating discrepancies between the two to measures of firm performance. It may also reflect a presumption that diversification of input bundles makes input use more heterogeneous at the industry level but not at the firm level.

Feenstra, Markusen, and Zeile (1992) provide the only exception I am aware of.[21] They argue that because Korean conglomerates (*chaebols*) are vertically integrated, when new intermediate producers join a conglomerate they effectively diversify the input menu for its final goods producers. Regressions confirm that, over a four-year period, TFP growth among final goods producers in 45 *chaebols* was positively correlated with the fraction of input expenditure going to new intra-*chaebol* intermediate goods suppliers.

This innovative study provides tantalizing evidence that input diversification contributes to productivity gains. However, data limitations prevent the authors from observing the connection between input variety and productivity as directly as one would like. Simultaneity bias is also an issue, since *chaebols* with high productivity growth are probably inclined to expand and incorporate new firms regardless of whether input diversification occurs.

The Evidence, Part 2: Import Discipline Effects It is much more common to relate firm-level productivity measures to proxies for the vigor of import competition. Most micro empirical studies that do so are based on first- or second-order approximations to the production function $q_{it} = A_{it}h(v_{it})$ and express the log of productivity, $\ln(A_{it})$, as a function of import competition proxies, I_{it}, and noise. In the first-order (Cobb-Douglas) case, this amounts to estimating:

$$\ln(q_i) = \sum_{j=1}^{J} \beta_i \ln(v_{ij}) + \gamma I_{it} + \varepsilon_{it} \tag{4.6}$$

Alternatively, the log of productivity can be thought of as a draw from a one-sided productivity distribution (for example, $\alpha_{it} < 0$) plus an orthogonal transitory shock beyond the control of managers: $\ln(A_{it}) = \alpha_{it} + \varepsilon_{it}$:[22]

$$\ln(q_i) = \sum_{j=1}^{J} \beta_i \ln(v_{ij}) + \alpha_{it} + \varepsilon_{it} \tag{4.7}$$

Treating the distribution of α_{it} as dependent upon import competition, one can investigate whether mean productivity levels or productivity dispersion respond to trade liberalization.

Regardless of whether one uses equation (4.6) or (4.7), one cannot measure import competition at the firm level. Thus its effect is identified by cross-industry or temporal variation in I_{it}. The former type of identification is problematic, because cross-industry regressions describe long-run equilibria, and all industry characteristics—including import penetration rates, protection rates, and concentration—are endogenous in the long run (Schmalensee 1989). Nonetheless, Caves and Barton (1990) use equation (4.7) to characterize the α_{it} distribution for each U.S. manufacturing industry and cross-industry variation in I_{it} to infer that "import competition (measured by imports' share of total supply) increases efficiency in industries whose domestic producers are concentrated" (p. 111).

Other studies use temporal variation in I_{it} to link import competition and productivity via equation (4.6) or (4.7). As I note in Tybout (2000), these studies (Harrison 1996; Nishimizu and Page 1982; Pavcnik 2002; Trefler 2001; Tybout, de Melo, and Corbo 1991; Tybout and Westbrook 1995) "tend to find that trade liberalization is associated with rising average efficiency levels" (p. 34). Similarly, liberalization drives down measured productivity dispersion relatively more in import-competing industries (Haddad and Harrison 1993; Pavcnik 2002; Tybout, de Melo, and Corbo 1991).[23] Both sets of findings are consistent with the import discipline hypothesis, but they also could reflect the kind of selection effects described by Bond (1986), Melitz (2003), and Bernard and others (2000).

The implications of these studies are further clouded by methodological problems. With the exception of Pavcnik (2002), they do not deal with the simultaneity bias that results from the dependence of factor inputs on productivity levels. Moreover, all of the studies use industrywide price deflators to convert plant-specific revenues to plant-specific measures of physical output. But since products within each industry are heterogeneous, this procedure attributes relative price fluctuation to physical output fluctuation, and it thus confounds efficiency with monopoly power. Trade-induced reductions in measured "productivity" dispersion may be no more than the reductions in mark-ups among firms with market power that I discussed before.

A general problem with this literature is that it tends to equate measured efficiency gains with welfare improvements. Thus when these gains are associated with trade liberalization, they are touted as a beneficial effect of foreign competition. But the costs of productivity gains are often embodied in overhead, license fees, training, and other items that do not get measured in the input vector. Furthermore, the benefits these expenditures generate are not fully reaped in the same periods in which they are incurred. I know of no study that attempts to measure the present value of firms' productivity-enhancing expenditures and compare them with the present value of the resulting productivity gains.

The Evidence, Part 3: Trade and Technology Diffusion Does trade serve as a conduit for technology diffusion? Many studies have established that exporters tend to be larger, more skill-intensive, and more productive than their domestically oriented counterparts (Aw and Hwang 1995; Aw, Chen, and Roberts 1997; Bernard and Jensen 1995, 1997; Bernard and Wagner 1997; Handoussa, Nishimizu, and Page 1986; Chen and Tang 1987). Furthermore, the case study literature on exporters documents instances in which technologically sophisticated buyers transmit blueprints and proprietary knowledge to the exporting firms.[24] However, there is some doubt as to whether the cross-sectional correlation between performance and exporting mainly reflects causality from performance to exporting. Firms may self-select into export markets, be sought out by foreign buyers, or both *because* they are high quality.

Several authors have attempted to resolve this issue by studying temporal changes in firms' performance and their relation to export market participation.

These studies amount to Granger causality tests based on variants of the autoregressive specification:

$$\ln(A_{it}) = \beta_0 + \sum_{j=1}^{J} \beta_j \ln(A_{it-j}) + \sum_{j=1}^{J} \gamma_j y_{it-j} + \cdots + \varepsilon_{it}, \tag{4.8}$$

where y_{it} is a dummy variable that indicates whether the ith firm exports in period t. Causality tests in this context establish whether exporting experience in the past helps explain productivity in the present, once other determinants of current productivity (including previous productivity) are controlled for. Given that y_{it} responds to productivity shocks, the distributed lag $\Sigma_{j=1}^{J}\gamma_j y_{it-j}$ will be orthogonal to ε_{it} only when ε_{it} is serially uncorrelated, so it is key to use a generous lag length (J) for the term $\Sigma_{j=1}^{J}\beta_j \ln(A_{it-j})$.

Fitting a version of equation (4.8) to plant-level panel data from Colombia, Mexico, and Morocco, Clerides, Lach, and Tybout (1998) find very little evidence that past exporting experience improves performance. Bernard and Jensen (1999) obtain similar results using U.S. data, finding that "exporting does not Granger-cause productivity but does Granger-cause employment, shipments and wages" (p. 14). In contrast, Kraay (chapter 7 in this volume) finds that lagged y_{it} values help explain current productivity among Chinese firms; Bigsten and others (1999) find evidence that exporting Granger-causes productivity among African firms. Aw, Chen, and Roberts (1997) obtain similar findings using census data from Taiwan (China) and the Republic of Korea.[25]

There are at least four problems with this literature. First, the contact between an exporting firm and its foreign client may occur well before export flows are actually observed in the data.[26] Second, as with the import discipline literature, there is a strong tendency to interpret productivity gains as good but no effort to quantify the costs of these gains. Third, the measures of performance are quite crude, as discussed in connection with the import discipline literature. Fourth, almost all of these studies focus on single conduits for technology transfer. But international activities such as exporting, importing intermediates, importing capital goods, and selling equity abroad are often complementary, so firms pursue them in bundles (Kraay, Soloaga, and Tybout in chapter 11 in this volume). Studies that focus on one activity at a time may generate misleading conclusions regarding channels of international technology diffusion.

Kraay, Soloaga, and Tybout (chapter 11 in this volume) tackle the third and fourth methodological problems using the same data sets that Clerides, Lach, and Tybout (1998) used to study learning by exporting. First, they document that international activities indeed come in bundles: exporting, importing intermediate goods, importing capital goods, and sales of equity to multinationals are clearly not independent activities.[27] Next, by using a nested logit representation of demand for the differentiated products and exploiting information on the market share of each product, they are able to separately measure product and process innovations at each firm.[28] Finally, they relate quality trajectories and average cost trajectories to firms' international activities, using generalized versions of equation (4.8). They find that activity histories do not usually help predict future product quality or reduce average production costs, once the histories of these performance variables are controlled for. Nonetheless, Colombian firms that engage in at least some international activities—especially those that import their intermediate goods—(tend to have higher product quality.[29] This finding suggests the kind of static efficiency effect that Ethier (1982) envisioned.

Summary

Measurement and methodological problems plague the literature I have reviewed here, but some findings seem robust. First, the evidence suggests that mark-ups fall with import competition. The most likely interpretation is that foreign competition increases the elasticity of demand that domestic firms face. It is not clear whether these trade-induced reductions in mark-ups reflect the elimination of market power or the creation of negative economic profits.

Second, contrary to the predictions of many simulation models, import-competing firms cut back their production levels when foreign competition intensifies. This is not consistent with the Helpman and Krugman (1985) monopolistic competition model, under which some domestic plants exit and the remaining plants either remain the same size (if their demand elasticities do not change) or expand (if their demand elasticities rise). Instead, it suggests that sunk entry or exit costs are important in most sectors.

Third, trade does seem to rationalize production in the sense that markets for the most efficient plants are expanded. Furthermore, if the methodological problems with measuring productivity are discounted, most studies suggest that exposure to foreign competition improves intraplant efficiency (an unknown cost).

Fourth, firms that engage in international activities do tend to be larger and more productive. But it is not obvious that the activities caused these characteristics and not vice versa.

Transition Dynamics

The theories I have mentioned describe static or steady-state equilibria; the regressions that give them empirical content deal mostly with patterns of contemporaneous correlation. But some important issues are inherently dynamic. For example, when a developing country dismantles its trade barriers and devalues its currency, as the World Bank often recommends, the effect of the new regime on the central bank's foreign currency reserves depends on the resulting changes in the export trajectory. Political support for a given reform package depends on the associated changes in firms' market values and employment trajectories that business representatives and workers anticipate. All of the literature I have reviewed thus far is silent on these high-profile issues.

The dynamic effects of policy reforms are difficult to characterize, because they reflect complex decisions on the part of firms. Faced with an uncertain future, some managers find themselves weighing the earnings effects of shutting down plants, firing workers, or both against the associated severance costs and the option value of retaining plants or workers for possibly better days. Others weigh the sunk costs of breaking into foreign markets, building new plants, or hiring workers against the net revenue streams these activities may generate. Their decisions are further complicated by the need to anticipate decisions by managers producing competing products. Below I discuss a nascent literature that tackles the relation between commercial policy reforms and industrial responses in settings with these features.

Export Dynamics

Theory In the past 15 years, several theoretical contributions to the trade literature have incorporated sunk costs and uncertainty in dynamic models. Among the first researchers to do so were Dixit (1989), Baldwin (1988), and Baldwin and Krugman (1989), who investigated the role of sunk costs and expectations in driving exporters' behavior. Generalizing their specification in anticipation of the discussion to follow, I specify an export profit function for the ith firm that depends on the exchange rate (e_t), marginal production costs (c_{it}), a foreign demand shifter (x_{it}), and serially uncorrelated noise (ε_{it}): $\pi^f(e_t, c_{it}, x_t) + \varepsilon_{it}$.[30] I assume that firms without prior exporting experience must establish distribution channels, repackage their products, and learn bureaucratic procedures and call the sum of these entry costs for new exporters Γ_S. Then, defining the indicator variable y_{it} to take a value of 1 in periods when the ith firm exports and 0 otherwise, the pay-off from being an exporter in year t may be written as:

$$u(e_t, c_{it}, x_{it}, \varepsilon_{it}, y_{it}, y_{it-1}) = \begin{cases} \pi^f(e_t, c_{it}, x_{it}) + \varepsilon_{it} & \text{if } y_{it} = 1; y_{it-1} = 1 \\ \pi^f(e_t, c_{it}, x_{it}) - \Gamma_S + \varepsilon_{it} & \text{if } y_{it} = 1; y_{it-1} = 0 \\ 0 & \text{if } y_{it} = 0; y_{it-1} = 0 \end{cases}$$

Presuming that the vector (e_t, c_{it}, x_t) follows a first-order Markov process, risk-neutral managers do best to choose a sequence of decision rules, $y_{it} = g_t(e_t, c_{it}, x_{it}, \varepsilon_{it}, y_{it-1})$, that maximizes their expected profit stream from export market participation: Equivalently, their patterns of export market participation should satisfy the following Bellman equation:

$$V(e_t, x_{it}, c_{it}, \varepsilon_{it}, y_{it-1}) = \max_{y_{it}} \{\pi^f(e_t, c_{it}, x_{it}) - (1 - y_{it-1})\Gamma_S$$
$$+ \varepsilon_{it} + \delta E_t V(e_{t+1}, c_{it+1}, x_{it+1}, \varepsilon_{it+1}, y_{it})\} \quad (4.9)$$

Expectations are taken conditioned on (e_t, c_{it}, x_t) and the Markov process that govern this vector's evolution.

This framework implies that seemingly identical policies and macro conditions can lead to different levels of exports, depending on how many firms have a history of export market participation, for several reasons. First, when firms have no exporting experience, they weigh the sunk costs of entry against the expected profit stream. But when most firms are already exporters, the aggregate response to export incentives reflects volume adjustments and has little to do with entry costs. Second, firms that begin exporting in response to a shock—say, a large devaluation—may not cease exporting when that shock is reversed. Third, expectations about future exchange rate trajectories and commercial policies may play a critical role in determining whether firms invest in becoming exporters today. Fourth, export responsiveness to any shock or regime switch depends critically on the amount of cross-firm heterogeneity in marginal costs and foreign demand, x_{it}. Many firms or just a scattered few may be on the verge of exporting.

Evidence Several studies have explored the empirical relevance of the sunk-cost export model sketched above. Roberts and Tybout (1997) begin from the implication of equation (4.9) that firms will find it optimal to export whenever

$$\pi^f(e_t, c_{it}, x_{it}) - (1 - y_{it-1})\Gamma_S + \varepsilon_{it}$$
$$+ \delta[E_t V(e_{t+1}, c_{it}, x_{it+1}, \varepsilon_{it+1} \mid y_{it} = 1) - E_t V(e_{t+1}, c_{it}, x_{it+1}, \varepsilon_{it+1} \mid y_{it} = 0)] > 0$$

The second bracketed term describes the option value of being an exporter in period t, that is, the expected current value of being able to export in period t+1 without having to pay sunk entry costs. Accordingly, its magnitude depends on expectations about the future operating profits one might generate by exporting. Combining terms that depend on current values of the state variables, the ith firm will do best to export whenever

$$f(e_t, c_{it}, x_{it}) + y_{it-1}\Gamma_S + \varepsilon_{it} > 0, \text{ where}$$
$$f(e_t, c_{it}, x_{it}) = \pi^f(e_t, c_{it}, x_{it}) + \delta[E_t V(e_{t+1}, c_{it+1}, x_{it+1}, \varepsilon_{it+1} \mid y_{it} = 1)$$
$$- E_t V(e_{t+1}, c_{it+1}, x_{it+1}, \varepsilon_{it+1} \mid y_{it} = 0)].$$

Using a reduced form approximation to $f(\theta)$ and assuming a particular distribution for the error term, ε_{it}, this equation implies a dynamic discrete choice model of export market participation. Roberts and Tybout (1997), Bernard and Jensen (1999), Campas (1999), and Sullivan (1997) fit this model as a dynamic probit or logit and test whether sunk entry costs affect export market participation. This amounts to testing whether lagged exporting status affects current status, once the other sources of persistence in behavior $(x_{it}, c_{it}, e_t, \varepsilon_{it})$ have been controlled for. If their sources of persistence are not completely controlled for, this approach to inference misattributes serial correlation in exporting status to sunk costs. It is therefore important to treat ε_{it} as a serially correlated disturbance when estimating the equation.

The universal finding of these studies is that sunk costs are important. Even after serial correlation in ε_{it} is treated, the probability that a firm will export, given $(x_{it}, c_{it}, e_t, \varepsilon_{it})$, can be up to 0.70 higher if it exported the previous period. From this, researchers have typically concluded that export aggregates are subject to important hysteresis effects and that sunk costs matter.

Das, Roberts, and Tybout (2001) revisit the question of how sunk costs shape export responsiveness among Colombian chemical producers. Instead of using a reduced-form version of the decision rule, they fit a structural model that explicitly describes the profit function and the autoregressive processes that govern the vector $(x_{it}, c_{it}, e_t, \varepsilon_{it})$. Using their estimates, they then examine the option value of export market participation for each firm: $\delta[E_t V(e_{t+1}, x_{it+1}, \varepsilon_{it+1} \mid y_{it} = 1) - E_t V(e_{t+1}, x_{it+1}, \varepsilon_{it+1} \mid y_{it} = 0)]$. This expression measures the importance of expectations about the future in shaping exporting decisions. They find that it is quantitatively important for small-scale exporters, whose foreign demand is relatively limited. However, the firms that supply the bulk of total exports earn operating profits that far exceed the option value term. Hence hysteresis effects are only important to fringe players in the export markets; aggregate exports are relatively insensitive to history or expectations. Put differently, if one is interested only in the aggregates, sunk entry costs and the subtleties they introduce can be ignored for many industries.[31]

One robust finding concerning exporters is that they tend to sell very small fractions of their output abroad (Aw, Chen, and Roberts 1997; Roberts, Sullivan, and Tybout 1995; Campas 1999). In principle, this could mean that foreign demand for each firm's product is very limited and inelastic, but this is not the way most people view foreign markets. A second explanation is that firms export just enough to exploit duty drawback schemes and purchase the imported intermediates or capital goods at duty-free prices. To my knowledge, this hypothesis has not been pursued, although it would be easy to do so. A third hypothesis is that firms export partly to diversify their earnings stream, exploiting the imperfect correlation between foreign and domestic shocks. Small stable shares in foreign markets may be rational under these assumptions.

Maloney and Azevado (1995) develop a simple model of this diversification motive for exports and fit it to firm-level panel data from Mexico. They find, among other things, that exchange rate volatility and the covariance between domestic and international demand shocks are significant determinants of export volumes. Hence, for example, when an overvalued exchange rate is allowed to float, the export response may be counterintuitive.

In sum, the initiation of exports appears to invoke some sunk start-up costs. These costs matter a good deal for marginal exporters but are unimportant relative to the operating profits that large exporters earn. Thus their effect on aggregate export responses to regime shifts or exchange rate shocks may not be large. Other determinants of export responsiveness that may be relevant include risk diversification considerations and domestic market demand shocks (when marginal costs are not flat). There is some evidence that risk diversification matters; domestic market demand shocks remain largely unexplored.

Industrial Evolution

Theory Theoretical models of industrial evolution demonstrate how the combination of sunk entry costs with imperfect foresight and cross-firm heterogeneity can lead to continual flux in the population of active firms (Jovanovic 1982; Hopenhayn 1992; Ericson and Pakes 1995). They also describe the implications of this flux in terms of job turnover patterns and productivity growth. Very little theoretical work has been done on the effects of commercial policy in an economy with these features.

Two exceptions merit note. The first is Melitz (2003), who focuses on the relation between openness and the steady state distribution of firm types. The other is Albuquerque and Rebelo (2000), who abstract from intraindustry heterogeneity to derive some analytical results about dynamic responses to trade liberalization. Only Albuquerque and Rebelo (2000) deals with transition issues, so I focus on it here.

Albuquerque and Rebelo consider an open economy with homogeneous firms in each of two sectors. New firms must pay a sunk fee to initiate production, so incumbents may earn positive profits in the steady state without inducing entry, and multiple equilibria are possible. Responses to policy shocks depend on the pre-reform equilibrium. When profits net of entry costs are zero in the exportable goods sector and profits before entry costs are zero in the import-competing sector, small reductions in the rate of protection should generate entry in the exportable goods sector and exit in the import-competing sector. Unanticipated reforms also induce intersectoral reallocations of variable factors in the period before entry and exit occur. Pre-announcing eliminates this short-run adjustment period. If, however, the economy begins from an interior steady state and reforms are too mild to trigger entry or exit, the effects of policy reforms are limited to variable factor movements and capital gains or losses for the owners of incumbent firms.

The dichotomy between responses beginning from zero-profit versus interior profits is an artifact of the assumption that firms within each sector are homogeneous. Intraindustry heterogeneity will generally mean that operating profits are close to zero for the marginal incumbent and profits net of entry costs will be close to zero for the marginal entrant. Nonetheless, the results mentioned above suggest how responses to reforms should depend on the density of incumbents and potential entrants near the zero-profit margin.

The Evidence, Part 1: Descriptive Studies It is well established that, even within narrowly defined industries, plants are heterogeneous in terms of their size and measured productivity (see the references cited in the first section of this chapter). Simultaneous plant entry and exit are the norm, as are market share reallocations and job creation and destruction among incumbent firms (see, for example, Dunne, Roberts, and Samuelson 1989; Davis, Haltiwanger, and Schuh 1996; Baldwin, Dunne, and Haltiwanger 1998; Roberts and Tybout 1996). These stylized facts inspired the modern theory of industrial evolution and are commonly cited as evidence of its relevance.

Much less is known about the effects of commercial policy shocks on industrial evolution patterns or how these effects depend on the initial population of firms. A number of studies document patterns of contemporaneous correlation between openness, firm size distributions, and entry/exit or market share–based efficiency gains (see the first section of this chapter). There is also a small amount of evidence on openness to patterns of job turnover (Levinsohn 1999).[32] However, these studies tell us little about the dynamic responses to reforms when threshold costs and uncertainty make firms' adjustments forward looking, gradual, or dependent on initial conditions.

The Evidence, Part 2: A Structural Model Lu and Tybout (2000) attempt to go beyond patterns of contemporaneous correlation and quantify these dynamic relationships. Drawing heavily on Ericson and Pakes (1995) and Pakes and McGuire (1994), they develop an empirical model with sunk costs, heterogeneity, and uncertainty. Their model portrays an import-competing industry populated by a finite number of potential entrepreneur/owners, including those already in the industry (incumbents) and those contemplating entry (potential entrants). Each incumbent is characterized by a unique product and a time-varying productivity index that summarizes both the product's appeal and the unit production costs.

Imports are represented by a single foreign variety whose price responds to exchange rate shocks and commercial policy reforms but not to domestic producers' behavior.

Entrepreneurs in this industry play a Markov-perfect dynamic game against one another. Each period, each entrepreneur attempts to maximize his or her discounted net profit stream, given the available information set. Potential entrants choose whether to enter the market, given their privately observed entry costs. Incumbents decide whether to remain in the market or exit, given the privately observed scrap value of their firms. The incumbents who remain active engage in Bertrand-Nash product market competition with one another, given the current price of the import-competing good and a simple logit demand system.

At the beginning of each period, all entrepreneurs learn the productivity level of each incumbent firm (industry structure), as well as the current realizations on the number of consumers and the real effective exchange rate (market conditions). If an incumbent firm remains in the industry, its productivity evolves from period to period according to a common knowledge exogenous Markov process, as do the exchange rate and the number of consumers. Firms solve for their optimal strategies and make their exit or entry decisions simultaneously. From period to period, the industry structure evolves with entry, exit, and random shocks to each firm's productivity.

Using Colombian panel data on the pulp and paper industry, Lu and Tybout (2000) estimate the demand parameters of their model. Combined with observed market shares these allow them to impute product quality trajectories for each producer and to estimate the associated Markov processes. Given these primitives, they calibrate the entry and exit cost distributions, so that simulated plant turnover rates approximate the industry's actual figures.

Lu and Tybout's main computational experiment is to simulate responses to a change in the exchange rate process that gradually intensifies import competition. The impact effect of this regime switch is to squeeze price-cost mark-ups, just as the econometric evidence suggests. However, the new exchange rate regime also discourages entry (but not exit), so over time the number of domestic producers gradually shrinks. With the menu of varieties falling, elasticities of demand for each variety fall, allowing the remaining incumbents to restore their mark-ups and cover their operating costs. This transition path suggests that the robust margin squeeze effects and output contraction effects identified by contemporaneous correlation patterns may not be permanent.

Although consumers initially benefit from cheaper imported goods and cheaper domestic goods, they are ultimately left with fewer domestic varieties at prices close to pre-appreciation levels. Hence in the scenario that Lu and Tybout analyze, the present value of consumer welfare actually falls with heightened import competition.[33] Producers suffer capital losses, of course, so they are worse off, too.

Extra costs are also imposed on workers, who endure higher job destruction rates during the transition period. Indeed, the job turnover effects predicted by this model are implausibly high, suggesting that it should be generalized to including severance and screening costs, as in Hopenhayn and Rogerson's (1993) simulations. By the same token, the apparent importance of hiring and firing costs means that firms' expectations are critical and suggests that static calculations of the employment effects of trade policies can be very inaccurate.

This framework provides a conceptually rigorous way to address the question of how changes in the intensity of import competition affect the market share–based efficiency changes described by the second term in equation (4.5). Lu and Tybout (2000) find that this type of efficiency gain is small, for two reasons. First, most of the adjustment in varieties comes from less entry rather than more exit. Incumbent firms that are relatively inefficient do not increase the rate at which they jump out of the market, because their entry costs are already sunk, their scrap values are small, and they perceive a possibility that conditions will improve in the future. Second, the firms that do enter or exit account for a relatively small fraction of total production. This is consistent with what is observed in the data.

These simulations are subject to several criticisms. Most fundamentally, they are partial equilibrium and thus do not document the capital gains and growing number of product varieties in sectors that benefit from exchange rate appreciation. Second, they do not permit the number of imported varieties to adjust. If foreign firms face sunk entry costs when breaking into the domestic market, some new firms will probably be induced to enter by the change in the exchange rate regime. Third, the model is highly stylized in many respects, including the demand system, the productivity growth process (which is presumed exogenous), and the distributions for entry costs and scrap values. Nonetheless, at a minimum the model demonstrates that conclusions based on contemporaneous patterns of correlation can be very misleading, and it brings together in a unified framework the phenomena that firms, workers, and consumers care about.

An Agenda for Future Research

I close with a few observations on directions for future trade research using firm- or plant-level data. First, as the previous section suggests, I am enthusiastic about the new insights that may be gained from dynamic structural models that link trade regimes and industrial evolution. These models suffer from some serious limitations, but they integrate many pieces of the response story that were heretofore treated in disjoint literatures. They also provide a basis for counterfactual simulations in the presence of threshold costs, uncertainty, and heterogeneous firms. As computers become more powerful and solution algorithms improve,

I am hopeful that econometrically estimated industrial evolution models can be made more realistic and used for applied policy work.

Second, despite the large volume of research on the link between trade and productivity, there are several ways in which this literature might be improved. One would be to stop pretending that firms in manufacturing industries produce homogeneous products and instead deal with pricing, output, and productivity measurement in unified frameworks (see, for example, Melitz 2000). Another would be to tighten the link between theory and tests. Theory has emphasized the effects of enhanced input variety (including both capital and intermediate goods) and, more recently, efficiency gains due to geographic agglomeration. But there is very little direct micro evidence on the importance of either. These are difficult topics to tackle, but creative empiricists should be able to make progress on both fronts.

Third, although the relationship between trade and wages has attracted considerable attention, there is only limited evidence on the micro details of worker displacement, job search processes, and reemployment patterns triggered by changes in the trade regime. The census bureaus of several countries (including the United States) have recently devoted some resources to matching household survey data with establishment survey data, so the characteristics of plants and workers can now be analyzed together and workers can be tracked as they change jobs. These matched data sets should provide a much better basis for inference on the employment effects of commercial policy reforms or changes in the exchange rate regime.

Notes

1. I ignore the empirical literature on multinationals and foreign direct investment, which is treated in chapter 3.

2. In these models protection takes the form of institutional arrangements for antidumping measures.

3. See Head and Ries (1999) for discussion and references.

4. This measure presumes that intermediate input and labor use are proportional to output and that the proportions are fixed across plants. See Schmalensee (1989) for further discussion of the limitations of price-cost margin as a performance measure.

5. See also Lee (1991) and Roberts and Tybout (1996,) for surveys of the literature on developing countries.

6. As Roberts and Tybout (1996, p. 196) note, "Efficient plants should be larger and have higher profits, so a positive correlation is generally expected between market shares and price-cost margins, regardless of whether firms have market power . . . ."

7. Generalizations have included allowing for nonconstant returns to scale and letting η and the mean productivity growth rate vary across firms. See, for example, Harrison (1994).

8. Pakes and Griliches (1984) estimate that it may take several years for newly installed capital to reach full productivity.

9. Klette and Griliches (1996) and Melitz (2003) discuss the consequences of this measurement problem for estimates of production function parameters.

10. The most common form of entry and exit barrier is sunk start-up costs. Firms continue to operate as long as their expected earnings stream covers their expected future expenditures, even if ex post they discover they cannot also recoup the sunk costs they paid to enter (see, for example, Albuquerque

and Rebelo 2000). Uncertainty about future market conditions is likely to increase the option value of remaining in operation, effectively compounding persistence in status. Firms that enjoyed excess profits before import competition intensified will also fail to exit.

11. I do not treat external returns to scale, because these are nearly impossible to measure.

12. The cases he analyzes are autarky versus free trade, more versus fewer countries in a customs union, and high versus low nontariff barriers (at home and abroad).

13. These studies span a wide range of countries. See Scherer and others (1975), Muller and Owen (1985), Baldwin and Gorecki (1986), Caves (1984), and Schwalbach (1988). Tybout (1993) provides further details on these studies.

14. This pattern is less apparent when size is measured with output or value-added, suggesting that efficiency gains occurred in the import-competing industries.

15. They find no significant cross-industry correlations between firm size and effective protection rates or import penetration rates, however.

16. See, for example, Harris (1984), Smith and Venables (1988), Norman (1990), and Brown, Deardorff, and Stern (1991). Tybout and Westbrook (1996) provide a more detailed discussion.

17. The size-productivity linkage is common in models with heterogeneous firms. See, for example, Hopenhayn (1992), Melitz (2000), and Bernard and others (2000).

18. His decomposition does not distinguish intraplant productivity gains due to scale efficiency from other sources of intraplant gains. Bernard and others (2000) show that revenue per unit output is a monotonic function of true total factor productivity if firms compete.

19. Trefler's (2001) intraindustry data are grouped by plant size, so he cannot rule out the possibility that the Free Trade Agreement generated productivity gains through reallocations within size classes or through entry and exit.

20. Bernard and Jensen (2000) link entry and exit patterns to trade indirectly by arguing that, with output prices pinned down by international arbitrage, Rybczynski effects should induce net entry in the sectors that intensively use the factors that are growing relatively rapidly. They confirm this conjecture using data from the United States, first with cross-industry regressions at the national level, then with similar regressions at the regional level. They find that where human capital and physical capital have grown relative to unskilled labor, exit rates have been low among skill-intensive goods and high among low-skill goods.

21. Feenstra and others (1999) use detailed data on trade flows, albeit not at the firm level, to link sectoral productivity to the diversity of final good and upstream exports.

22. Detailed discussions of this approach to productivity analysis may be found in the stochastic frontier literature (see, for example, Greene 1993).

23. There is also evidence that innovative activities are stimulated by import competition (see Blundell, Griffith, and van Reenen 1999).

24. Much of this literature focuses on East Asia. Pack (chapter 2 in this volume) and Westphal (2001) provide recent surveys.

25. Both Kraay (chapter 7 in this volume) and Bigsten and others (1999) are based on annual data with short lag lengths, J, and do not provide tests for serial correlation. Hence they may be picking up a spurious correlation. Aw, Chen, and Roberts (1997) compare censuses at five-year intervals, so their study is also likely to suffer from this problem.

26. For interesting discussions of the case study literature on pre-exporting contacts with buyers, see Pack (chapter 2 in this volume) and Westphal (2001).

27. Using multinomial probit models, Aw, Roberts, and Winston (2001) document similar dynamic complementarities among worker training, R&D, and exporting.

28. One unappealing feature of their approach is that it assumes that the ratio of physical output to intermediate input use is constant across all producers in a given four-digit industry and geographic region.

29. Given the way Kraay, Soloaga, and Tybout impute quality, this is almost a corollary to the finding that firms engaging in international activities have large domestic market shares.

30. Domestic product market conditions are kept out of the analysis by assuming flat marginal cost schedules with respect to output.

31. Using a reduced-form econometric model and descriptive statistics, Campas (1999) draws similar conclusions from Spanish data.

32. Levinsohn (1999) finds that job turnover patterns in Chile during the 1980s were not closely linked to commercial policy or exchange rate shocks. He argues that turnover rates were higher among tradable goods than among nontradables. Thus liberalization in economies like Chile's should reduce job security and may meet resistance, for the political economy reasons detailed by Fernandez and Rodrik (1991).

33. This result is partly an artifact of the demand system they use, which probably overstates the value consumers place on goods with small market shares.

References

Abbott, Thomas, Zvi Griliches, and Jerry Hausman. 1989. "Short-Run Movements in Productivity: Market Power versus Capacity Utilization." Department of Economics, Harvard University, Cambridge, MA.

Albuquerque, Rui, and Sergio Rebelo. 2000. "On the Dynamics of Trade Reform." *Journal of International Economics* 51: 21–48.

Aw, Bee-Yan, and Amy Hwang. 1995. "Productivity and the Export Market: A Firm-Level Analysis." *Journal of Development Economics* 47: 313–32.

Aw, Bee-Yan, Xiaomin Chen, and Mark Roberts. 1997. "Firm-Level Evidence on Productivity Differentials and Turnover in Taiwanese Manufacturing." Working Paper 6235, National Bureau of Economic Research, Cambridge, MA.

Aw, Bee-Yan, Mark Roberts, and Tor Winston. 2001. "Investment in Knowledge and the Evolution of Firm Productivity in the Taiwanese Electronics Industry." Department of Economics, Pennsylvania State University, University Park, PA.

Baldwin, John, and Paul Gorecki. 1986. *The Role of Scale in Canada–U.S. Productivity Differences in the Manufacturing Sector: 1970–79.* Toronto: University of Toronto Press.

Baldwin, John, Timothy Dunne, and John Haltiwanger. 1998. "A Comparison of Job Creation and Job Destruction in Canada and the United States." *Review of Economics and Statistics* 80: 347–56.

Baldwin, Richard. 1988. "Hysteresis and the Beachhead Effect." *American Economic Review* 78: 773–85.

Baldwin, Richard, and Paul Krugman. 1989. "Persistent Trade Effects of Large Exchange Rate Shocks." *Quarterly Journal of Economics* 104: 635–54.

Bernard, Andrew, and J. Bradford Jensen. 1995. "Exporters, Jobs, and Wages in U.S. Manufacturing 1976–87." *Brookings Papers on Economic Activity: Microeconomics*, Washington, DC.

———. 1997. "Why Some Firms Export: Experience, Entry Costs, Spillovers, and Subsidies." School of Management, Yale University, New Haven, CT.

———. 1999. "Exceptional Exporter Performance: Cause, Effect, or Both?" *Journal of International Economics* 47: 1–26.

———. 2000. "Who Dies? International Trade, Market Structure and Plant Closures." Tuck School of Business, Dartmouth College, Hanover, NH.

Bernard, Andrew, and Joachim Wagner. 1997. "Exports and Success in German Manufacturing." *Weltwirtschaftliches Archiv* 133: 134–57.

Bernard, Andrew, Jonathan Eaton, J. Bradford Jensen, and Samuel Kortum. 2000. "Plants and Productivity in International Trade." Working Paper 7688, National Bureau of Economic Research, Cambridge, MA.

Bhagwati, Jagdish. 1978. *Foreign Trade Regimes and Economic Development: Anatomy and Consequences of Exchange Control Regimes.* Lexington, MA: Ballinger for the National Bureau of Economic Research.

Bigsten, Arne, Paul Collier, Stefan Dercon, Bernard Gauthier, Jan Gunning, Anders Isaksson, Abena Oduro, Remco Ooostendorp, Cathy Patillo, Mans Soderbom, Michel Sylvain, Francis Teal, and

Albert Zeufack. 1999. "Exports and Firm-Level Efficiency in the African Manufacturing Sector." School of Business, University of Montreal.

Blundell, Richard, Rachel Griffith, and John van Reenen. 1999. "Market Share, Market Value and Innovation in a Panel of British Manufacturing Firms." *Review of Economic Studies* 66: 529–54.

Bond, Eric. 1986. "Entrepreneurial Ability, Income Distribution, and International Trade." *Journal of International Economics* 20: 343–56.

Brown, Drucilla, Alan Deardorff, and Robert Stern. 1991. "A North American Free Trade Agreement: Analytical Issues and a Computational Assessment." Department of Economics, University of Michigan, Ann Arbor, MI.

Campas, Jose. 1999. "Exchange Rates and Trade: How Important Is Hysteresis in Trade?" Stern School of Business, New York University, New York.

Caves, Richard. 1984. "Scale, Openness and Productivity in Manufacturing." In *The Australian Economy: A View from the North*, ed. Richard Caves and Lawrence Krause. Washington, DC: Brookings Institution.

Caves, Richard, and David Barton. 1990. *Efficiency in U.S. Manufacturing Industries*. Cambridge, MA: MIT Press.

Chen, Tain-Jy, and De-Piao Tang. 1987. "Comparing Technical Efficiency between Import-Substituting and Export-Oriented Foreign Firms in a Developing Country." *Journal of Development Economics* 36: 277–89.

Clerides, Sofronis, Saul Lach, and James Tybout. 1998. "Is Learning by Exporting Important? Micro-Dynamic Evidence from Colombia, Mexico and Morocco." *Quarterly Journal of Economics* 113: 903–47.

Corden, W. Max. 1974. *Trade Policy and Economic Welfare*. Oxford: Clarendon Press.

Das, Sanghamitra, Mark Roberts, and James Tybout. 2001. "Market Entry Costs, Producer Heterogeneity, and Export Dynamics." Working Paper 8629, National Bureau of Economic Research, Cambridge, MA.

Davis, Steven, John Haltiwanger, and Scott Schuh. 1996. *Job Creation and Destruction*. Cambridge, MA: MIT Press.

de Long, J. Bradford, and Lawrence Summers. 1991. "Equipment Investment and Economic Growth." *Quarterly Journal of Economics* 56: 445–502.

Devarajan, Shanta, and Dani Rodrik. 1991. "Pro-Competitive Effects of Trade Reforms: Results from a CGE Model of Cameroon." *European Economic Review* 35: 1157–84.

Dixit, Avinash. 1989. "Entry and Exit Decisions under Uncertainty." *Journal of Political Economy* 97: 620–38.

Dunne, Timothy, Mark Roberts, and Larry Samuelson. 1989. "Plant Turnover and Gross Employment Flows in the U.S. Manufacturing Sector." *Journal of Labor Economics* 7: 48–71.

Dutz, Mark. 1996. "'Oligopolistic Firms' Adjustment to Quota Liberalization: Theory and Evidence." In *Industrial Evolution in Developing Countries*, ed. Mark Roberts and James Tybout. New York: Oxford University Press.

Ericson, Richard, and Ariel Pakes. 1995. "Markov-Perfect Industry Dynamics: A Framework for Empirical Work." *Review of Economic Studies* 62: 53–82.

Ethier, Wilfred. 1982. "National and International Returns to Scale in the Modern Theory of International Trade." *American Economic Review* 72: 950–59.

Feenstra, Robert, James Markusen, and William Zeile. 1992. "Accounting for Growth with New Inputs: Theory and Evidence." *American Economic Review: Papers and Proceedings* 82: 415–21.

Feenstra, Robert, Dorsati Madani, Tzu-Han Yang, and Chi-Yuan Liang. 1999. "Testing Endogenous Growth in South Korea and Taiwan." *Journal of Development Economics* 60: 317–41.

Fernandez, Raquel, and Dani Rodrik. 1991. "Resistance to Reform: Status Quo Bias in the Presence of Individual Specific Uncertainty." *American Economic Review* 81: 1146–55.

Goh, Ai-Ting. 2000. "Opportunity Cost, Trade Policies and the Efficiency of Firms." *Journal of Development Economics* 62: 363–83.

Greene, William. 1993. "Frontier Production Functions." Working Paper EC-93-20, Stern School of Business, New York University, New York.

Grether, Jean-Marie. 1996. "Mexico 1985–90: Trade Liberalization, Market Structure, and Manufacturing Performance." In *Industrial Evolution in Developing Countries,* ed. Mark Roberts and James Tybout. New York: Oxford University Press.

Grossman, Gene, and Elhanan Helpman. 1991. *Innovation and Growth in the Global Economy.* Cambridge, MA: MIT Press.

Haddad, Mona, and Ann Harrison. 1993. "Are There Positive Spillovers from Direct Foreign Investment? Evidence from Panel Data for Morocco." *Journal of Development Economics* 42: 51–74.

Hall, Robert. 1988. "The Relation between Price and Marginal Cost in U.S. Industry." *Journal of Political Economy* 96: 921–47.

Handoussa, Heba, Mieko Nishimizu, and John Page. 1986. "Productivity Change in Egyptian Public Sector Industries after the 'Opening,' 1973–1979." *Journal of Development Economics* 20: 53–73.

Harris, Richard. 1984. "Applied General Equilibrium Analysis of Small Open Economies with Scale Economies and Imperfect Competition." *American Economic Review* 74: 1016–32.

Harrison, Ann. 1994. "Productivity, Imperfect Competition, and Trade Reform: Theory and Evidence." *Journal of International Economics* 36: 53–73.

———. 1996. "Determinants and Effects of Foreign Direct Investment in Côte d'Ivoire, Morocco, and Venezuela." In *Industrial Evolution in Developing Countries,* ed. Mark Roberts and James Tybout. New York: Oxford University Press.

Hart, Oliver. 1983. "The Market Mechanism as an Incentive Structure." *Bell Journal of Economics* 14: 366–82.

Head, Keith, and John Ries. 1999. "Rationalization Effects of Tariff Reductions." *Journal of International Economics* 47: 295–320.

Helpman, Elhanan, and Paul Krugman. 1985. *Market Structure and Foreign Trade: Increasing Returns, Imperfect Competition, and the International Economy.* Cambridge, MA: MIT Press.

Hopenhayn, Hugo. 1992. "Entry, Exit, and Firm Dynamics in Long-Run Equilibrium." *Econometrica* 60: 1127–50.

Hopenhayn, Hugo, and Richard Rogerson. 1993. "Job Turnover and Policy Evolution: A General Equilibrium Analysis." *Journal of Political Economy* 101: 915–38.

Jovanovic, Boyan. 1982. "Selection and the Evolution of Industry." *Econometrica* 50: 649–70.

Klette, Tor J., and Zvi Griliches. 1996. "The Inconsistency of Common Scale Estimators When Output Prices Are Unobserved and Endogenous." *Journal of Applied Econometrics* 11: 343–61.

Krishna, Pravin, and Devashish Mitra. 1998. "Trade Liberalization, Market Discipline, and Productivity Growth: New Evidence from India." *Journal of Development Economics* 46: 447–52.

Krugman, Paul. 1979. "Increasing Returns, Monopolistic Competition, and International Trade." *Journal of International Economics* 9: 469–79.

———. 1987. "The Narrow Moving Band, the Dutch Disease, and the Consequences of Mrs. Thatcher: Notes on Trade in the Presence of Scale Economies." *Journal of Development Economics* 27: 41–55.

Lee, Norman. 1991. "Market Structure and Trade in the Developing Countries." In *Trade Policy, Industrialization and Development: New Perspectives,* ed. Gerald K. Helleiner. Oxford: Clarendon Press.

Levinsohn, James. 1993. "Testing the Imports-as-Market-Discipline Hypothesis." *Journal of International Economics* 35: 1–12.

———. 1999. "Employment Responses to International Liberalization in Chile." *Journal of International Economics* 47: 321–44.

Liu, Lili. 1993. "Entry-Exit, Learning and Productivity Change: Evidence from Chile." *Journal of Development Economics* 42: 217–42.

Liu, Lili, and James Tybout. 1996. "Productivity Growth in Colombia and Chile: Panel-Based Evidence on the Role of Entry, Exit and Learning." In *Industrial Evolution in Developing Countries,* ed. Mark Roberts and James Tybout. New York: Oxford University Press.

Lu, Shihua, and James Tybout. 2000. "Import Competition and Industrial Evolution: A Computational Experiment." Department of Economics, Pennsylvania State University, University Park, PA.

Maloney, William, and Rodrigo Azevado. 1995. "Trade Reform, Uncertainty and Export Promotion: Mexico 1982–88." *Journal of Development Economics* 48: 67–89.

Melitz, Marc. 2000. "Estimating Productivity in Differentiated Product Industries." Department of Economics, Harvard University, Cambridge, MA.

———. 2003. "The Impact of Trade on Intra-industry Reallocations and Aggregate Productivity." *Econometrica* 71: 1695–1725.

Miyagawa, Kaz, and Yuko Ohno. 1995. "Closing the Technology Gap under Protection." *American Economic Review* 85: 755–70.

Muller, Jurgen, and Nicholas Owen. 1985. "The Effects of Trade on Plant Size." In *Industry, Structure, and Performance*, ed. Joachim Schwalbach. Berlin: Edition Sigma.

Nishimizu, Mieko, and John Page. 1982. "Total Factor Productivity Growth, Technological Progress, and Technical Efficiency Change: Dimensions of Productivity Change in Yugoslavia." *Economic Journal* 92: 920–36.

Norman, Victor. 1990. "Assessing Trade and Welfare Effects of Trade Liberalization." *European Economic Review* 34: 725–51.

Pakes, Ariel, and Zvi Griliches. 1984. "Estimated Distributed Lags in Short Panels with an Application to the Specification of Depreciation Patterns and Capital Stock Constructs." *Review of Economic Studies* 51: 243–62.

Pakes, Ariel, and Paul McGuire. 1994. "Computing Markov-Perfect Nash Equilibria: Numerical Implications of a Dynamic Differentiated Product Model." *RAND Journal of Economics* 25: 555–89.

Pavcnik, Nina. 2002. "Trade Liberalization, Exit, and Productivity Improvements: Evidence from Chilean Plants." *Review of Economic Studies* 69: 245–76.

Prusa, Thomas J. 1992. "Why Are So Many Antidumping Petitions Withdrawn?" *Journal of International Economics* 33: 1–20.

Roberts, Mark, and James Tybout. 1991. "Size Rationalization and Trade Exposure in Developing Countries." In *Empirical Studies of Commercial Policy*, ed. Robert Baldwin. Chicago: University of Chicago Press for the National Bureau of Economic Research.

———, eds. 1996. *Industrial Evolution in Developing Countries*. New York: Oxford University Press.

——— 1997. "The Decision to Export in Colombia." *American Economic Review* 87: 545–65.

Roberts, Mark, Theresa Sullivan, and James Tybout. 1995. "Micro-Foundations of Export Booms." World Bank, Washington, DC.

Rodrik, Dani. 1992. "Closing the Technology Gap: Does Trade Liberalization Really Help?" In *Trade Policy, Industrialization and Development: New Perspectives*, ed. Gerald Helleiner. Oxford: Clarendon Press.

Scharfstein, David. 1988. "Product Market Competition and Managerial Slack." *Rand Journal* 19: 47–55.

Scherer, Frederic M., Alan Beckenstein, Erich Kaufer, Dennis R. Murphy, and Francine Bougeon-Massen. 1975. *The Economics of Multi-Plant Operation: An International Comparison Study*. Cambridge, MA: Harvard University Press.

Schmalensee, Richard. 1989. "Inter-Industry Studies of Structure and Performance." In *Handbook of Industrial Organization*, ed. Richard Schmalensee and Robert Willig. Amsterdam: North-Holland.

Schwalbach, Joachim. 1988. "Economies of Scale and Intra-Community Trade." In *Research on the 'Cost' of Non-Europe: Basic Findings*, Vol. 2. Brussels: Commission of the European Communities.

Smith, Alistair, and Anthony Venables. 1988. "Completing the Internal Market in the European Community." *European Economic Review* 32: 1501–25.

Staiger, Robert W., and Frank A. Wolak. 1989. "Strategic Use of Antidumping Law to Enforce Tacit International Collusion." Working Paper 3016, National Bureau of Economic Research, Cambridge, MA.

Sullivan, Theresa. 1997. *Estimating the Manufactured Export Supply Function for Morocco*. Ph.D. dissertation, Department of Economics, Georgetown University, Washington, DC.

Trefler, Daniel. 2001. "The Long and Short of the Canada-U.S. Free Trade Agreement." Working Paper 8293, National Bureau of Economic Research, Cambridge, MA.

Tybout, James. 1991. "Linking Trade and Productivity: New Research Directions." *World Bank Economic Review* 6: 189–212.

———. 1993. "Internal Returns to Scale as a Source of Comparative Advantage." Georgetown University Working Paper 93-01, Washington, DC.

———. 2000. "Manufacturing Firms in Developing Countries: How Well Do They Do, and Why?" *Journal of Economic Literature* 38: 11–44.

Tybout, James, and M. Daniel Westbrook. 1995. "Trade Liberalization and Dimensions of Efficiency Change in Mexican Manufacturing Industries." *Journal of International Economics* 39: 53–78.

———. 1996. "Scale Economics as a Source of Efficiency Gains." In *Industrial Evolution in Developing Countries,* ed. Mark Roberts and James Tybout. New York: Oxford University Press.

Tybout, James, Jaime de Melo, and Vittorio Corbo. 1991. "The Effects of Trade Reforms on Scale and Technical Efficiency: New Evidence from Chile." *Journal of International Economics* 31: 231–59.

Voustden, Neil, and Neil Campbell. 1994. "The Organizational Cost of Protection." *Journal of International Economics* 37: 219–38.

Westphal, Larry. 2001. "Technology Strategies for Economic Development in a Fast-Changing Global Economy." Department of Economics, Swarthmore College, Swarthmore, PA.

Young, Alwyn. 1991. "Learning by Doing and the Dynamic Effects of International Trade." *Quarterly Journal of Economics* 106: 369–405.

PART II

FOREIGN TRADE AND PRODUCTIVITY

5

ON THE QUANTITY AND QUALITY OF KNOWLEDGE: THE IMPACT OF OPENNESS AND FOREIGN R&D ON NORTH-NORTH AND NORTH-SOUTH TECHNOLOGY SPILLOVERS

Maurice Schiff and Yanling Wang

The theory of endogenous growth based on increasing returns to knowledge accumulation originated with Romer (1986, 1990). One of the implications of this theory is that policies affecting knowledge accumulation can have a permanent effect on the rate of economic growth.

Knowledge is assumed to differ from traditional inputs in two ways. First, it has public good characteristics. Second, new knowledge complements existing knowledge, so that the marginal product of additional units of knowledge increases. A widely available new idea raises productivity and increases market size, raising the return to additional ideas. A high-knowledge economy is likely to be able to make productive use of an advanced piece of knowledge, while a knowledge-scarce economy may not.

The assumption that knowledge is a public good means that, once generated, it diffuses at no cost and is available to the entire economy. Though knowledge clearly possesses public goods characteristics, most knowledge is privately produced and is rarely a pure public good, diffusion of which is instantaneous or free. Much new

The authors would like to thank Aaditya Mattoo, Marcelo Olarreaga, and participants at a World Bank International Trade Seminar for their useful comments.

knowledge is embedded in new products or in improved qualities of existing products and does not diffuse either freely or instantaneously. This is especially true in the face of additional barriers, including tariffs and quantitative restrictions on imports, different standards and regulations, and higher communication costs (including those related to language differences).

In the case of domestic knowledge diffusion, Griliches (1957) shows that the adoption of hybrid corn in the United States was gradual and that the cumulative adoption process followed a logistic or *S*-function. His work spawned a number of studies that found *S*-shaped patterns in other technology diffusion processes, implying that it may take a long time before most firms adopt a new technology. Greenwood (1997) finds that it took 54 years for the rate of adoption of steam locomotives to rise from 10 percent to 90 percent of existing firms in the United States. In the case of diesels, the corresponding figure was five years. Manuelli and Seshadri (2003) find that the process took 35 years for tractors.[1]

This chapter examines the process of international rather than intranational technology diffusion. Keller (2002a) shows that the effect of research and development (R&D) performed in the G-5 countries on productivity in smaller Organisation for Economic Co-operation and Development (OECD) countries declines with distance—that is, international technology diffusion is costly, and its cost rises with distance. The studies cited above imply that access to knowledge is not instantaneous. Rather, its rate of adoption is subject to the usual cost-benefit considerations and typically leads to gradual adoption. Consequently, knowledge accumulation occurs through an increase in quantity (increased diffusion of existing knowledge), an increase in quality (new knowledge), or both.

In an international context, trade-related knowledge diffusion can occur through an increase in a country's level of exposure to that knowledge through trade (quantity), an increase in the knowledge-content of that trade (quality), or both. This chapter investigates how these two components of knowledge diffusion affect productivity. Given the higher cost of international relative to domestic knowledge diffusion, examining the differential impact of the quantity and quality of knowledge diffusion in an international context seems particularly promising.

A recent literature has examined the impact of trade on knowledge diffusion by constructing measures of access to foreign knowledge and estimating its effect on productivity.[2] The seminal paper is Coe and Helpman (1995). It estimates the impact on total factor productivity (TFP) of foreign R&D, defined as the sum of trading partners' R&D stocks (that is, a measure of knowledge quality), weighted by the bilateral trade shares (a measure of knowledge quantity). Using aggregate data, Lumenga-Neso, Olarreaga, and Schiff (forthcoming) find that foreign R&D has a significantly positive impact on TFP in developed countries; Coe, Helpman, and Hoffmaister (1997) find similar results for developing countries. TFP rises with the degree of a country's openness and the trading partners' R&D stocks. Similar findings are obtained at the industry level by Keller (2002b) for developed countries and by Schiff, Wang, and Olarreaga (2002) for developing countries.

These studies treat the two components of trade-related knowledge diffusion—openness and trading partners' R&D stocks—symmetrically in their empirical analysis. This chapter subjects the symmetry assumption to rigorous testing and concludes that the impact of the two components is asymmetric. It finds that openness plays a more (less) important role than trading partners' R&D stocks in North-South (North-North) knowledge diffusion and has a greater (smaller) impact on productivity in the South (North) than found in the existing literature.

Analytical Framework

This chapter investigates the relative contributions to TFP in a developed or a developing country of R&D performed in OECD countries (quality of knowledge) and the degree of access to this knowledge through trade (quantity of knowledge). TFP is assumed to be given by

$$TFP = TFP(T, Z), \tag{5.1}$$

where T denotes technological knowledge and Z is a vector of other factors affecting TFP, including, for instance, education. T in a given country is assumed to be given by

$$T = T(DRD, NRD);\ T_1, T_2 > 0, \tag{5.2}$$

where DRD is the stock of R&D produced in that country, NRD is access to the trading partners' R&D stocks, and T_1 and T_2 are the first-order derivatives with respect to DRD and NRD (what is referred to in the literature as foreign R&D), respectively. Access to the foreign stock of R&D, NRD, is assumed to be given by

$$NRD = NRD(OPEN, RDC);\ NRD_1, NRD_2 > 0,\ NRD_{11} < 0. \tag{5.3}$$

NRD is thus a function of $OPEN$, the degree of a country's openness, and RDC, a measure of trading partners' R&D stocks (that is, the R&D content of the country's trade). The second derivative, NRD_{11}, is assumed to be negative to reflect the fact that the additional knowledge a country obtains from the imports of a given machine is likely to diminish with the number of units of that machine it imports.

Past studies that have examined trade-related technology diffusion have assumed that openness and trading partners' R&D stocks enter symmetrically in NRD, that is, that equation (5.3) takes the form

$$NRD = NRD(OPEN * RDC) \tag{5.4}$$

and that equation (5.1) takes the form

$$TFP = TFP(T(DRD, OPEN * RDC), Z). \tag{5.5}$$

This chapter investigates whether the variables *OPEN* and *RDC* actually enter symmetrically into the TFP equation. The hypothesis is tested for both North-North trade (trade among OECD countries) and North-South trade (trade between OECD and developing countries).

North-North Trade

Empirical Implementation

For North-North trade, we use the same aggregate data used by Coe and Helpman (1995) and Keller (1998). The data set consists of 22 developed countries (21 OECD countries and Israel) and 20 years (1971–90). Using aggregate data allows a more precise comparison of our results with those of Coe and Helpman (1995) and Keller (1998). It is also useful because a third of the sample of OECD countries does not report R&D data at the industry level.[3]

Coe and Helpman (1995) define *NRD* as the sum over all trading partners of their R&D stocks multiplied by the trade shares. They define trade shares as the imports from trading partners divided by the sum of these imports. The fact that the trade shares add up to 1 implies that the level of total imports does not affect *NRD* (as long as the shares are unchanged).[4] Here trade shares are defined relative to GDP, and the level of imports does affect *NRD*. Thus *NRD* for country *c* is given by

$$NRD_c \equiv \sum_{k \neq c} \left(\frac{M_{ck}}{GDP_c} \right) RD_k, \tag{5.6}$$

where *M* is imports, *RD* is the stock of R&D, and *k* stands for the trading partner countries.

We define two transformations of *NRD*, one that measures openness and one that measures the R&D content of imports. Openness is defined as

$$OPEN_c \equiv \sum_{k \neq c} \frac{M_{ck}}{GDP_c}, \tag{5.7}$$

which is obtained by setting $RD_k = 1, \forall k$, in equation (5.6). The R&D content of trade, *RDC*, is defined as

$$RDC_c \equiv \sum_{k \neq c} RD_k, \tag{5.8}$$

which is obtained by setting $M_{ck}/GDP_c = 1, \forall c, k,$ in equation (5.6).

Two alternative equations are estimated:

$$\log TFP_{ct} = \beta_0 + \beta_1 \log DRD_{ct} + \beta_N \log NRD_{ct} + \beta_T \log OPEN_{ct} + \sum_t \beta_t D_t + \sum_c \beta_c D_c + \varepsilon_{ct}, \tag{5.9}$$

and

$$\log TFP_{ct} = \beta_0' + \beta_1' \log DRD_{ct} + \beta_N' \log NRD_{ct} + \beta_R' \log RDC_{ct} + \sum_t \beta_t' D_t + \sum_c \beta_c' D_c + \varepsilon_{ct}', \tag{5.10}$$

where *NRD*, *OPEN*, and *RDC* are defined in equations (5.6), (5.7), and (5.8); DRD_c is the home country's stock of R&D; and D_t (D_c) are time (country) dummies. Equation (5.10) is estimated as a check on robustness of the estimation results of equation (5.9).

Assume that when imports or the R&D content of imports change, the proportional change is the same across all trading partners. Then in equation (5.9) the elasticity of TFP with respect to openness is $\beta_N + \beta_T$ and the elasticity with respect to the R&D content of trade is β_N. In equation (5.10) the elasticity of TFP is β_N' with respect to openness and $\beta_N' + \beta_R'$ with respect to the R&D content of trade. If openness and the R&D content of trade enter symmetrically in TFP (as they do in *NRD* in equation [5.6]), the following results should hold: $\beta_T = 0$ in equation (5.9), $\beta_R' = 0$ in equation (5.10), and $\beta_N = \beta_N'$. If, however, the impact of openness on TFP is larger (smaller) than the impact of the R&D content, the following results should hold: $\beta_T > (<)\ 0$ and $\beta_R' < (>)\ 0$.

Note that the value of these coefficients is likely to depend on the level of development of the importing country. In OECD countries in which communication and information systems are more highly developed, new knowledge is likely to spread more rapidly, and the marginal impact of additional imports is likely to be smaller. In contrast, knowledge is likely to spread less rapidly in developing countries, where communication and information systems are less developed, and the marginal impact of additional imports is likely to be larger.

Empirical Results

Before turning to the econometric analysis, we need to consider the fact that two or more variables may be trended and contain unit roots, making the regression results spurious (unless the variables are co-integrated). As noted earlier, we use the same data used by Coe and Helpman (1995). They found that their variables exhibited a clear trend but were co-integrated, justifying the estimation of a relationship in the levels of the variables without having to adjust them.

TABLE 5.1 Determinants of Total Factor Productivity in OECD Countries

Variable	Model (1)	Model (2)	Model (3)
log*DRD*	0.112* (14.1)	0.119* (14.77)	0.045* (4.25)
log*NRD*	0.045* (4.9)	0.055* (5.82)	–0.004 (–0.43)
log*OPEN*		–0.071* (23.67)	
log*RDC*			0.208* (8.63)
Adjusted R^2	0.60	0.61	0.66

Source: Authors.

Note: $N = 440$. Figures in parenthesis are *t*-statistics. Results on time and country dummies are not reported. *DRD* is the domestic R&D stock; *NRD* is the trade-related foreign R&D stock, as defined in equation (5.6); *OPEN* is the ratio of a country's total imports over its GDP; and *RDC* is the sum of the R&D stock for the rest of the sample.

*Significant at the 1 percent level.

The empirical results are presented in table 5.1. Regression (1) indicates that the elasticities of both *DRD* (own R&D stock) and *NRD* are positive and significant at the 1 percent level, with the elasticity of *DRD* (0.112) more than twice as large as that of *NRD* (0.045). Regression (2) adds the variable *OPEN*. The elasticities with respect to *DRD* (0.119) and *NRD* (0.055) are similar to those in regression (1), and the elasticity with respect to *OPEN* is –0.071 (all significant at the 1 percent level). This implies that the elasticity is $\beta_N = 0.055$ with respect to the R&D content of trade, and $\beta_N + \beta_T = -0.016$ with respect to openness and is not significantly different from zero. These results imply that symmetry is rejected and that the R&D content of trade (the quality of trade) has a greater impact on TFP than the degree of openness (the quantity of trade).

In regression (3), $\beta'_N = -0.004$ (is not significant) and $\beta'_R = 0.208$ (is significant at the 1 percent level). Thus the elasticity of TFP with respect to openness is $\beta'_N = -0.004$ (not significant), and the elasticity with respect to the R&D content of trade is $\beta'_N + \beta'_R = 0.204$. Although the elasticities in regression (3) differ from those in regression (2), the qualitative result is the same: symmetry is rejected, and the R&D content of trade has a greater impact on TFP than the degree of openness.[5]

Why does the R&D content of trade have a greater impact on TFP than the degree of openness? First, most OECD countries have low trade barriers, and openness is typically high. Consequently, the marginal impact of trade is likely to be small ($NRD_{11} < 0$ in equation [5.3]). Second, alternative means of diffusing knowledge are available, including FDI, licensing, the Internet and other telecommunications technology, scientific journals, and international meetings, all of which reduce the importance of trade as a means of knowledge diffusion.

Note that conditions with respect to openness and other channels of technology diffusion differ greatly in the South. Over the estimation period 1976–98, tariffs averaged 6.3 percent for 15 OECD countries in the sample and 20.5 percent (more than three times more) in 24 developing countries in the sample. As for alternative channels of technology diffusion, over the period 1990–98, the average number of Internet users was 52.8 per 1,000 people in the 15 OECD countries and 6.8 (about one-eighth as many) in the 24 developing countries. FDI averaged $95.6 billion in the 15 OECD countries and $33.8 billion (about a third as much) from the 15 OECD countries to all developing countries. Unsurprisingly, as shown below, the empirical results for North-South and North-North trade differ.

Comparison with the Literature

In their seminal paper, Coe and Helpman (1995) estimate regression (1) and obtain an elasticity of TFP with respect to *NRD* of 0.092 and of 0.060 when they add a dummy variable to capture the additional effect of domestic R&D in the G-7 countries.[6] In a much cited paper, Keller (1998) uses the same data and estimates alternative versions of TFP equations. In one specification, he regresses TFP on *DRD* and *RDC*, with results that are as good as or better than those of Coe and Helpman. He obtains an elasticity of TFP with respect to *RDC* of 0.161 and of 0.129 when the dummy variable is added to capture the additional effect of domestic R&D in the G-7 countries. He concludes that a country's trading partners' R&D does not necessarily diffuse through trade and that nontrade channels should be investigated as well.

Keller's point seems well taken. His results are close to those in regression (3), though our interpretation differs somewhat from his. By incorporating both *NRD* and *RDC* (in addition to *DRD*) in regression (3), we allow the data to determine which of the two effects—openness or trading partners' R&D stocks—dominates. Our interpretation is that the effect of trading partners' R&D stocks on TFP is dominant in North-North trade but that this reflects both the influence of nontrade channels and the fact that OECD economies are typically very open, so that the marginal effect of openness is small.

This result does not generalize to North-South trade, because of the different conditions in the South described above. As is shown below, opposite findings are obtained for North-South trade, with openness having a significant impact on TFP.

North-South Trade

Empirical Implementation

For North-South trade, we make use of a data set of industry-level, trade-related technology diffusion used in Schiff, Wang, and Olarreaga (2002). The data set consists of 16 manufacturing industries, 24 developing economies, 15 OECD trading partners, and 22 years (1977–98).[7] The 16 industries are divided into high and low R&D-intensity groups, with R&D intensity defined as the ratio of expenditures on R&D to value-added. The average R&D intensity is 1.3 percent for the "low" group and 11 percent for the "high" group.[8] As in Coe, Helpman, and Hoffmaister (1997), domestic R&D was not included because of the lack of data.

Schiff, Wang, and Olarreaga (2002) define North-foreign R&D (that is, knowledge obtained through trade) in industry i of developing country c, NRD_{ci}, as

$$NRD_{ci} \equiv \sum_j a_{cij} RD_{cj} = \sum_j a_{cij} \left[\sum_k \left(\frac{M_{cjk}}{VA_{cj}} \right) RD_{jk} \right], \tag{5.11}$$

where c (k) indexes developing (OECD) countries, j indexes industries, M denotes imports, VA denotes value-added, and RD denotes R&D. a_{cij} is the import input-output coefficient, which measures for country c the share of imports of industry j sold to industry i.

The first part of equation (5.11) says that in developing country c, North-foreign R&D in industry i, NRD_{ci}, is the sum, over all industries j, of RD_{cj}, the industry-j foreign R&D obtained through imports, multiplied by a_{cij}, the share of imports of industry j that is sold to industry i. The second part of equation (5.11) says that RD_{cj} is the sum, over OECD countries k, of M_{cjk}/VA_{cj}, the imports of industry-j products from OECD country k per unit of industry-j value-added (that is, the bilateral openness share), multiplied by RD_{jk}, the stock of industry-j R&D in OECD country k.

We define an openness variable as

$$OPEN_{ci} = \sum_j a_{cij} \left[\sum_k \left(\frac{M_{cjk}}{VA_{cj}} \right) \right], \tag{5.12}$$

which is derived from equation (5.11) by setting $RD_{jk} = 1, \forall j, k$.

We define an R&D variable as

$$RDC_{ci} = \sum_j a_{cij} \left[\sum_k RD_{jk} \right], \tag{5.13}$$

which is derived from equation (5.11) by setting $M_{cjk}/VA_{cj} = 1, \ \forall c, j, k$.

Education is included in the regression as a control variable, as it is in Coe, Helpman, and Hoffmaister (1997); Schiff, Wang, and Olarreaga (2002); and others. Two alternative equations are estimated:

$$\begin{aligned} \log TFP_{cit} &= \beta_0 + \beta_N \log NRD_{cit} + \beta_T \log OPEN_{cit} + \beta_E E_{ct} \\ &\quad + \sum_t \beta_t D_t + \sum_c \beta_c D_c + \sum_i \beta_i D_i + \varepsilon_{cit}, \end{aligned} \tag{5.14}$$

and

$$\begin{aligned} \log TFP_{cit} &= \beta_0' + \beta_N' \log NRD_{cit} + \beta_L' \log RDC_{cit} + \beta_E' E_{ct} \\ &\quad + \sum_t \beta_t' D_t + \sum_c \beta_c D_c + \sum_i \beta_t' D_i + \varepsilon_{cit}', \end{aligned} \tag{5.15}$$

where E denotes education, and D_t, (D_c), and (D_i) represent time, country, and industry dummies. The effects for high and low R&D-intensity industries are estimated by introducing a dummy variable, DR, with $DR = 1$ for high R&D-intensity industries and $DR = 0$ otherwise.

Empirical Results

As with the analysis of North-North trade, we need to consider the possibility that two or more variables might be trended and contain unit roots, making the regression results spurious (unless the variables are co-integrated). The unit root hypothesis was rejected at the 1 percent significance level for log *TFP*, log *NRD*, log *OPEN*, and log *RDC*. The education variable *E* is significant at the 1 percent level in all six regressions, with a 1 percentage point increase in education raising TFP by 6.8–7.5 percent (table 5.2).

Regressions (1) and (2) are reproduced from Schiff, Wang, and Olarreaga (2002). They impose symmetric effects of openness and R&D on TFP. Regression (1) shows a positive impact of *NRD* on TFP (significant at the 1 percent level), with an elasticity of about 0.19. Regression (2) shows an elasticity of about 0.14 for low R&D-intensity industries and 0.28 for high R&D-intensity industries, both significant at the 1 percent level. As might be expected, foreign R&D has a greater impact on the productivity of R&D-intensive industries.

TABLE 5.2 Determinants of Total Factory Productivity in Developing Countries

Variable	Model					
	(1)	(2)	(3)	(4)	(5)	(6)
log *NRD*	0.188** (6.11)	0.138** (4.03)	–0.012 (–0.19)	–0.065 (–0.99)	0.369** (8.82)	0.294** (6.31)
log *NRD***DR*		0.141** (3.52)		0.285** (4.02)		0.156** (3.5)
log *OPEN*			0.251** (3.89)	0.295** (4.23)		
log *OPEN***DR*				–0.244** (–3.13)		
log *RDC*					–0.408** (–6.38)	–0.248** (–3.89)
log *RDC***DR*						–0.189* (–2.15)
E	6.831** (4.34)	6.823** (4.34)	6.9** (4.37)	6.866** (4.37)	7.462** (4.74)	7.411** (4.72)
Adjusted R^2	0.23	0.23	0.23	0.24	0.24	0.24

Source: Authors.

Note: $N = 5{,}721$. Figures in parentheses are *t*-statistics. The constant and regression results on country, year, and industry dummies are not reported. *NRD* is the trade-related North-foreign R&D defined in equation (5.11), *OPEN* is defined in equation (5.12), and *RDC* is defined in equation (5.13). *E* is the secondary school completion ratio for the population 25 and above. *DR* = 1 for R&D-intensive industries and 0 for low R&D-intensity industries.

*Significant at the 5 percent level.

**Significant at the 1 percent level.

Columns (3) and (4) regress TFP on *NRD* and *OPEN* (see equation [5.14]). Regression (3) shows that the elasticity of TFP with respect to R&D is –0.012 (not significantly different from zero), and the elasticity with respect to openness is about 0.24 (0.251–0.012), significant at the 1 percent level. Regression (4) shows that for low R&D-intensity industries, the elasticity of TFP with respect to R&D is –0.065 (not significantly different from zero), and the elasticity with respect to openness is 0.23 (0.295–0.065), significant at the 1 percent level. For high R&D-intensity industries, the elasticity of TFP with respect to R&D is 0.22 (0.285–0.065), significant at the 1 percent level, and the elasticity with respect to openness is about 0.27(0.22 + 0.295 – 0.244), significant at the 1 percent level. The results from regression (4) imply that R&D has no impact on the TFP of low R&D-intensity industries and a significant impact on the TFP of high R&D-intensity industries. Openness has a significant impact on the TFP of both low and high R&D-intensity industries. The impact of openness is larger than that of R&D, significantly so for low R&D-intensity industries (where only openness matters), somewhat less so for high R&D-intensity industries. The results on the importance of R&D are quite plausible: one would expect the embodied technology or R&D content of imports to matter more in industries in which technology plays a more important role (that is, in R&D-intensive industries).[9]

Columns (5) and (6) correspond to equation (5.15). Regression (5) shows an elasticity of TFP of about 0.37 with respect to openness (significantly different from zero at the 1 percent level) and of about (–0.039) with respect to R&D (not significantly different from zero). These results confirm those of regression (3).

Regression (6) shows an elasticity of TFP with respect to openness equal to about 0.29 for low R&D-intensity industries and 0.45 for high R&D-intensity industries, both significant at the 1 percent level. The elasticity of TFP with respect to R&D is not significantly different from zero for low R&D-intensity industries (0.046 = 0.294 – 0.248) or high R&D-intensity industries (0.013 = 0.046 + 0.156 – 0.189). These results confirm those of regression (4), though the elasticities with respect to openness in both industry groups are larger in this case and the elasticity with respect to R&D in high R&D-intensity industries is smaller.

Comparison with the Literature

The results obtained here imply that the impact of openness on TFP in developing countries is greater than that obtained in Schiff, Wang, and Olarreaga (2002), where the effects of openness and R&D were constrained to be symmetric. For all industries taken together, the elasticity of TFP with respect to openness is 0.19 in the presence of symmetry and 0.24–0.37 in its absence. When industries are split between high and low R&D-intensity industries, the elasticity of TFP with respect to openness is 0.14 for low R&D-intensity industries given symmetry and 0.23–0.29 in the unconstrained case. For high R&D-intensity industries, the elasticity is 0.28 given symmetry and 0.27–0.45 in the unconstrained case. Thus the openness elasticity for the low R&D-intensity industries is 60–100 percent larger than when symmetry between the R&D and openness effects is imposed and 0–60 percent larger for the high R&D-intensity industries.

Coe, Helpman, and Hoffmaister (1997) estimated the impact of North-South R&D spillovers at the aggregate level. They tried a variety of specifications, including some with an openness variable. In their preferred specification, they obtain an elasticity of TFP with respect to *NRD* of 0.058 and an elasticity of TFP with respect to the share of imports to GDP (openness) of 0.0279, both significant at the 1 percent level.[10] The sum of weights for *NRD* in Coe, Helpman, and Hoffmaister (1997) is 1, so the elasticity of TFP is 0.058 with respect to R&D and 0.279 with respect to openness. Coe, Helpman, and Hoffmaister (1997) estimate 10 different specifications. The average value of the elasticity of TFP is negative with respect to *NRD* and 0.304 with respect to openness. These results support our findings that the elasticity of TFP with respect to openness is larger than that with respect to R&D and that the values obtained fall within the range of our estimates.

The conclusions of both Falvey, Foster, and Greenaway (2002) and Keller (2000) lend support to our results. Falvey, Foster, and Greenaway (2002) estimate North-South R&D spillovers at the aggregate level and use various definitions of *NRD*, including that of Coe and Helpman (1995) and Lichtenberg and van Pottelsberghe de la Potterie (1998). They conclude that only the specifications that include the level of imports result in positive coefficients for the effect of knowledge spillovers. Keller (2000) concludes that openness should play a greater role for technology diffusion and productivity growth in developing countries than in developed ones.

Conclusion

A recent literature has examined the impact of trade-related technology diffusion on TFP. That literature imposed symmetry between the impact of openness and that of the R&D content of trade. This chapter shows that the assumption of symmetry is not warranted in the case of either North-North or North-South technology diffusion.

For North-North trade, the R&D content of trade has a greater impact on TFP than openness. Unlike studies imposing symmetry between the TFP effects of the R&D content of trade and openness, which find a positive impact of openness, our analysis indicates that the effect of openness on TFP is not significantly different from zero. For North-South trade, openness has a greater impact on TFP than the R&D content of trade. The impact of openness on TFP is greater than that obtained when symmetry is imposed. The impact of the R&D content of trade on TFP is not significantly different from zero in low R&D-intensity industries and seems positive in R&D-intensive industries. These results imply that the gains from trade liberalization in developing countries are larger than earlier literature suggested and that the liberalization of trade in those industries is likely to have a greater impact on growth than previously thought.

Notes

1. Gradual adoption is typically attributed to some market imperfection, including lobbying (Parente and Prescott 1994), imperfect information (Jovanovic and MacDonald 1994), and learning by doing (Jovanovic and Lach 1989; Jovanovic and Nyarko 1996). Manuelli and Seshadri (2003) obtain the same diffusion pattern for tractors in a frictionless model, with gradual adoption due to the change in exogenous variables including labor costs, over time.

2. Recent interest in the relationship between trade and growth and in international technology spillovers is based on the development of endogenous growth theories (Romer 1986, 1990) and their application to the open economy case (Grossman and Helpman 1991).

3. Aggregate data are also used because of the difficulty of producing reasonable empirical results at the industry level for G7 countries plus Sweden. These countries were selected to compare the results with those of Keller (2002b), who examines the impact of domestic R&D and trading partners' R&D (that is, *RDC*) but not of openness. Keller (2002b) covers the years 1970–91. The sample here was restricted because industry-level trade data are available only from 1976. The results are not reported, because the domestic R&D variable was never significant in the restricted sample, even though *NRD* was positive and significant.

4. Coe and Helpman (1995) do add the share of imports in GDP in their third TFP regression, though in a different way from trade shares and R&D stocks (see Lumenga-Neso, Olarreaga, and Schiff [forthcoming] for more details).

5. We estimated TFP as a function of *NRD, OPEN,* and *RDC* but did not obtain satisfactory results because of multicollinearity problems between these variables.

6. The trade shares used in the definition of *NRD* in the first two equations estimated by Coe and Helpman differ from ours. In Coe and Helpman these shares add up to 1, and the effect of an overall increase in openness cannot be simulated. For given shares, the elasticity of TFP with respect to *NRD* reflects the impact of an equiproportionate increase in trading partners' R&D stocks. Thus their results are not so far from those obtained here. Coe and Helpman do include a measure of openness in a third regression by multiplying it by log NRD. However, they model a log-log relationship with respect to NRD but a semi-log relationship with respect to openness.

7. The 24 developing economies are Bangladesh, Bolivia, Chile, Cameroon, Colombia, Cyprus, Ecuador, Egypt Arab Republic, Guatemala, Hong Kong (China), Indonesia, India, the Islamic Republic of Iran, Jordan, the Republic of Korea, Kuwait, Mexico, Malawi, Malaysia, Pakistan, the Philippines, Poland, Trinidad and Tobago, and República Bolivariana de Venezuela. The 15 OECD countries are Australia, Canada, Denmark, Finland, France, Germany, Ireland, Italy, Japan, the Netherlands, Norway, Spain, Sweden, the United Kingdom, and the United States.

8. The 16 manufacturing industries are food, beverages, and tobacco (code 31 in the International Standard Industrial Classification); textiles, apparel, and leather (32); wood products and furniture (33); paper, paper products, and printing (34); *chemicals, drugs, and medicines (351/2); petroleum refineries and products (353/4)*; rubber and plastic products (355/6); nonmetallic mineral products (36); iron and steel (371); nonferrous metals (372); metal products (381); *nonelectrical machinery, office and computing machinery (382); electrical machinery and communication equipment (383); transportation equipment (384); professional goods (385)*; and other manufacturing (39). (High R&D-intensity industries are shown in italics.) For the "high" group, the average R&D intensity minus two standard deviations is 3.8 percent, which is larger than the average plus two standard deviations of the "low" group (3.1 percent). Assuming a normal distribution, the hypothesis that any of the industries in the "high" R&D intensity cluster belongs to the "low" cluster is rejected at the 1 percent significance level.

9. This result may also partly reflect the fact that greater openness has a disciplining effect by increasing the level of contestability and competitiveness of the domestic industry.

10. Coe, Helpman, and Hoffmaister (1997) use imports of machinery and equipment rather than total imports.

References

Atkeson, Andrew, and Patrick Kehoe. 2001. "The Transition to a New Economy after the Second Industrial Revolution." NBER Working Paper No. W8676, National Bureau of Economic Research, Cambridge, MA.

Coe, David T., and Elhanan Helpman. 1995. "International R&D Spillovers." *European Economic Review* 39 (5): 859–87.

Coe, David T., Elhanan Helpman, and Alexander W. Hoffmaister. 1997. "North-South R&D Spillovers." *Economic Journal* 107 (440): 134–49.

Falvey, R., N. Foster, and D. Greenaway. 2002. "North-South Trade, Knowledge Spillovers and Growth." Research Paper 2002/23, Leverhulme Centre for Research on Globalisation and Economic Policy, University of Nottingham, United Kingdom.

Greenwood, Jeremy. 1997. "The Third Industrial Revolution: Technology, Productivity and Income Inequality." American Enterprise Institute Studies on Understanding Economic Inequality. Washington, DC: AEI Press.

Griliches, Zvi. 1957. "Hybrid Corn: An Exploration in the Economics of Technological Change." *Econometrica* 25 (4): 501–22.

Grossman, M. Gene, and Elhanan Helpman. 1991. *Innovation and Growth in the Global Economy.* Cambridge, MA: MIT Press.

Jovanovic, Boyan, and Saul Lach. 1989. "Entry, Exit, and Diffusion with Learning by Doing." *American Economic Review* 79 (4): 690–99.

Jovanovic, Boyan, and Glenn M. MacDonald. 1994. "Competitive Diffusion." *Journal of Political Economy* 102 (1): 24–52.

Jovanovic, Boyan, and Yaw Nyarko. 1996. "Learning by Doing and the Choice of Technology." *Econometrica* 64 (6): 1299–310.

Keller, Wolfgang. 1998. "Are International R&D Spillovers Trade-Related? Analyzing Spillovers among Randomly Matched Trade Partners." *European Economic Review* 42 (8): 1469–81.

———. 2000. "Do Trade Patterns and Technology Flows Affect Productivity Growth?" *World Bank Economic Review* 14 (1): 17–47.

———. 2002a. "Geographic Localization of International Technology Diffusion." *American Economic Review* 92 (1): 120–42.

———. 2002b. "Trade and the Transmission of Technology." *Journal of Economic Growth* 7 (1): 5–24.

Lichtenberg, Frank R., and Bruno van Pottelsberghe de la Potterie. 1998. "International R&D Spillovers: Comment." *European Economic Review* 42 (8): 1483–91.

Lumenga-Neso, Marcelo Olarreaga, and Maurice Schiff. Forthcoming. "On 'Indirect' Trade-Related R&D Spillovers." *European Economic Review.* (Earlier version published as World Bank Policy Research Working Paper 2580 (April 2001), Washington, DC.)

Manuelli, Rodolfo E., and Ananth Seshadri. 2003. "Frictionless Technology Diffusion: The Case of Tractors." Working Paper 9604, National Bureau of Economic Research, Cambridge, MA.

Parente, Stephen L., and Edward C. Prescott. 2004. "Unified Theory of the Evolution of International Income Levels." Research Department Staff Report 333, Federal Reserve Bank of Minneapolis, Minneapolis, MN.

Romer, Paul M. 1986. "Increasing Returns and Long-Run Growth." *Journal of Political Economy* 94 (5): 1002–37.

———. 1990. "Endogenous Technical Change." *Journal of Political Economy* 98 (5): S71–S102.

Schiff, Maurice, Yanling Wang, and Marcelo Olarreaga. 2002. "Trade-Related Technology Diffusion and the Dynamics of North-South and South-South Integration." Policy Research Working Paper 2861, World Bank, Washington, DC.

6

THE KNOWLEDGE CONTENT OF MACHINES: NORTH-SOUTH TRADE AND TECHNOLOGY DIFFUSION

Giorgio Barba Navaretti, Maurice Schiff, and Isidro Soloaga

This chapter examines the impact of imported technologies on productivity growth in developing countries. Many studies analyze trade-related channels for transferring technologies and knowledge and their effects on productivity.[1] In contrast to this earlier literature, we focus on the technological content of imported factors of production rather than on imports per se. We also explicitly model the choice of technology and therefore deal with the endogenous nature of the relationship between imported technologies and productivity growth.

We study the effects of knowledge, the "weightless" good (Quah 1999), when it is embodied in machines, the most physical of all factors of production. We use an indirect measure to capture the amount of knowledge contained in machines: the average unit value per ton of machine imported.[2] Does this measure, relating the value of a weightless good to the weight of its container, make sense? Federal Reserve Chairman Alan Greenspan once noted that through the second half of the

Earlier drafts of this chapter were circulated with the title "Weightless Machines and Costless Knowledge: An Empirical Analysis of Trade and Technology Diffusion." The authors wish to thank Francis Ng and Alessandro Nicita for invaluable help in constructing the database; Gani Aldashev, Alessandra Tucci, and Paolo Zagaglia for their research assistance; and Bernard Hoekman, Aart Kraay, Riccardo Lucchetti, James Tybout, Jonathan Eaton, and the anonymous referees of the *Journal of International Economics* for providing useful suggestions. Giorgio Barba Navaretti is also grateful to David Tarr and Tony Venables for arranging and supporting his stay at the World Bank while working on earlier drafts of this chapter.

twentieth century, the United States tripled the real value of its output with no increase in the weight of the material produced (*Washington Post* 2000). Indeed, at any point in time, the price of machines reflects their relative productivity. The price of computers grows with their megahertz or other embodied features.

Over time, however, the relative price of equipment falls, and the increasing productivity of increasingly weightless (knowledge-intensive) machines is not mirrored in their prices (Gordon 1990; Eaton and Kortum 2001). The cost of computers has been declining for years, although their capacity to process information has skyrocketed. Thus the evolution of unit values of machines over time fails to capture their technological content. To overcome this problem, we normalize the unit values of the machines imported by our sample countries by the unit value of the same machines imported at the same point in time by the United States, which is assumed to be the technological frontier. In this way, we derive an index, measuring the distance of imported machines from their technological frontier at any point in time, that can be used for comparisons across countries, industries, and time.

We analyze the machines exported by the European Union to a sample of neighboring developing countries and transition economies in Central and Eastern Europe and the Southern Mediterranean. We find that although developing economies buy increasingly productive machines over time, the technology embodied in these machines persistently lags behind those purchased by the United States.

What drives these choices, and how do they affect productivity and growth? We develop a theoretical model that analyses the choice of technology and relates it to the expected productivity outcomes. The model is then tested, using industry-specific data for our sample countries. We find that the choice of lower technologies is optimal for developing countries, given local skills and factor prices. However, an increase in the level of complexity of the machines imported has a positive impact on total factor productivity (TFP) growth, which turns out to be larger than an increase in the share of imported machines in total investments.

This chapter contributes in several ways to the literature on trade-related technology spillovers, including Coe and Helpman (1995) and other papers building on their framework, such as Coe, Helpman, and Hoffmaister (1997); Keller (2000); and Schiff, Wang, and Olarreaga (2002). First, it clearly distinguishes between the quantity and quality of imports by measuring the knowledge content of imported machines. The impact of the technological content of imported machines has rarely been examined in this literature.[3] Imagine a country, A, that imports the same total value of goods (or capital goods) from two countries, B and C, that have the same R&D stock. The impact of imports from both countries on domestic productivity is the same in the Coe and Helpman framework. However, the composition of goods imported from B and C may be different. For instance, imports from B may consist of fewer but more knowledge-intensive (more productive) machines than those from C. Whether imports from B or C have a greater impact on A's productivity will depend on the elasticity of TFP with respect to the knowledge-intensity of machines and the quantity of machines. This is one of the issues examined in this chapter.

Second, studies based on the Coe-Helpman framework treat the choice of technology as exogenous. This chapter explicitly models the choice of technology—the knowledge-intensity of machines—by relating it to its potential effect on productivity growth.[4]

Third, this chapter extends earlier works analyzing the choice of the vintage of imported machinery (Barba Navaretti, Soloaga, and Takacs 2000) and the impact of imported machines on export performance (Barba Navaretti, Galeotti, and Mattozzi 2000).

A Model of Knowledge Production and Choice of Machines

This section develops a simple model of supply of knowledge-embedded machines by developed countries and (import) demand by developing countries. Among other things, the model tries to explain two empirical observations. The first is the fact that even though machines with greater knowledge content are more expensive at a given moment in time, they are cheaper over time. The second is the fact that countries with higher levels of human capital import more knowledge-intensive machines.

Supply

Assume a developed country industry in which each firm's cost of producing additional knowledge at time t, ΔS_t (its expenditure on R&D), is c_t. Knowledge is transmitted by embedding it in a machine (for example, manufacturing equipment or a CD-ROM). The cost of a knowledge-free machine is $m_t = m, \forall t$.

The production of knowledge and machines takes place in a competitive setting. Each firm produces only one machine. In other words, there are no contemporaneous economies of scale in knowledge production at the firm level.[5] The cost of a knowledge-embedded machine is

$$C_t = m_t + c_t = m + c_t. \tag{6.1}$$

Machines can absorb only a fixed amount of additional knowledge $\Delta S_t = \Delta S, \forall t$ (for example, fixed space in a CD-ROM), and they depreciate after one period, while knowledge does not. Knowledge at time t is thus

$$S_t = \sum_{1}^{t} \Delta S_\tau = \sum_{1}^{t} \Delta S = t * \Delta S. \tag{6.2}$$

All firms know S_{t-1} at the start of period t. In other words, private knowledge in period $t-1$ becomes public knowledge at time t (possibly due to reverse engineering). S_t is not known publicly at time t. Any firm that wants to obtain and sell S_t at time t has to produce ΔS_t.

Additional knowledge, ΔS_t, increases with c_t, the expenditure on R&D, and with S_{t-1}, the stock of publicly available knowledge at the start of period t. The assumption is that privately produced additional knowledge is complementary to the existing stock of publicly available knowledge. For instance, a high-knowledge economy is likely to more easily produce an advanced piece of knowledge than a knowledge-scarce economy. Thus knowledge production benefits from increasing returns at the industry level:

$$\Delta S_t = f(c_t, S_{t-1}), f_1, f_2, f_{12} > 0. \quad (6.3)$$

For simplicity, assume

$$\Delta S_t = c_t * S_{t-1}. \quad (6.4)$$

From equations (6.2) and (6.4):

$$c_t = \Delta S / S_{t-1} = 1/(t-1), \quad t > 1. \quad (6.5)$$

Thus the cost of producing additional knowledge, c_t, falls over time, and so does the cost of machines, C_t (equation 6.1), despite the fact that their knowledge content increases. The reason is that the cost of machines is related to the cost of producing *additional* knowledge, not to the stock of knowledge, because the stock of knowledge in the previous period is publicly available at zero cost and the increasing stock of publicly available knowledge reduces the cost of developing additional knowledge.

From equations (6.1) and (6.5):

$$C_t = m + 1/(t-1). \quad (6.6)$$

At time t, firms can build "new" machines embedded with the latest knowledge, S_t, at cost C_t, or they can build "old" machines with publicly available knowledge, S_{t-1}, at cost $C_t^* = m$.[6] Thus in equilibrium, (at most) two types of machines will be built.

Note that the difference in the cost of the two types of machines is

$$C_t - C_t^* = 1/(t-1). \quad (6.7)$$

The cost difference declines over time. Consequently (as shown formally below), the proportion of new machines used increases over time.

Demand

Downstream firms in developing countries need to buy a machine in every period in order to be able to produce. They demand either an old or a new machine. A downstream firm's choice of type of machine depends on its knowledge-absorption capacity, ϕ, and on country-specific human capital, H. The higher a firm's ϕ, the more effectively it can make use of the latest knowledge in its production process and the greater the likelihood that it will buy the latest (new) machines. Similarly, the higher the level of country-specific human capital, *H*, the easier it is for a firm to productively absorb new knowledge.

Even though S_{t-1} is public knowledge for all firms in the knowledge-producing industry of the developed country, it is not public knowledge for the downstream firms in developing countries that produce other goods and services. Their knowledge-absorption capacity at time *t* is assumed to depend on a firm-specific exogenous component, ε, and on whether they are using old or new machines at $t-1$. If firms use the latest machines at $t-1$, they acquire a greater capacity to adopt new knowledge and be more effective in using the latest machines at time *t*. Thus

$$\phi_t = \phi_t\left(\varepsilon, M_{t-1}^i\right), \phi_1, \phi_2 > 0, \tag{6.8}$$

where M_{t-1}^i denotes the effect of a machine of type *i* (*i* = *OLD*, *NEW*) at $t-1$ on a firm's knowledge-absorption capacity at time *t*, and $M_{t-1}^{NEW} > M_{t-1}^{OLD}$. For simplicity, let

$$\phi_t = \varepsilon * M_{t-1}^i. \tag{6.9}$$

Assume that at time *t*, the productivity gain downstream firms obtain from additional knowledge, ΔS_t, is ΔR_t. Normalizing firm output and sales price to unity, the added profits from ΔS_t are (approximately) ΔR_t, given by

$$\Delta R_t^i = g(\phi_t, H_t) = \phi_t\left(\varepsilon, M_{t-1}^i\right) * H_t = \varepsilon * M_{t-1}^i * H_t, \tag{6.10}$$

where the subscript *i* of ΔR_t^i refers to the type of machine used by the firm at $t-1$.

Downstream firms choose a new (old) machine at time *t* as long as the productivity gain is larger (smaller) than the additional cost, that is, as long as $\Delta R_t^i >$ $(<) C_t - C_t^* = 1/(t-1)$.

The critical value ε_{ct}, the exogenous component of a firm's knowledge-absorption capacity, where the firm is indifferent between the two types of machines at time *t*, is given by the condition $\Delta R_t^i = 1/(\mathrm{t}-1)$ or

$$\varepsilon_{ct} = 1\Big/\left[(t-1) * M_{t-1}^i * H_t\right]. \tag{6.11}$$

The higher a country's level of human capital, the lower the critical firm-specific value of ε_{ct} at which firms switch from old to new machines, that is, the lower the share of old machines and the lower the average age of machines. If a downstream firm's $\varepsilon < \varepsilon_{ct}$, then $\Delta R_t < 1/(t-1)$ and it buys an old machine at time *t*. A firm using an old machine at $t-1$ has a higher value of ε_{ct} (equation 6.11) and is thus more likely to continue using an old machine at time *t*. Given that the cost of additional knowledge falls over time (equation 6.5), ε_{ct} falls over time (equation 6.11). When ε_{ct} has fallen to the point at which the inequality is reversed ($\varepsilon > \varepsilon_{ct}$), the firm switches to new machines forever (assuming no sudden fall in H_t). If at first $\varepsilon > \varepsilon_{ct}$, the firm buys new machines from the start.

Thus the productivity of firms with low values of ε is hurt both because of the low value of ε and because firms with low values of ε tend to buy old machines. Similarly, a low level of human capital hurts a firm's productivity both because of the lower level of human capital itself and because it raises the value of ε_{ct} and increases the likelihood that the firm will buy old machines.

With this setting and the distribution of firm-specific knowledge-absorption capacity ε, we can determine the share of a given type of machine (old or new) imported by any country. Let the distribution of ε be uniform, with $\varepsilon \in [0,1]$. Then a given country's share of old machines is

$$\varepsilon_{ct} = 1/\left[(t-1) * H_t * M_{t-1}^{OLD}\right]. \tag{6.12}$$

Given that the age of new machines is zero and that of old machines is one, the average age of the stock of machines is also ε_{ct}. Thus the critical value ε_{ct} is equal to the share of old machines and to the average age of machines. It is inversely proportional to a country's level of human capital.[7]

The average cost of machines is given by

$$AC_t = C_t^* * \varepsilon_{ct} + C_t * (1-\varepsilon_{ct}) = m * \varepsilon_c + (m+c_t)(1-\varepsilon_{ct}) = m + c_t(1-\varepsilon_{ct}). \tag{6.13}$$

A country with a very high level of human capital (a country at the frontier) would have a very low critical value of ε_{ct}. From equation (6.12), $\varepsilon_c \to 0$ as $H \to \infty$. Thus in an economy at the frontier, all firms use new machines, and the average cost of machines relative to the average cost at the frontier, RAC_t, is

$$RAC_t = \frac{m + c_t(1-\varepsilon_{ct})}{m + c_t}. \tag{6.14}$$

RAC_t, which reflects the average quality of machines in a given country relative to that at the technological frontier, is used in the empirical analysis. It falls with ε_{ct}—and thus rises with the level of human capital (equation 6.12)—and it falls with $c_t = 1/(t-1)$, the cost of producing additional knowledge. Given that c_t falls over time, RAC_t increases and so does the share of new machines.

In the empirical analysis that follows, we estimate a combination of equation (6.10), which relates productivity gains to the choice of machines, and equation (6.12), which shows the (average) choice of machines made by a given country.

Data and Sample Countries

The study focuses on the imports of machines from the European Union by three countries in Central and Eastern Europe (Bulgaria, Hungary, and Poland) and three in the Southern Mediterranean (Egypt, Israel, and Turkey) between 1989 and 1997. The sample countries differ in terms of their level of development, with GNP per capita ranging from $1,380 in Bulgaria to $16,180 in Israel in 1997.

Economic integration increased between the European Union and the sample countries in this period, with growing flows of trade and FDI. All of the sample countries have preferential trade agreements with the European Union, which is by far their major trading partner and source of imported technologies (60–90 percent of their machines are imported from the European Union) (Barba Navaretti, Galeotti, and Mattozzi 2000). Eaton and Kortum (2001) provide evidence that world production of machines is highly concentrated in a small number of countries and that developing countries are almost invariably net importers of such machines.

Measuring Technological Complexity

How should the level of technological complexity of imported machines be measured? The theoretical section showed that the average level of technological complexity of machines used by a country is measured by its average cost. We proxy the technological complexity of imported machines by their average unit values, which are constructed from trade statistics on imports. EU trade statistics (Comext-Eurostat) provide sufficiently disaggregated data in both values and quantities. Quantities of machines are measured in metric tons.[8]

The use of unit values as a proxy of technological complexity raises several concerns. First, how closely does this indirect measure capture differences in technological complexity? As shown in the theoretical model, in a competitive market, differences in the prices of similar machines should reflect differences in their knowledge content and productivity. Indeed, the unit values of the metalworking machines exported from the United States are highly correlated with the skill index of technological complexity discussed above. These correlations range from 0.60 to 0.95, depending on the level of aggregation of the categories of machines.

Second, how accurate are unit values for comparing different types of machines at any point in time and the same machines across time? Different types of machines have different prices because they are inherently different—and not only because they are more or less complex. Moreover, the price of a machine declines with time, due to obsolescence. To control for these effects, we construct a unit value index by normalizing the unit values of machines imported by a given country, classified at the six-digit level in trade statistics (harmonized code) by the unit value of the same machines imported by the United States, which is assumed to be the technological frontier. Specifically, the unit value index for a six-digit machine i imported by country c at time t is given by

$$UVI_{ict} = (UV_{ict}/UV_{iUSt}), \tag{6.15}$$

where the denominator is the unit value of the same machine *i* imported by the United States at time *t*. This is essentially the relative average cost of machines as defined in equation (6.14).

Third, unit values may capture market imperfections, such as market power or trade barriers, which also affect prices. However, the countries in the sample are small, and we can reasonably assume that the price of machines is given for them. Moreover, we use freight on board prices in current ECU at the EU border. These prices should not be distorted by trade or other policies in the importing country.

Once unit values indexes (UVIs) are computed, it is necessary to derive correspondences at the industry level between categories of machines imported and the industries using them in production. If, for example, we are interested in computing the UVI for the textile industry, we must aggregate it over all textile machines. We are able to do so at the three-digit ISIC industry level by matching data on productivity derived from industrial statistics (UNIDO) and data on imports of technology derived from trade statistics (Comext-Eurostat).[9] The industry matching, reported in annex A, is available for 13 sectors. Thus the average UVI of the machines used in the three-digit ISIC industry *j* in country *c* at time *t* is given by

$$UVI^{6}_{jct} = \sum_{i=1}^{n}\left(UVI_{ict}\frac{V_{ict}}{V_{jct}}\right), \qquad (6.16)$$

where *n* is the number of six-digit categories *i* corresponding to the ISIC three-digit category *j*, V_{ict} is the value of machines *i* imported by country *c* at time *t*, and V_{jct} is the total value of machines used in sector *j* imported by country *c* at time *t*.[10]

Trends of Unit Value Indexes

It is useful to observe how the UVI^6 behave across countries and over time. Figure 6.1 reports the trends of the average of UVIs for the sample importing countries and some other Southern Mediterranean countries. The U.S. index is set at 100. For most years and countries, the index is lower than 100 and declining. These trends support the theoretical prediction that on average developing countries import less technology-intensive machines than does the United States. The country with the smallest gap is Israel, the country with the highest per capita income in the sample. The technology gap widens for all countries in the sample, quite dramatically so for Hungary and Poland, although their trend is affected by the dramatic turnaround of their economies after 1989.[11]

FIGURE 6.1 Average Unit Values in Select Countries, 1989–97

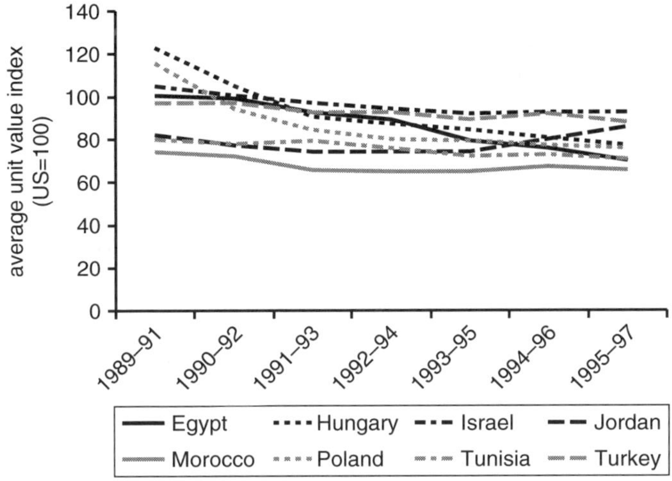

Source: Authors.
Note: Figures show three-year moving averages.

Annex D examines the persistence in the technology gap of imported machines in a different way. It compares UVIs with their values lagged one, two, and seven years and shows a striking persistence in the technology gap.

Do Embodied Imported Technologies Boost Productivity? Econometric Analysis

We have shown some descriptive evidence that relative to the United States, developing countries import machines embodying simpler technologies. The theoretical model shows that technology imports and productivity are two endogenous choices to be analyzed jointly. The use of more advanced technologies is expected to increase productivity, but firms will buy them only if the increase in productivity is worth the cost.

We examine the impact of machine imports on TFP in those industries (j) using them as factors of production. We work with a panel comprising 13 industries, 6 countries, and 8 years (1989–96). Annex B describes how we computed TFP; annex C reports the data sources and the basic descriptive statistics of the variables used in the estimations.

TFP at time t is assumed to depend on lagged productivity (which proxies the exogenous component of the knowledge absorption capacity), on productivity at the frontier, on the types of machines used in production at t and earlier, and on the overall level of development of the importing country. It can be empirically implemented as follows:

$$\ln(TFP_{cjt}) = \alpha_1 + \alpha_2\left(\sum_{\tau=1}^{n} \ln TFP_{cjt-\tau}\right) + \alpha_3 \ln TFP_{USjt-1} + \alpha_4 \ln IMP_{cjt-1}$$
$$+ \alpha_5 \ln UVI_{cjt-1} + \alpha_6 \ln DOMINV_{ct-1} + \alpha_7 \ln GDP_{ct-1}$$
$$+ \alpha_8 Dj + \alpha_9 Dc + \alpha_{10} Dt + \varepsilon_{cjt}, \quad (6.17)$$

where TFP_{cjt} is TFP for industry j in the importing country c at time t; TFP_{USjt} measures TFP for industry j in the United States, the technological leader, which captures the effects of technological progress at the frontier on TFP in c;[12] IMP is the share of imported machines in total investments in industry j, which controls for the relative importance of imported machines; UVI is the unit value index as defined in equation (6.15), which proxies the complexity of the machines imported and used in sector j in the importing country c at time t; and $DOMINV$ is a variable controlling for the technological content of domestic investments. As $DOMINV$ is not observable, we use domestic consumption of electricity as a proxy. GDP is per capita GDP, and measures aggregate demand and the overall level of development of country c at time t; it proxies human capital, as well as infrastructure and institutional development. Dj, Dc, and Dt are industry, country, and time dummies.

We also need to analyze the choice of embodied technology, as a function, among other things, of expected productivity, given that the two are jointly determined. Equation (6.14) in the model shows that the proximity to the technological frontier of the machines imported each year by country c—and consequently their relative average cost—is affected by the ability of importers to use high-tech technologies efficiently. This term is made up of firm-specific components (that is, the firm's absorption capacity) and of country-specific components. A firm's absorption capacity, and thus the gains in productivity achievable through the new technologies, depends on its past productivity (TFP_{t-1}); on the technologies used in the past (UVI_{t-1}); and on other factors, such as its relationship to foreign firms (OPT_{t-1}). As we do not have consistent sector-specific data on FDI, we measure it indirectly by looking at the share of exports of sector j from country c classified as outward processing trade (OPT). OPT captures flows of temporary trade between subcontractors and between parent companies and subsidiaries.[13] The country-specific components include the overall level of development, as proxied by GDP per capita (GDP_{t-1}). We also include relative factor prices (w/r)

to control for the relative labor intensity of different types of technologies.[14] We thus have

$$\ln UVI_{cjt} = \beta_0 + \beta_1(\ln UVI_{cjt-1}) + \beta_2 \ln\left(\frac{w_{cjt}}{r_{ct}}\right) + \beta_3 \ln TFP_{cjt-1}$$
$$+\beta_4 GDP_{ct-1} + \beta_5 \ln OPT_{cjt-1} + \beta_6 Dj + \beta_7 Dc + \beta_8 Dt + v_{cj}. \quad (6.18)$$

As for productivity, all technological choices are observed at the sector level. Equations (6.17) and (6.18) define a system that jointly determines the choice of imported technologies and productivity.

We face several econometric problems. First, our results may be driven by spurious correlation, in that unobserved time-invariant factors may affect both productivity and the choice of technology. Such factors could be the share of foreign investors in the industry or the degree of export orientation. Although we can control for some of these variables, others may remain unobservable. Second, as discussed in the theory, there is persistence over time in both productivity and the choice of technology, not necessarily related to the learning process associated with high-tech machines. Third, an endogeneity problem arises from the simultaneity among productivity, the choice of technology, and potentially most of the explanatory variables.

To eliminate the effect of time-invariant unobservable factors, we carry out the estimations in first differences. To isolate the impact of technological choices on productivity from trend effects, we estimate both productivity and the choice of technology on their lagged values. As for the endogeneity problem, we run two independent regressions in which all explanatory variables can be instrumented using the appropriate lagged variables, using the GMM-Instrumental Variable method (GMM-IV) developed by Arellano and Bond (1991) for dynamic panels.[15]

The two first-difference equations to be estimated are obtained by transforming equations (6.17) and (6.18) as follows:

$$\Delta\ln(TFP_{cjt}) = a_1\Delta\ln(TFP_{cjt-1}) + a_2\Delta\ln TFP_{USjt-1} + a_3\Delta\ln UVI_{cjt-1}$$
$$+a_4\Delta\ln IMP_{cjt-1} + a_5\Delta\ln DOMINV_{ct-1}$$
$$+a_6\Delta\ln GDP_{ct-1} + a_7 Dt + \Delta\varepsilon_{cjt}. \quad (6.19)$$

$$\Delta\ln UVI_{cjt} = b_1\Delta\ln UVI_{cjt-1} + b_2\Delta\ln\left(\frac{w_{cjt}}{r_{ct}}\right) + b_3\Delta\ln TFP_{cjt-1}$$
$$+ b_4\Delta\ln OPT_{cjt-1} + b_5\Delta\ln GDP_{ct-1} + b_6 Dt + \Delta v_{cjt}. \quad (6.20)$$

The results of our estimations are reported in table 6.1 for productivity (equation 6.19) and table 6.2 for the choice of technology (equation 6.20). Both regressions perform well. The nonsignificant Sargan tests indicate that the instruments are appropriate. The estimations also successfully take care of the serial autocorrelation of disturbances, given that there is no second-order autocorrelation.

We first focus on the determinants of TFP in table 6.1. Regressions (1) and (2) include the import share. As this variable has many missing observations (see annex C), to test the robustness of our results we also report regression (3), which does not include the import share and has many more observations, a number similar to those used in estimating the choice of technology. We find that embodied technologies, as measured by the UVI of the machines imported, have a positive effect on TFP. The coefficient is significant and robust in the three regressions (and in alternative specifications not reported). Import shares have a positive but not significant coefficient (and are not robust to changes in this specification), thus confirming our presumption that what matters for productivity and growth is the quality of the machines imported and that once we control for quality, quantities are not important. TFP in the United States also has a positive effect on TFP in the importing country, indicating that technical progress in the leader generates spillovers onto the laggard, independently of the two countries' trade relations. GDP has a positive effect in regression (1), confirming the prediction that country-specific factors related to the level of development enhance the efficient use of advanced technologies. However, this variable is sector invariant and almost time invariant and thus not very robust in our estimations. It is no longer significant in (2), where we include electricity consumption per capita, a proxy of the quality of domestic investments. Electricity consumption per capita is not significant in regression (2), probably because it is highly correlated to GDP per capita. GDP per capita is also not significant in regression (3).

We now move on to the analysis of the choice technology. Results are reported in table 6.2. Lagged TFP has a positive and significant effect on UVI, which proxies for the choice of technology. Firms buy high-tech machines if they have enough skills to use them in a sufficiently productive way. Factor prices have no effects. Alternatively, we use the ratio of real wages on real interest and the wage rental ratio, as described in footnote 14. The wage rental ratio is the best measure of relative factor costs, but it includes the price of U.S. machines, which is also in the denominator of the dependent variable, thereby causing spurious correlation. Results, though, are unaffected by the use of these alternative measures.

OPT, which captures the involvement of foreign firms, is also nonsignificant, even when we check for nonlinearity by introducing *OPT* squared. As expected, GDP has a positive impact on the choice of technology, confirming that frontier technology is purchased only when the overall level of development, and implicitly

TABLE 6.1 Determinants of Total Factor Productivity

Dependent variable: diff ln total factor productivity			
	Regression		
Variable	(1)	(2)	(3)
Lag diff ln total factor productivity	−0.008 (0.075)	0.076 (0.588)	0.277*** (3.34)
Lag diff ln unit value index	0.228*** (2.59)	0.166** (2.14)	0.169** (2.06)
Lag diff ln import shares	0.024 (1.20)	0.018 (0.999)	
Lag diff ln gross domestic product per capita	1.021** (2.00)	0.503 (1.04)	0.181 (0.538)
Lag diff ln total factor productivity US	0.556** (2.47)	0.516*** (2.53)	0.339* (1.85)
Lag diff ln electricity consumption per capita		0.471 (1.07)	
Number of observations	153	153	345
Wald (joint)[1]	23.99 (0.000)	57.91 (0.000)	26.81 (0.000)
Sargan test[2]	37.36 (1.000)	36.36 (1.000)	53.59 (0.708)
AR (1) test	−2.078 (0.038)	−2.220 (0.026)	−2.560 (0.010)
AR (2) test	1.638 (0.101)	1.836 (0.066)	1.741 (0.082)

Note: Table includes results from the first step of two-stage GMM-Instrumental Variables estimates plus the Sargan test derived from the second step. "Diff" indicates first-order differencing. Time dummies are included in all equations. The absolute values of *t*-statistics are given in parentheses. All explanatory variables except *TFP US* are treated as endogenous. GMM–type level instruments are their second or earlier lags. The GMM estimations were performed using the program DPD for OX (Doornik, Arellano, and Bond 2002).

1. Wald statistics are a test of the joint significance of the independent variable asymptotically distributed as χ^2 with k degrees of freedom (k is the number of coefficients estimated excluding time dummies), under the null hypothesis of no relationship.

2. The Sargan test of overidentifying restrictions is distributed as a χ^2 with as many degrees of freedom as the number of overidentifying restrictions, under the null hypothesis of the validity of the instruments. The test based on the two-step GMM estimator is heteroskedasticity-consistent.

*Significant at the 10 percent level. **Significant at the 5 percent level. ***Significant at the 1 percent level.

TABLE 6.2 Choice of Technology

Dependent variable: Diff ln *UVI*[6]	Regression (1)	Regression (2)
Lag diff ln unit value index	–0.081 (0.60)	0.041 (0.662)
Lag diff ln total factor productivity	0.259* (1.93)	0.256** (2.10)
Lag diff ln wage rental	0.048 (0.84)	
Lag diff ln real wage over real interest rate		–0.091 (1.12)
Lag diff ln outward processing trade	0.925 (0.47)	–0.877 (1.11)
Lag diff ln outward processing trade squared	–2.553 (0.853)	
Lag diff ln gross domestic product per capita	0.744** (1.97)	0.035 (0.09)
Number of observations	379	379
Wald (joint)	17.80 (0.007)	11.16 (0.048)
Sargan test	55.71 (1.000)	62.21 (1.000)
AR (1) test	–4.571 (0.000)	–4.312 (0.000)
AR (2) test	–0.465 (0.642)	1.203 (0.229)

Source: Authors.

Note: See note to table 6.1.

*Significant at the 10 percent level.

**Significant at the 5 percent level.

human capital, is high. However, as in the estimation of the determinants of TFP, this result is not robust to changes in the specifications of the model: the coefficient of GDP is not significant when we measure relative factor costs as real wages over real interest rates.

Conclusion

This chapter explores the impact of imported technologies on productivity in manufacturing sectors for a sample of developing countries and transition economies in Central and Eastern Europe and the Southern Mediterranean. These countries have recently integrated their economies with the European Union.

The analysis, which is based on a theoretical model of the choice of technology, departs from earlier studies of international technology diffusion by focusing on the technology embodied in imported machines. The technological level of the imported machines is proxied by an index relating the unit value of the machines imported by a given country to the unit value of the same machines imported by the United States. We find very strong regularities in the pattern of imported machines. Unit values are generally stable across time, except for countries facing dramatic shocks in the period observed. There is a constant and even increasing gap between the unit value of the machines imported by the United States and the machines imported by our sample of developing countries. The increasing gap may be due partly to the fact that productivity grew unusually rapidly in the United States in the 1990s.

The technology gap reflects two inherent characteristics of technological progress in the past decade. On the one hand, the price of machines has been stable over time, despite the rise in their technological content. On the other hand, at any point in time, the prices of machines increases with the machines' technological content. Therefore, as the price of technology has fallen over time, developing countries have imported increasingly more advanced machines, while the gap with respect to technological leaders has remained roughly constant. We show that this gap is persistent and that it is higher the lower the level of development of the importing country.

We also show that productivity in manufacturing depends on the type of machines imported in a given industry. Thus the cheaper and less sophisticated machines that developing countries import result in a lower TFP than frontier technologies, even though the choice may well be optimal given their relative factor prices and their reduced ability to use frontier technologies.

In contrast with earlier studies we find that importing machines per se does not enhance importers' productivity. Once we control for the quality of imports, quantity does not seem to matter.

Chapter 6 Annex A: Matching between Machines and Products

TABLE 6A.1 Matching between Machines and Products

Machines		
Harmonized	SITC code	Description
8437/38 (excluding 84384/79)	727	Food machinery, nondomestic
84384, 842121, 842122, 8435	727	Food machinery, nondomestic
847810, 84790	72843	Tobacco working machines
8444–8451	7244, 7245, 7246, 7247	Textile machinery
8452	7243	Sewing machines
8453	7248	Skin, leather working machines
84793,8465, 8466	72812, 72819, 72844	Machine tools for working woods and wood treating machines
8439–8441	725	Paper mill machinery
8440–8442	726	Printing and binding machinery
8456–8463, 8466	731, 733, 735	Machine tools for metal
8454, 8455, 8468, 8515	737	Metalworking machinery
8475, 8464, 2019	72841	Glass working machinery
8477	72842	Rubber and plastic working machines

TABLE 6A.1 (*Continued*)

Products		
NACE	ISIC rev. 2	Description
411–423	311	Food
424–428	313	Beverages
429	314	Tobacco
431–439	321	Textiles
453–456	322	Clothing
441, 442, 451, 452	323, 324	Shoes and leather
461–467	331, 332	Wood and wood furniture
471, 472	341	Paper and paper products
473	342	Printing
312–319, 321–328, 351–353, 361–365	381, 382, 384	Metal products and machines (including transport but excluding electrical machines)
221–233, 311	371	Iron and steel
247	362	Glass
481–483	355,356	Rubber and plastic

Source: Authors.

Chapter 6 Annex B: Empirical Derivation of Total Factor Productivity

Measuring Changes in Total Factor Productivity

The estimation procedures used are very straightforward. We assume that sectoral GDP (Yj) is produced using two factors, physical capital (K) and labor (L), using a Cobb-Douglas production function:

$$Y_{jt} = A_{jt}(0)e^{\lambda_{jt}}\left(K_{jt}^{\alpha_{jt}} L_{jt}^{1-\alpha_{jt}}\right), \tag{6B.1}$$

where j indicates sector, $A_j(0)$ represents initial conditions, λ_j is the rate of technological progress in sector j, α_j measures the importance of physical capital in output, and $1-\alpha_j$ measures the importance of labor. Taking logs and differentiating with respect to time yields

$$d\ln(Y_{jt}) = \lambda_j + \alpha_j d\ln(K_{jt}) + (1-\alpha_j)d\ln(L_{jt}). \tag{6B.2}$$

We estimated equation (6.2) by sector j and time t. We pooled data for all c countries in our sample, added a time trend dummy (Dt) and a country dummy (Dc), and, by country, added a dummy for periods of recession in economic activity (DR_{cjt}) that takes the value 1 whenever $Y_{cjt} < Y_{cjt} - 1$. The final equation estimated is

$$d\ln(Y_{cjt}) = \lambda_{cjt} + \alpha_j d\ln(K_{cjt}) + (1-\alpha_j)d\ln(L_{cjt}) + Dc + Dt + DR_{cjt} + \varepsilon_{cjt}. \tag{6B.3}$$

To gain in efficiency, we take into account the simultaneous correlation between the disturbances in different sectors (due to, for instance, common shocks) by estimating all the sectors as a system, by SUR.

Changes in TFP by country and by sector were calculated as

$$\Delta TFP_{cjt} = d\ln(Y_{cjt}) - \hat{\alpha}_j d\ln(K_{cjt}) - (1-\hat{\alpha}_j)d\ln(L_{cjt}) - \hat{D}c - \hat{D}t - D\hat{R}_{cjt}. \tag{6B.4}$$

Values estimated for α (the contribution of capital) varied from 0.25 for the food sector to 0.75 for the machinery sector.

The Data

TFP was estimated for 13 sectors, disaggregated on the basis of the three-digit ISIC rev. 2 code (see annex C), for 1980–96. Because of lack of data availability, TFP at the sector level could be computed only for Bulgaria, Egypt, Hungary, Israel, Poland, Turkey, and the United States. Capital stocks were calculated according to the perpetual inventory method. The data came from the UNIDO Industrial Statistics database.

Chapter 6 Annex C: Description of Variables, Sources of Variables, and Descriptive Statistics

TABLE 6C.1 Description and Source of Data for Variables

Variable	Description	Data source
Total factor productivity (TFP)	Total factor productivity (see annex B)	Unido Industrial Statistics
Total factor productivity US	Total factor productivity in the United States (see annex B)	Unido Industrial Statistics
Unit value index (UVI)	Ratio between the unit value of machines imported by a country from the European Union and the unit value of machines imported by the United States	Comext-Eurostat
Wage rental rate	$\frac{w_{cjt}}{P_{USjt}(1+r_{ct}+\delta)}$, where P_{us} is the unit values of machines imported by the United States, r_{ct} is the real interest rate, and δ is a fixed 10 percent depreciation rate	Unido Industrial Statistics and Comext-Eurostat
Real wage	Interest rate-deflated wages	Unido Industrial Statistics and Comext-Eurostat
Import share (IMP)	Average share of imported machines in total investments	Comext-Eurostat and Unido Industrial Statistics
Outward-processing trade (OPT)	Shares of outward-processed exports in total exports of the sample country	Comext-Eurostat
GDP per capita	Real gross domestic product of the importing country	*World Development Indicators* (World Bank)

Source: Authors.

Note: All variables except GDP are measured for sector *j* in country *c* at time *t*.

TABLE 6C.2 Descriptive Statistics

Variable	Number of observations	Mean	Standard deviation
TFP	489	91.22	16.96
TFP US	504	107.39	13.33
Unit value index (UVI)	648	96.54	40.15
Wage rental rate	540	215.32	222.56
Real wage	573	249.47	382.54
Import share (IMP)	340	0.38	0.24
Outward-processing trade (OPT)	648	0.06	0.14
GDP per capita	648	6,609.63	4,326.49

Source: Authors.

Chapter 6 Annex D:

An alternative way to assess the persistency in the technology gap in machines imported is to compare sector- and country-specific unit value indexes with their lags. We plot the unit value indexes of machines imported at time *t* with the unit value indexes of the same machines imported one year earlier (figure 6D.1), two years earlier (figure 6D.2), and seven years earlier (figure 6D.3). The figures show a striking persistence in the gap, with the indexes positively correlated with their lagged values even with a seven-year lag. Moreover, a large share of the observations lie above the diagonal and this share increases the longer the lag, implying that for many sectors and countries the gap is increasing.

FIGURE 6D.1 Persistency of the Technology Gap with One-Year Lag

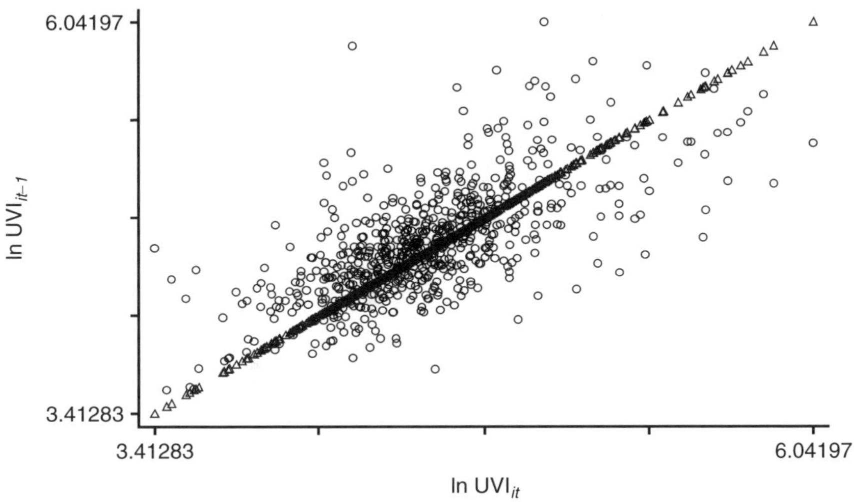

Source: Authors.

Note: Δ denotes ln UVI_{it-1} = ln UVI_{it}.

FIGURE 6D.2 Persistency of the Technology Gap with Two-Year Lag

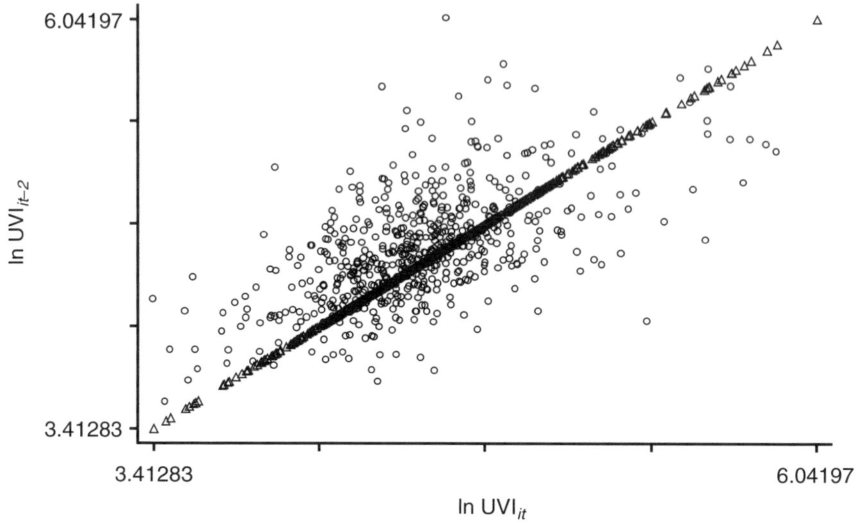

Source: Authors.

Note: Δ denotes $\ln \text{UVI}_{it-2} = \ln \text{UVI}_{it}$.

FIGURE 6D.3 Persistency of the Technology Gap with Seven-Year Lag

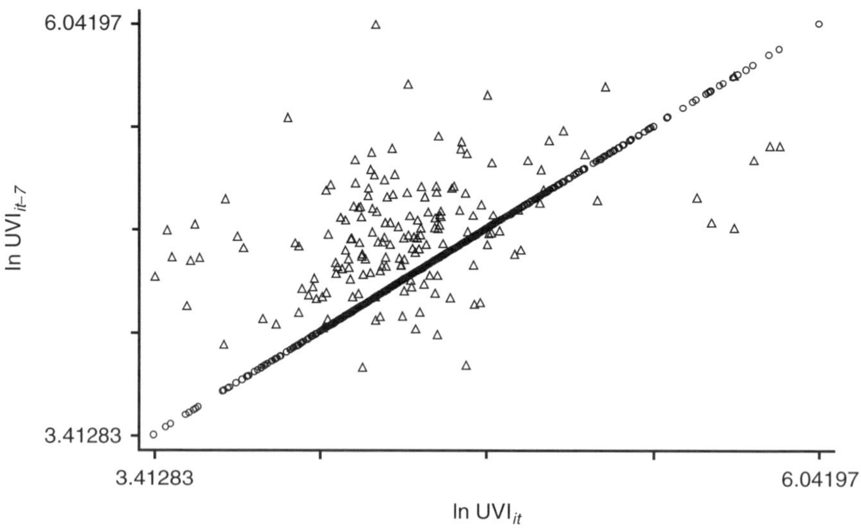

Source: Authors.

Note: ∘ denotes $\ln \text{UVI}_{it-7} = \ln \text{UVI}_{it}$.

Notes

1. These channels are imports (Coe and Helpman 1995; Coe Helpman, and Hoffmaister 1997; Keller 2000; Djankov and Hoekman 1996, Mody and Yilmaz 2002; Schiff, Wang, and Olarreaga 2002); foreign direct investments (Blomström and Kokko 1998); and exports (Clerides, Lach, and Tybout 1998; Bernard and Jensen 1999; Aw, Chung, and Roberts 2000; Kraay [chapter 7 of this volume]). See also Barba Navaretti and Tarr (2000) for a review.

2. Unit values per ton of machine are very highly correlated to unit values per number of machines, but unit values per number of machines are available only for a limited number of machines and countries.

3. Eaton and Kortum (2001) deal with the choice of capital equipment and its impact on productivity. Caselli and Wilson (2003) classify imports of capital equipment according to the R&D intensity of the capital equipment–producing industry. However, in their framework all capital equipment imported of a given type is technology invariant, whereas our framework specifically takes into account differences in technological content within specific categories of machines.

4. Eaton and Kortum (2001) analyze both the choice of imported capital equipment and its impact on productivity. Caselli and Wilson (2003) analyze the choice of technology and its link to per capita income. Studies looking at the link between productivity and exports explicitly recognize that the choice of exporting is endogenous with the respect to the effects of productivity on exporting (Clerides, Lach, and Tybout 1998; Bernard and Jensen 1999; Aw, Chung, and Roberts 2000; Kraay (chapter 7 of this volume).

5. However, as shown below, the model incorporates increasing returns at the industry level over time.

6. The "old" machines are not old, but the knowledge embedded in them is. Firms can also build machines with older knowledge at cost m, but doing so is never optimal.

7. In reality, the negative link between human capital and the age of machines may be even stronger, because countries with lower levels of income and human capital will tend to have more low-productivity firms.

8. For some countries, the quantity of machines is also measured in terms of the number of machines, but these data are not as widely available as metric ton data. Unit values computed using the two quantity units are very highly correlated.

9. General machines, such as computers, which are used by all industries and cannot therefore be attributed to any, are omitted from the index.

10. This index is subject to a composition effect. The index can increase with time, either because countries buy the same bundle of machines and the value of each or some of them increases or because bundles change toward machines with a higher average unit value. To avoid this problem, it is possible to construct Tornqvist price indexes in which weights are fixed over time, normally the period average weights (Aw and Roberts 1986). However, our unit values are already normalized across machines. Thus an increase in the index due to a composition effect captures the process of technological upgrading that we want to observe.

11. A possible explanation for the abrupt decline of the index in Hungary and Poland that is consistent with earlier findings based on the skill index (Barba Navaretti, Galeotti, and Mattozzi 2000) is as follows. Eastern European countries used to buy most of their machines within the Soviet Bloc; they imported only very sophisticated machines from Western Europe. The first years observed in our data may capture this earlier distortion. Once trade was liberalized with the European Union, a geographical reorientation of imports took place and most machines were later imported from Europe. Consequently, the average quality of the machines imported from the European Union fell.

12. Given how UVI is constructed, if both machines imported by the United States and c improve at the same pace with time, UVI_c remains constant even if c imports increasingly productive machines. To partially control for this effect, we include industry-specific TFP in the United States.

13. It is likely that local subcontractors or local subsidiaries of Western companies use more advanced machines, for a variety of reasons, including the standards imposed on them by their foreign partners or parents.

14. In some specification of our empirical estimations, we use the wage rental ratio $w_{cjt}/[P_{USjt}(1+r_{ct}+\delta)]$. This measure allows us to control for the effects of δ, a fixed yearly depreciation rate of 10 percent, and P_{USjt}, the price of the machines imported by the United States in sector j at time t (a proxy of the price of the highest technology machines). Unfortunately, this variable is also in the denominator of *UVI* and is thus a source of spurious correlation. We also used an alternative measure of labor cost, controlling for the skill composition of the labor force using the ILO's Labour Statistics Database. Unfortunately, these data are sector invariant and are of little value in cleaning sector-specific labor cost data. Using these alternative measures of labor costs did not change the results.

15. We also estimated a system of simultaneous equations in which productivity and technology are jointly determined. These estimations give similar results as the GMM-IV ones, but the endogeneity of other variables except for *TFP* and *UVI* cannot be controlled for, a major shortcoming.

References

Arellano, Manuel, and Stephen Bond. 1991. "Some Tests of Specification for Panel Data: Monte Carlo Evidence and an Application to Employment Equations." *Review of Economics Studies* 58 (194): 277–97.

Aw, Bee Yan, and Mark J. Roberts. 1986. "Measuring Quality Change in Quota-Constrained Import Markets: The Case of U.S. Footwear." *Journal of International Economics* 21 (1/2): 45–60.

Aw, B. Y., S. Chung, and M.J. Roberts. 2000. "Productivity and Turnover in the Export Market: Micro-Level Evidence from Taiwan (China) and the Republic of Korea." *World Bank Economic Review* 14 (1): 65–90.

Barba Navaretti, Giorgio, and David G. Tarr. 2000. "International Knowledge Flows and Economic Performance: An Introductory Survey." *World Bank Economic Review* 14 (1): 1–16.

Barba Navaretti, Giorgio, Marzio Galeotti, and Andrea Mattozzi. 2000. "Moving Skills from Hands to Heads." Working Paper 2525, Centre for Economic Policy Research, London.

Barba Navaretti, Giorgio, Isidro Soloaga, and Wendy Takacs. 2000. "Vintage Technology and Skill Constraints: Evidence from U.S. Exports of Used and New Machines." *World Bank Economic Review* 14 (1): 91–109.

Bernard, Andrew B., and J.B. Jensen. 1999. "Exceptional Exporter Performance: Cause, Effect or Both?" *Journal of International Economics* 47 (1): 1–25.

Blomström, M., and A. Kokko. 1998. "Foreign Investment as a Vehicle for International Technology Transfer." In *Creation and Transfer of Knowledge: Institutions and Incentives*, ed. G. Barba Navaretti, P. Dasgupta, K.G. Maler, and D. Siniscalco, 279–311. Heidelberg and Berlin: Springer Verlag.

Caselli, Francesco, and Daniel Wilson. 2003. "Importing Technology." Department of Economics, Harvard University, Cambridge, MA.

Clerides, S., S. Lach, and J. Tybout. 1998. "Is Learning by Exporting Important? Micro-Dynamic Evidence from Colombia, Mexico and Morocco." *Quarterly Journal of Economics* 113 (3): 903–47.

Coe, D., and E. Helpman. 1995. "International R&D Spillovers." *European Economic Review* 39 (5): 859–87.

Coe, D., E. Helpman, and W. Hoffmaister. 1997. "North-South R&D Spillovers." *Economic Journal* 107 (440): 134–49.

Dijankov, S., and B. Hoekman. 1996. "Intra-Industry Trade, Foreign Direct Investments and the Reorientation of East European Exports." Discussion Paper 1377, Centre for Economic Policy Research, London.

Doornik, Jurgen A., Manuel Arellano, and Stephen Bond. 2002. "Panel Data Estimation Using DPD for Ox." Nuffield College, Oxford.

Eaton, J., and S. Kortum. 2001. "Trade in Capital Goods." *European Economic Review* 45 (7): 1195–1235.

Gordon, R.J. 1990. *The Measurement of Durable Goods Prices*. Chicago: University of Chicago Press.

Keller, W. 2000. "Do Trade Patterns, Technology Flows Affect Productivity Growth?" *World Bank Economic Review* 14 (1): 17–48.

Mody, A., and K. Yilmaz. 2002. "Imported Machines for Export Competitiveness." *World Bank Economic Review* 16 (3): 23–48.

Quah, Danny. 1999. "The Weightless Economy in Economic Development." Discussion Paper 417, Centre for Economic Policy Research, London.

Schiff, M., Y. Wang, and M. Olarreaga. 2002. "Trade-Related Technology Diffusion and the Dynamics of North-South and South-South Integration." Policy Research Working Paper 2861, World Bank, Washington, DC.

7

EXPORTS AND ECONOMIC PERFORMANCE: EVIDENCE FROM A PANEL OF CHINESE ENTERPRISES

Aart Kraay

This chapter investigates whether firms learn from exporting. Using a panel data set of 2,105 Chinese industrial enterprises between 1988 and 1992, it finds that, controlling for past firm performance and unobserved firm characteristics, past exports are a significant predictor of current enterprise performance. These learning effects can be quite large: the estimated coefficients indicate that a 10 percentage point increase in a firm's export to output ratio in a given year leads to a 13 percent increase in labor productivity, a 2 percent increase in total factor productivity (TFP), and a 6 percent decrease in unit costs the following year. Interestingly, these learning effects are most pronounced among established exporters. For new entrants to export markets, learning effects are insignificant and occasionally negative.

Although the superior economic performance of exporting firms relative to nonexporters has been extensively documented in a number of developed and developing countries,[1] the question of causality between exports and firm performance has only recently begun to receive attention. After all, the better average

The author would like to thank David Dollar, William Easterly, Mary Hallward-Driemeier, Pascal Mazodier, Barry Naughton, Jakob Svensson, James Tybout, Colin Xu, and seminar participants at the World Bank and the Université d'Auvergne for helpful comments and Jean Imbs for diligent research assistance. This chapter appeared in French as "Exportations et performances économiques: étude d'un panel d'entreprises chinoises," *Revue d'économie du développement* 1–2/1999, 183–207.

performance of exporters may simply be due to exporters self-selecting into export markets precisely because they are more efficient. It is also possible that firms learn from exporting through a variety of channels. Somewhat informally, it is often argued that exposure to global markets elicits greater entrepreneurial effort on the part of managers, forcing them to become more "competitive." More concretely, there is evidence that developing country exporters benefit from a range of trade-related linkages with their developed country customers, such as production or managerial advice embedded in supplier specifications.[2]

Work by Bernard and Jensen (1995, 1999a, 1999b) and Clerides, Lach, and Tybout (1998) attempts to disentangle the direction of causation between exports and firm performance, using panel data on samples of U.S. firms and Colombian, Moroccan, and Mexican firms, respectively. In contrast with this chapter, both of these studies find little evidence that past exports are associated with improvements in future firm performance, casting doubt on the existence of learning effects.

This chapter extends the work of these authors in two directions. First, I consider a different panel data set of 2,105 large and medium-size Chinese industrial enterprises between 1988 and 1992. The experience of these firms is of considerable independent interest, given China's swift growth and the rapid expansion of its trade from a very low base during the 1980s. Although China's opening to world markets is often cited as one of the key factors responsible for China's growth, there is little evidence on the effects of trade on economic performance at the firm level.[3]

Second, I employ a somewhat different empirical methodology than these earlier studies. For a partial correlation between past exports and current performance to constitute evidence of learning from exporting, it is necessary to employ a methodology that rules out two alternative sources of this correlation: unobserved firm-specific factors that are correlated with both exports and firm performance and the confluence of persistence in firm performance and self-selection of better firms into export markets. Accordingly, I employ a dynamic panel specification in which firm performance depends on lagged performance and lagged exports. I address the difficulty of unobserved firm-specific effects by working with a first-differences specifications; the presence of lagged performance and an appropriate choice of instruments rule out the second explanation.[4] Finally, unlike these earlier studies, I allow the coefficient on lagged exports to vary with the export history of the firm; I thus allow learning effects to depend on how long firms have been in export markets. I find that there are large differences in learning effects between recent entrants and established exporters. This raises the possibility that the failure of earlier studies to find learning effects may simply be because these studies pool information across firms with different export histories.

Are Exporters and Nonexporters Different?

In this section, I briefly discuss the data set and document the characteristics of the sample of firms. I then summarize some of the features of the exporting firms in the sample and confirm that in China, as in other countries, exporters tend to be larger and more efficient than nonexporters. The question of causation between exports and firm performance is taken up in the following section.

Data and Sample Characteristics

This study exploits data from a rich panel data set of more than 7,000 large and medium-size Chinese industrial enterprises between 1988 and 1992 compiled by China's State Statistical Bureau. The data set is a subsample of the annual Chinese industrial survey on which published aggregate industrial statistics are based; it contains consistent time series for a large number of physical and financial indicators of enterprises, including exports. Unfortunately, many values are missing in the data, especially for the exports variable. This forces consideration of a much smaller subsample of 2,105 enterprises for which it is possible to construct a balanced panel of observations on all relevant variables. (Details of how the variables were constructed from the underlying data are given in the annex.)

Tables 7.1 and 7.2 present an overview of key variables for the sample of 2,105 firms. Although the sample represents only a tiny fraction of the hundreds of

TABLE 7.1 Summary Statistics for Enterprise Sample

Statistic	1988	1989	1990	1991	1992
Gross output value					
Sample (billions of yuan)	185.1	220.4	232.9	263.5	320.3
Population[1] (billions of yuan)	1,822.4	2,201.7	2,392.4	2,824.8	3,706.6
Sample/population (percent)	10.2	10.0	9.7	9.3	8.6
Employment					
Sample (thousands)	6,502.6	6,684.8	6,724.1	6,827.2	6,978.3
Population[1] (thousands)	61,580.0	62,280.0	63,780.0	65,510.0	66,210.0
Sample/population (percent)	10.6	10.7	10.5	10.4	10.5
Exports					
Sample (billions of yuan)	19.7	24.3	27.9	31.9	31.8
Population[2] (billions of yuan)	176.7	195.6	298.6	382.7	467.6
Sample/population (percent)	11.2	12.4	9.3	8.3	6.8

Source: Author.

1. Industry only.

2. Economywide exports.

TABLE 7.2 Distribution of Sample by Ownership and Sector, 1990 (percent)

Item	Gross output value		Employment	
	Sample	Population	Sample	Population
Ownership[1]				
State-owned units	95.92	54.60	96.07	68.42
Collectively owned units	2.49	35.62	2.63	29.41
Others	1.59	9.77	1.30	2.16
Sector[2]				
Textiles	25.69	14.55	29.29	14.05
Garments	1.20	2.63	0.92	3.11
Basic chemicals	5.59	9.47	4.11	7.20
Pharmaceuticals	5.14	2.26	3.56	1.53
Chemical fibers	3.77	1.73	2.08	0.64
Construction materials	1.36	n.a.	1.51	n.a.
Ferrous metals	32.11	8.25	27.15	5.66
Machinery	13.67	10.63	21.30	18.65
Automobiles	6.94	4.53	6.25	3.79
Electronics	4.52	3.71	3.84	2.96
Other[3]	—	42.24	—	42.42

Source: Author.

Note: — = not applicable. Sectoral distribution for population is approximate, due to imperfect concordance between documentation of the data set and published national tools.

1. All industry, including mining and utilities.
2. Manufacturing only.
3. Sectors not represented in sample.

thousands of Chinese industrial enterprises, the large firms in the sample accounted for about 10 percent of industrial gross output value and employment and 11 percent of exports in 1988. Although the share of the sample in total employment is roughly constant over the sample period, the share of the sample in gross output falls to 8.6 percent in 1992 and the share in exports falls precipitously to 6.8 percent. This is due primarily to biases in the composition of the sample of firms relative to the population of Chinese industrial enterprises, as documented in the first panel of table 7.2. The sample of firms is heavily skewed toward state-owned enterprises, which have experienced slower output and export growth on average than the rest of the economy (World Bank 1996, 1997). The sample is also somewhat skewed in its sectoral composition, with sectors such as textiles and ferrous metals overrepresented and many sectors entirely unrepresented. These compositional biases, as

TABLE 7.3 Summary Statistics on Exporters, 1988–92

Statistic	1988	1989	1990	1991	1992
Number of exporting firms	1,348	1,410	1,490	1,509	1,492
Exporters' percentage share of					
Number of firms	64.04	66.98	70.78	71.69	70.88
Gross output value	73.24	77.14	81.32	83.57	82.14
Employment	72.09	76.52	80.38	82.22	81.27
Exports/gross output value					
Mean (weighted by gross output value)	0.145	0.143	0.147	0.145	0.121
Standard deviation	0.218	0.217	0.217	0.225	0.216

Source: Author.

well as any additional selection biases caused by limiting the sample to a balanced panel, suggest that some caution is in order in extending the results to all firms.

Characteristics of Exporting Firms

Table 7.3 presents summary statistics on the exporting firms in the sample. The first panel reveals that 64–72 percent of all firms reported positive exports over the sample period. Exporters account for a somewhat larger share of employment and gross output, reflecting the larger average size of exporters relative to nonexporters. The second panel shows that on average, exporting firms export 12–15 percent of their gross output, but the large standard deviation indicates that there is substantial cross-sectional variation in this ratio across firms. Table 7.4 highlights the contribution to this variation of differences in export ratios across sectors and forms of ownership. In 1990 state-owned enterprises, which constitute the majority of the sample, exported about 14 percent of their output, considerably less than collectively owned enterprises, which exported 25 percent of their output. Textiles, garments, and pharmaceuticals are the most export-oriented sectors, exporting 30 percent, 21 percent, and 19 percent of the gross value of industrial output, respectively, while automobiles (3 percent) and ferrous metals (5 percent) are the least export-oriented sectors.

The large sectoral variation in the incidence of exporters, documented in the second column of table 7.4, suggests that the larger average size of exporters implied by table 7.3 could simply be an artifact of the sectoral composition of exporting firms, with exporters being more prevalent in sectors populated by larger firms. However, table 7.5, which reports average firm size controlling for the sectoral, regional, and ownership composition of exporters, shows that this is not

TABLE 7.4 Summary Statistics on Exporters by Ownership and Sector, 1990

Statistic	Number of exporters	Exporter percentage share of			Exports/gross output	
		Number of firms	Gross output value	Employment	Mean	Standard deviation
Ownership						
State-owned units	1,353	70.36	81.45	80.49	0.14	0.21
Collectively owned units	99	74.44	80.16	77.65	0.25	0.26
Others	38	77.55	75.10	77.86	0.16	0.19
Sector						
Textiles	540	87.95	93.47	92.31	0.30	0.24
Garments	19	63.33	62.79	60.13	0.21	0.35
Basic chemicals	86	68.80	79.81	77.42	0.11	0.13
Pharmaceuticals	139	81.29	90.27	86.71	0.19	0.18
Chemical fibers	24	40.00	58.71	62.31	0.08	0.13
Construction materials	22	46.81	53.34	49.12	0.08	0.05
Ferrous metals	65	60.19	77.95	76.64	0.05	0.12
Machinery	414	62.73	76.27	75.41	0.11	0.17
Automobiles	58	45.67	78.01	72.51	0.03	0.08
Electronics	123	75.46	80.47	80.33	0.11	0.19

Source: Author.

the case.[5] Even within these groups, exporters tend to be nearly twice as large as nonexporters, in terms of both gross output value and employment, and this difference is highly significant.

Table 7.6 presents some indicators of within-firm variation in exports between 1988 and 1992. The upper panel documents the fact that a firm's status as an exporter or nonexporter is highly persistent over time by reporting the probability that an exporting (nonexporting) firm continues to export (not export) from one year to the next, averaging over all firms in all periods. Given that a firm exported in a particular year, the probability that it continues to export the following year is 0.905. Similarly, the probability that a firm does not export in a given year, conditional on not having exported the previous year, is 0.836, and the probability that a nonexporter breaks into export markets is only 0.164. This is not to say, however, that there is little time-series variation in the export ratio itself.

TABLE 7.5 Size of Exporters and Nonexporters, 1988–92 (unweighted averages of selected variables)

Item	1988	1989	1990	1991	1992
Gross output value					
Exporters (thousands of yuan)	100,545	120,553	127,133	145,916	176,370
Nonexporters (thousands of yuan)	65,421	72,497	70,747	72,643	93,309
p-value[1]	0.00	0.00	0.00	0.00	0.00
Employment					
Exporters	3,478	3,628	3,627	3,720	3,801
Nonexporters	2,397	2,259	2,145	2,036	2,133
p-value[1]	0.00	0.00	0.00	0.00	0.00

Source: Author.

Note: Unweighted averages, controlling for ownership, sector, and province effects.

1. *p*-value is for test of the null hypothesis that the means for exporters and nonexporters are equal.

The lower panel of table 7.6 presents the mean and standard deviation of year-over-year changes in the export to gross output value ratio. The large standard deviations indicate that there is considerable variation around the average change in exports. For example, between 1990 and 1991, these figures indicate that fully half of the firms in the sample experienced a change in their export ratio of more than 7 percent in absolute value, and one-quarter experienced changes of more than 12 percent in absolute value.

TABLE 7.6 Persistence and Volatility of Export Status

Item	Exports at time $t + 1$		Does not export at time $t + 1$	
Exports at time t	0.905		0.095	
Does not export at time t	0.164		0.836	
	1988–89	1989–90	1990–91	1990–92
Change in exports/gross output value[1]				
Mean	0.010	0.017	0.007	–0.020
Standard deviation	0.093	0.106	0.122	0.121

Source: Author.

1. Refers to firms with positive exports in both years over which change is calculated.

Do Exporters Perform Better Than Nonexporters?

In order to investigate differences in enterprise performance between exporters and nonexporters, I require indicators of economic performance over time for all firms. I consider three measures: labor productivity, TFP, and unit costs. I measure labor productivity as gross output per worker in constant 1990 prices, I measure unit costs as the current price ratio of cost of goods sold to sales revenue. TFP is constructed as the residual from a three-factor constant returns Cobb-Douglas gross output production function using capital, labor, and materials as inputs. Capital is measured as the net value of fixed assets, labor input is measured as the annual average number of workers, and materials are measured in constant 1990 prices. The output elasticities are measured as the current-price shares of factor payments in gross output and are allowed to vary across sectors and over time. In particular, the output elasticity of labor in each sector and year is estimated as the sectoral average of wage payments (including bonuses) to gross output, while the output elasticity of materials is measured as materials consumption divided by gross output. Given the assumption of constant returns to scale, the capital output elasticity is one minus the other two elasticities.[6]

It is worth noting in passing that considerable controversy surrounds various measures of enterprise productivity in China.[7] Much of this controversy results from efforts to reconcile the anomalous behavior of various published price indices for output and material inputs and the consequences of different reconciliations for the relative productivity performance of state and nonstate industry. Since I am not interested in comparisons between state- and nonstate industry but rather between exporters and nonexporters in a sample consisting primarily of state-owned enterprises, I am able to sidestep much of the controversy surrounding the choice of appropriate deflators. Moreover, in the empirical work in the following sections, I enter the firm performance variables in logarithms and include ownership dummies interacted with period dummies, which has the effect of sweeping out any variation in deflators across ownership forms.

Table 7.7 summarizes the differences in firm performance between exporting and nonexporting firms. The first panel reports the simple averages of labor productivity, TFP, and unit costs in exporting and nonexporting firms, pooling the data for all firms and years. Both measures of productivity are significantly higher in exporting firms, with exporters enjoying a productivity advantage of 2–9 percent over nonexporters. In terms of unit costs, exporters appear to have slightly worse performance than nonexporters, but this is an artifact of the sectoral composition of exporters. The lower panel presents the same averages as the upper panel, controlling for the sectoral, regional, and ownership distribution of exporters. The productivity

TABLE 7.7 Performance of Exporters and Nonexporters

Item	Exporters	Nonexporters	*p*-value[1]
Unconditional			
Labor productivity	36.652	33.694	0.00
Total factor productivity	2.014	1.965	0.00
Unit costs	0.922	0.912	0.001
Conditional[2]			
Labor productivity	45.819	42.844	0.00
Total factor productivity	2.105	2.057	0.00
Unit costs	0.949	0.961	0.00

Source: Author.

1. *p*-value is for test of the null hypothesis that the means for exporters and nonexporters are equal.
2. Controlling for sector, ownership, and regional effects.

advantage of exporters persists, and exporters now also enjoy slightly lower unit costs than nonexporters.

In summary, the empirical regularities observed in this sample of predominantly state-owned Chinese industrial enterprises are consistent with those documented in several other countries. Exporting firms tend to be larger than nonexporting firms and to enjoy higher productivity and lower unit costs. Although a firm's status as an exporter is very persistent over time, there is substantial time-series variation in the export to gross output ratio of exporting firms. This last feature of the data is important, because in the next section I use the within-firm time-series variation in exports and performance measures to identify learning effects.

Do Exporters Learn from Exporting?

In this section I formally test whether the better average performance of exporting firms documented in the previous section can be attributed to exporters learning from exporting. In contrast to other work, I find a statistically and economically significant effect of lagged exports on current firm performance, suggesting the presence of learning from exporting.

Specification and Identification

The empirical strategy in this section is to test whether a firm's performance, as measured by labor productivity, TFP, and unit costs, depends on its past export experience. For such a test to constitute evidence in favor of learning by exporting, it is necessary to rule out two alternative explanations for any observed correlation between past exports and current enterprise performance.

First, current enterprise performance may depend on past export experience, due to unobserved enterprise characteristics that affect both performance and exports. For example, certain firms may have more energetic managers who run efficient operations with lower unit costs than their competitors and also aggressively seek out foreign markets, while other firms may be run by more conservative managers who are unwilling to implement efficiency-enhancing reforms and prefer to rely on traditional domestic markets. Such unobservable firm characteristics may give rise to spurious correlations between lagged exports and current enterprise performance.

Second, firm performance is itself likely to be persistent over time, and it is jointly determined with a firm's export performance. For example, if production is characterized by scale economies, an expansion of plant size today may result in lower unit costs for many periods in the future. If firms with better performance self-select into export markets and firm performance is correlated over time, then current performance will be correlated with past export behavior even in the absence of learning effects.

To distinguish the learning hypothesis from these two alternative explanations, I estimate a series of regressions of firm performance on lagged exports and lagged firm performance. The estimation technique is selected so as to yield consistent estimates of the coefficient on lagged firm performance and lagged exports even in the presence of unobserved firm-specific, time-invariant effects that are correlated with the explanatory variables.[8] To the extent that unobserved firm characteristics, such as managerial ability, do not vary over time, the first explanation for the correlation between lagged exports and current firm performance is ruled out. The inclusion of lagged firm performance in the regression controls for serial dependence in this variable, and the appropriate choice of lagged variables as instruments addresses the problem that exports and firm performance are jointly determined. Hence I can also distinguish the learning hypothesis from the second explanation mentioned above.

Specifically, I estimate variants on the following equation:

$$y_{ijt} = \beta \cdot y_{ij,t-1} + \gamma \cdot x_{ij,t-1} + \eta_i + \lambda_{jt} + \varepsilon_{ijt}, \tag{7.1}$$

where y_{ijt} and x_{ijt} denote the logarithm of firm performance and the export to gross output value ratio of firm i of ownership form j at time t, η_i is an unobserved firm-specific effect, and λ_{jt} is an unobserved period-specific effect that may also vary across ownership forms. Finally, ε_{ijt} is a well-behaved zero-mean disturbance term.

To eliminate the individual- and period-specific effects from equation (7.1), I take deviations of all variables from period- and ownership-specific means and then difference these deviations to obtain:

$$\Delta y_{ijt} = \beta \cdot \Delta y_{ij,t-1} + \gamma \cdot \Delta x_{ij,t-1} + \Delta \varepsilon_{ijt}, \tag{7.2}$$

where Δ is the first difference operator and y_{ijt}, $x_{ij,t-1}$ and ε_{ijt} denote deviations from period- and ownership-specific averages.[9]

Two assumptions on the structure of the disturbances are required to identify the parameters of the model. First, I assume that there is no serial dependence in ε_{ijt}, that is, $E[\varepsilon_{ijt} \cdot \varepsilon_{ijs}] = 0$ for all $s \neq t$. Second, I assume that, although firm performance and exports are jointly determined, exports are predetermined with respect to ε_{ijt}, that is, $E[x_{ijt} \cdot \varepsilon_{ijs}] = 0$ for all $s > t$. Although the first-differencing introduces a correlation between the transformed residual and the first difference of lagged firm performance, the assumption of no serial dependence in the untransformed residual ensures that second and higher lags of firm performance are available as instruments for $\Delta y_{ij,t-1}$. The assumptions of no serial correlation and predeterminacy of exports imply that second and higher lags of exports are valid instruments for $\Delta x_{ij,t-1}$.

How valid are these identifying assumptions? If firm performance exhibits higher than first-order serial dependence, the residual term in equation (7.1) will also be serially correlated, invalidating the first identifying assumption. Since the short available time span of the data set makes it difficult to include several lags of the dependent variable, the only alternative is to test whether the estimated differenced residuals in equation (7.2) exhibit second-order serial dependence. It turns out that they do not, providing some comfort for the first identifying assumption. The assumption that exports are predetermined with respect to ε_{ijt} would arise naturally in any model in which both exports and firm performance depend on their own and the other's lagged values (as well as other exogenous variables).[10]

Basic Results

The results of this basic specification are presented in table 7.8. The model is estimated using one lag of firm performance and one lag of exports over the period from 1990 to 1992, since the differencing and choice of twice-lagged variables as instruments eliminates the first two years from the estimation period. As instruments, I use only the second lags of firm performance and exports. Results using the full set of all available second and higher lags as instruments are quite similar and are not reported for brevity.[11]

The highly significant and positive coefficients on lagged firm performance in the first row of the table confirm that firm performance is persistent over time.

TABLE 7.8 Basic Model Results

Item	Labor productivity	Total factor productivity	Unit costs
Lagged performance	1.037** (0.087)	0.541** (0.032)	0.221** (0.031)
Lagged exports	1.321**(0.252)	0.233** (0.079)	–0.647*(0.108)
p-value for no serial correlation test	0.003**	0.947	0.264

Source: Author.

Note: Model estimated for 1990–92, using second lags of firm performance and exports as instruments. Standard errors are in parentheses.

*Significant at the 5 percent level.

**Significant at the 1 percent level.

More interestingly, there is statistically significant evidence of learning from exporting. Past exports are positively associated with current labor productivity and TFP and negatively associated with unit costs. The magnitude of these effects is also economically significant. Recalling that the performance measure is expressed in logarithms, the magnitudes of the coefficients on lagged exports imply that an increase of 0.1 in a firm's export ratio (roughly one sample standard deviation) causes a 13 percent increase in labor productivity, a 2 percent increase in TFP, and a 6 percent reduction in unit costs the following year.

The last row of table 7.8 reports the p-value associated with a test of the null hypothesis that there is no second-order serial correlation in the differenced residuals in equation (7.2). In the case of unit costs and TFP, this null hypothesis is not rejected at the 5 percent level, suggesting that the identifying assumptions are indeed valid. In the case of labor productivity, the null hypothesis is rejected at the 5 percent level. However, if I augment this specification with an additional lag of labor productivity, the null is no longer rejected, and the coefficient on lagged exports actually becomes much larger.

The rather large estimates of learning using these measures should be interpreted with some caution. Both exports and some measures of firm performance are quite persistent over time. As a result, lagged levels of these variables will not be very highly correlated with contemporaneous changes in these variables. This suggests that lagged levels may be weak instruments for the right-hand-side variables in equation (7.2). In the case of weak instruments, it is well known that in finite samples the instrumental variables estimates are biased toward the probability limits of the ordinary least squares estimates of equation (7.2). Therefore, the instrumental variables procedure I employ may only imperfectly address the endogeneity problems that motivate its use, and hence the learning effects may be somewhat overstated.[12]

Controlling for Export Histories

One shortcoming of the above regressions is that they do not control for the export history of a firm. The relationship between lagged exports and current firm performance may depend on the export history of a firm, for a number of reasons. Suppose, for example, that learning from exporting is a one-shot affair, in the sense that breaking into export markets is associated with a one-time improvement in firm performance. In this case, lagged exports should be positively correlated with firm performance only in the first few years after a firm enters the export market, not if a firm has been exporting for many years.

If, however, learning is an ongoing process, then lagged exports will be positively associated with current performance even in firms that have been exporting for many years. Moreover, if there are high entry costs that must be incurred before a firm can begin to export, then measured firm performance may initially deteriorate as the firm breaks into export markets. Only as the firm recoups these start-up costs will exports eventually be associated with better firm performance. Finally, it is even possible that firms choose to exit from export markets because they have exhausted all the learning benefits from exporting.

This raises the possibility that the basic regressions of the previous subsection are misspecified, since they impose the same coefficient on lagged exports for all firms, regardless of their export history. I address this concern by re-estimating equation (7.1), allowing the coefficient on lagged exports to vary with the export history of the firm. Specifically, the export history of a firm may be thought of as a sequence of indicator variables for each firm in each year that take on the value 1 if a firm exports in that year and 0 if it does not. Even with only five years of data, a very large number of distinct export histories is represented in the sample of firms, which would result in a very large number of different coefficients on lagged exports to be estimated. In order to conserve on degrees of freedom and impose some structure on the problem, I consider five types of export histories: firms that export over the entire sample period (established exporters), firms that initially do not export but at some point during the sample begin exporting and continue to export through the end of the sample period (entrants), firms that initially export but leave the export market for the duration of the sample period (exiters), firms that switch between exporting and not exporting more than once over the sample (switchers), and firms that never export (nonexporters).[13] As shown in table 7.9, slightly more than half of the firms in the sample fall in the first category. Relatively few firms are classified as either entrants or exiters (252 and 113, respectively, representing 12 and 5 percent of the total sample). The remainder are switchers (279) or firms that never export (343).

Next I specify a more general form of equation (7.1), in which the coefficients on lagged exports are allowed to vary with the export history of the firm. In particular,

TABLE 7.9 Export Histories

History	Number of firms	Share of total
Always exported	1,118	53.1
Entrants	252	12.0
Entered in 1989	79	3.8
Entered in 1990	67	3.2
Entered in 1991	63	3.0
Entered in 1992	43	2.0
Exiters	113	5.4
Exited in 1989	26	1.2
Exited in 1990	15	0.7
Exited in 1991	36	1.7
Exited in 1992	36	1.7
Switchers	279	13.3
Never exported	343	16.3

Source: Author.

for $t = 1991$ and $t = 1992$, the following pair of equations relates performance to lagged performance and lagged exports:

$$y_{i91} = \beta \cdot y_{i90} + \Big[\gamma^{ENT} \cdot D_{i89}^{ENT} + \gamma^{EXT} \cdot \left(D_{i89}^{EXT} + D_{i90}^{EXT}\right) + \gamma^{ALW} \cdot D_i^{ALW} + \gamma^{SWI} \cdot D_i^{SWI}\Big] \cdot x_{i90} + \eta_i + \lambda_{91} + \varepsilon_{i91}$$

$$y_{i92} = \beta \cdot y_{i91} + \Big[\gamma^{ENT} \cdot \left(D_{i89}^{ENT} + D_{i90}^{ENT}\right) + \gamma^{EXT} \cdot D_{i91}^{EXT} + \gamma^{ALW} \cdot D_i^{ALW} + \gamma^{SWI} \cdot D_i^{SWI}\Big] \cdot x_{i91} + \eta_i + \lambda_{92} + \varepsilon_{i92}, \tag{7.3}$$

where $\gamma^j, j =$ ENT, EXT, ALW, SWI is the coefficient on lagged exports for entrants, exiters, established exporters, and switchers, respectively; D_{it}^{ENT} (D_{it}^{EXT}) is a dummy variable that takes on the value 1 if an entrant (exiter) firm *i* enters (exits) the export market in year *t*; and D_i^{ALW} and D_i^{SWI} are dummy variables that take on the value 1 if the firm always exports or is a switcher. Firms that do not export are excluded.

Although equation (7.3) differs slightly from equation (7.2) in the sense that there are different right-hand-side variables in the two time periods, it can be estimated in the same manner as before. That is, I first take deviations of all variables from period- and ownership-specific means to eliminate period effects and then take first differences to eliminate firm-specific effects. Then the assumptions that exports are predetermined and that there is no serial correlation in the residual

TABLE 7.10 Basic Model Controlling for Export Histories

Variable	Labor productivity	Total factor productivity	Unit costs
Restricted model			
Lagged performance	0.811*** (0.134)	0.474*** (0.046)	0.195*** (0.039)
Lagged exports	1.188*** (0.254)	0.227** (0.101)	–0.687*** (0.150)
Unrestricted model			
Lagged performance	0.778*** (0.145)	0.469*** (0.046)	0.196*** (0.039)
Lagged exports in entrants	–5.428* (3.153)	–1.082 (0.667)	–0.275 (0.579)
Lagged exports in exiters	0.982 (2.974)	0.236 (1.353)	0.094 (1.758)
Lagged exports in established exporters	1.477*** (0.295)	0.293*** (0.110)	–0.688*** (0.164)
Lagged exports in switchers	0.735 (0.743)	0.151(0.309)	–.801* (0.414)

Source: Author.

Note: Model estimated for 1991–92, using second lags of firm performance and exports as instruments.

*Significant at the 10 percent level.

**Significant at the 5 percent level.

***Significant at the 1 percent level.

again imply that second and higher lags of firm performance and exports interacted with the various indicator variables are available as instruments in each period. Note that there is no variation in twice-lagged exports of entrants for $t = 1990$ (since exports in 1988 are by definition zero for this group of firms). For this reason, I can estimate this specification only for $t = 1991$ and $t = 1992$.

The results of this more general specification are presented in table 7.10 for the three measures of firm performance. For comparison purposes, the first panel presents the results of re-estimating the restricted model in equation (7.1) over the period 1991–92. These results are quite similar to those obtained for the full period. The lower panel allows the coefficient on lagged exports to vary with the export history of the firm. The most striking result that emerges is that learning effects are consistently positive and significant only for established exporters. Among entrants, there is weakly significant evidence that lagged exports are associated with lower current productivity, consistent with the view that entry into export markets is initially costly.

It is somewhat puzzling that learning effects seem to be most pronounced among established exporters—firms that report positive exports in all five years of the sample. This runs counter to the common intuition that new entrants to export markets should benefit most from exposure to the competitive pressures of global markets. One possible explanation for this relies on the structure of China's trade and foreign exchange institutions, which may have given the exporting firms in the sample preferential access to foreign exchange.[14] If this foreign exchange was used to purchase better quality imported capital goods and other inputs, it may account for the association between exports and enterprise performance documented above. However, this explanation is not entirely convincing during the sample period of the late 1980s and early 1990s, at which time a foreign exchange swap market was relatively well established.

Another explanation may simply be that the classification of "established exporters" is misleading. If many of these firms began exporting shortly before the beginning of the sample period in 1988 and if learning from exporting occurs for several years after breaking in to export markets, then the learning effects observed in the sample may be representative of those experienced by new entrants into export markets.

Conclusions

This chapter examines whether firms learn from exporting, in the sense that past exports lead to improvements in current firm performance. In contrast to the findings of other authors for different countries, I find that in this sample of Chinese industrial enterprises, past exports are significantly associated with higher labor productivity, higher TFP, and lower unit costs, suggesting that firms do reap efficiency benefits from exporting. It is particularly interesting to note that these sizeable learning effects are present in a sample consisting primarily of state-owned enterprises, which have been widely viewed as being slow to respond to and benefit from China's move to a more market-oriented economy.

These results raise a number of questions for future research. Although the correlation between past exports and current firm performance provides interesting and suggestive evidence of learning from exporting, it is less clear what exactly firms are learning, from whom, and how. Do firms become better at producing their outputs (perhaps thanks to supplier specifications provided by their clients), or do firms simply learn to be better exporters (as they develop familiarity with export markets and distribution channels)? Do the benefits of learning accrue solely to the exporting firm, or are there external effects from one firm's exports to another's performance? How long does it take for the benefits of exporting to appear? Answers to these questions will require more detailed investigation and better data with longer time-series coverage on the precise nature of exporting firms' relationships with their clients and with one another.

Chapter 7 Annex A: Data Construction

The variables used in the chapter were constructed as follows. Gross output value and materials consumption in current prices are taken directly from the original data set, and their constant 1990 values are obtained using the gross output and materials deflators for state industry reported in table 3 of Jefferson, Rawski, and Zheng (1996). Employment is defined as the period-average workforce, and no attempt is made to adjust for nonproduction workers. Capital input is measured as the net value of fixed assets, as reported in the data set. This constitutes a very rough measure of capital, particularly because it is likely that firms simply cumulate undeflated annual investments to arrive at this stock.

Total factor productivity is constructed as

$$tfp_{ijt} = q_{ijt} \cdot \left(k_{ijt}^{\alpha_{kjt}} \cdot n_{ijt}^{\alpha_{njt}} \cdot m_{ijt}^{\alpha_{mjt}} \right)^{-1}, \tag{7A.1}$$

where tfp_{ijt}, q_{ijt}, k_{ijt}, n_{ijt}, and m_{ijt} represent TFP, gross output, capital input, labor input, and materials input of firm *i* in sector *j*, respectively. The output elasticities for labor and materials in sector *j* at time *t*, α_{njt}, and α_{mjt}, are measured as the current price ratios of wages (including bonuses) and materials to gross output in that sector and year. The capital output elasticity is measured as one minus the other two elasticities. Unit costs are constructed as sales revenues divided by cost of goods sold, in current prices.

Constructing export to gross output ratios is complicated by the fact that the data report exports in current prices for certain years and in constant prices for other years. For 1988–90 I take the ratio of exports to gross output at 1980 prices, while for 1991–92 I use the current price ratio. While this splicing of two different measures is undesirable, it may not be too serious. For 1990 it is possible to construct both the current and constant 1980 price ratios, and the correlation across firms in this ratio is very high (0.97).

The original data set consists of 7,252 firms. Restricting the sample to those firms with a complete time series on all variables of interest reduces the sample by about two-thirds. The sample of remaining firms was reduced to the final sample of 2,105 firms by eliminating firms with obvious coding errors or other extreme outliers in the variables of interest.

Notes

1. See, among others, Chen and Tang (1987) (Taiwan [China]); Haddad (1993) (Morocco); Aw and Hwang (1995) (Taiwan [China]); Bernard and Jensen (1995) (United States); and Djankov and Hoekman (1997) (Bulgaria).

2. See Clerides, Lach, and Tybout (1998) for examples and a theoretical model.

3. There is a vast literature on estimating the productivity performance of Chinese enterprises (see Jefferson, Rawski, and Zheng 1992 for an overview). This literature is concerned primarily with the effects of China's incremental reforms on the productivity performance of state- and nonstate-owned enterprises. The papers by Perkins (1996, 1997) are among the few studies in this literature that explicitly consider the role of exports at the firm level. He concludes that exporting firms enjoy higher productivity. Using city-level data, Wei (1993) finds that exports are associated with higher growth.

4. The model estimated by Clerides, Lach, and Tybout (1998) allows for individual effects but requires them to be independent of explanatory variables, such as lagged exports. They address the problem of persistence and self-selection by jointly estimating a system that regresses firm performance on lagged performance and a sequence of lagged dummy variables indicating export market participation and using a probit equation to describe the decision to participate in export markets. Although the joint estimation of the performance and participation equations will yield more efficient estimates than the single-equation instrumental variables estimator employed here, it requires more restrictive distributional assumptions on the disturbances and individual effects. The evidence on causation from exports to firm performance in Bernard and Jensen (1995) consists of regressions of employment growth and wage growth on initial export status, controlling for observable firm characteristics. Initial export status is a poor predictor of wage growth, but it has some explanatory power for employment growth. This specification is subject to both of the concerns mentioned above.

5. Specifically, I report the coefficients on dummy variables for exporters and nonexporters in a cross-sectional regression of firm size on these variables and a set of industry, provincial, and ownership dummies for each year in the sample.

6. It is well known that measures of TFP constructed in this simple manner will overstate productivity if there are increasing returns or firms have market power. Since I will be identifying learning effects from the within-firm variation in firm performance, these concerns are relevant only if there are changes in the extent of increasing returns or firms' market power varies over time and these measures are correlated with export activity in the right way. It seems reasonable to think of the extent of increasing returns as a fairly stable feature of technology that is unlikely to vary much over time. It is, however, possible that firms' market power declines as firms break into export markets, as they are less able to charge mark-ups over unit costs. As a result, export market participation may be associated with declines in measured TFP. This will have the effect of obscuring, rather than exaggerating, any learning effects on this measure of firm performance.

7. See Jefferson, Rawski, and Zheng (1992, 1996) and Woo and others (1993, 1994) for a review of these issues. In this chapter I use deflators for output and materials advocated by Jefferson, Rawski and Zheng (1992, 1996). These deflators are based on indexes of factory-gate prices of industrial products and material inputs rather than the implicit gross output deflator.

8. I use a dynamic panel instrumental variables estimator proposed by Arellano and Bond (1991). For applications of this technique to cross-country growth regressions, see, for example, Caselli, Esquivel and Lefort (1996) and Easterly, Loayza, and Montiel (1996).

9. That is, I regress each variable on a set of time dummies interacted with a set of ownership dummies and retrieve the residuals from this regression as deviations from period- and ownership-specific means.

10. Only if exports depend on future performance will this identifying assumption be invalid. In this case, only sufficiently lagged exports will be valid instruments.

11. To allow for the possibility that exports depend on one-period ahead productivity, I also estimated the specification using thrice-lagged variables as instruments. This yielded similar results, with even larger and more significant learning effects.

12. See Blundell and Bond (1998) for a discussion of these issues and possible remedies.

13. Clearly, since I do not have a full export history for every firm but only five years of data, this categorization may misclassify some firms. For example, some firms that are identified as always having exported may have entered export markets in 1987 and hence should be classified as entrants.

14. During the 1980s, trade and foreign exchange continued to be subject to considerable regulation. Firms were subject to foreign exchange surrender requirements and had to rely on planned allocations of foreign exchange to finance imports. However, beginning in the early 1980s, firms were allowed to retain limited amounts of above-quota foreign exchange for their own use. See World Bank (1994) for details.

References

Arellano, M., and S. Bond. 1991. "Some Tests of Specification for Panel Data: Monte Carlo Evidence and Application to Employment Equations." *Review of Economic Studies* 58 (1): 277–97.

Aw, B.Y., and Hwang, A.R. 1995. "Productivity and the Export Market: A Firm-Level Analysis." *Journal of Development Economics* 47 (2): 313–32.

Bernard, A., and J.B. Jensen. 1995. "Exporters, Jobs, and Wages in U.S. Manufacturing: 1976–87." *Brookings Papers on Economic Activity Microeconomics* 67–119, Washington, DC.

———. 1999a. "Exceptional Exporter Performance: Cause, Effect or Both?" *Journal of International Economics* 47 (1): 1–25.

———.1999b. "Exporting and Productivity." School of Management, Yale University, New Haven, CT.

Blundell, Richard, and Stephen Bond. 1998. "Initial Conditions and Moment Restrictions in Dynamic Panel Data Models." *Journal of Econometrics* 87 (1): 115–43.

Caselli, F., G. Esquivel, and F. Lefort. 1996. "Reopening the Convergence Debate: A New Look at Cross-Country Growth Empirics." *Journal of Economic Growth* 1 (3): 363–89.

Chen, T., and D. Tang. 1987. "Comparing Technical Efficiency between Import-Substitution-Oriented and Export-Oriented Foreign Firms in a Developing Economy." *Journal of Development Economics* 26 (2): 277–89.

Clerides, S., S. Lach, and J. Tybout. 1998. "Is Learning by Exporting Important? Micro-Dynamic Evidence from Colombia, Mexico and Morocco." *Quarterly Journal of Economics* 103 (3): 903–48.

Djankov, Simeon, and Bernard Hoekman. 1997. "Trade Reorientation and Productivity Growth in Bulgarian Enterprises." Policy Research Working Paper 1708, World Bank, Washington, DC.

Easterly, W., N. Loayza, and P. Montiel. 1996. "Has Latin America's Post-Reform Growth Been Disappointing?" Research Department, World Bank, Washington, DC.

Haddad, M. 1993. "How Trade Liberalization Affected Productivity in Morocco." Policy Research Working Paper 1096, World Bank, Washington, DC.

Jefferson, G., T. Rawski, and Y. Zheng. 1992. "Growth, Efficiency, and Convergence in China's State and Collective Industry." *Economic Development and Cultural Change* 40 (2): 239–66.

———. 1996. "Chinese Industrial Productivity: Trends, Measurement Issues and Recent Developments." *Journal of Comparative Economics* 23 (1): 146–80.

Perkins, F.C. 1996. "Productivity Performance and Priorities of the Reform of China's State-Owned Enterprises." *Journal of Development Studies* 32 (3): 414–44.

———. 1997. "Export Performance and Enterprise Reform in China's Coastal Provinces." *Economic Development and Cultural Change* 45 (3): 501–39.

Wei, S.J. 1993. "Open Door Policy and China's Rapid Growth: Evidence from City-Level Data." Working Paper 4602, National Bureau of Economic Research, Cambridge, MA.

Woo, W.T, W. Hai, Y. Jin, and G. Fan. 1993. "The Efficiency and Macroeconomic Consequences of Chinese Enterprise Reform." *China Economic Review* 4 (2): 153–68.

———. 1994. "How Successful Has Chinese Enterprises Reform Been? Pitfalls in Opposite Biases and Focus." *Journal of Comparative Economics* 18 (3): 410–37.
World Bank. 1994. "China: Foreign Trade Reform." Washington, DC.
———.1996. "China: Fighting Inflation, Deepening Reforms." Washington, DC.
———.1997. "China: Development Challenges in the New Century." Washington, DC.

PART III

FOREIGN DIRECT INVESTMENT, TECHNOLOGY TRANSFER, AND PRODUCTIVITY

8

FOREIGN INVESTMENT AND PRODUCTIVITY GROWTH IN CZECH ENTERPRISES

Simeon Djankov and Bernard Hoekman

A rich case study literature documents how firms and industries adopt new technologies and know-how. It points to the vital role of imports and openness to trade, both for learning through re-engineering and direct inputs into production and through communications with and information from foreign partners (suppliers and buyers). Using aggregate data, a number of studies have concluded that trading with countries that are relatively R&D intensive leads to higher productivity growth of domestic industry (Coe and Helpman 1995; Coe, Helpman, and Hoffmaister 1997). While these findings are not inconsistent with the endogenous growth literature, they do not reveal much about how technology transfer occurs.

The microeconomic literature has emphasized three channels for the international transmission of technology: imports of new capital and differentiated intermediate goods (Feenstra, Markusen, and Zeile 1992; Grossman and Helpman 1995); learning by exporting (Clerides, Lach, and Tybout 1998); and foreign investment (Blomström and Kokko 1997). Attention has centered on the role of foreign investment as a channel of knowledge transfer and on the spillovers of know-how to other firms in the economy. Foreign investment should be associated with the transfer of knowledge, since by definition it is driven by the existence of intangible assets owned by the parent firm (Markusen 1995). The conventional wisdom is that foreign investment is a major channel of technology transfer to developing countries. Pack and Saggi (1997) note

This chapter originally appeared in the *World Bank Economic Review* 2000, 14 (1): 49–64. The authors are grateful to Magnus Blomström, Caroline Freund, Ann Harrison, Roberto Rocha, Jim Tybout, and three anonymous referees for helpful comments and suggestions.

that intrafirm transactions in royalties and license fees between parent firms and subsidiaries account for more than 80 percent of total global flows.

What matters for economic growth are the spillovers to firms within and across industries, evidence on which is much less robust. The case study literature has argued that positive spillovers are significant. It has also documented the importance of the availability of local skills and in-house technological capacity in adapting and using techniques developed elsewhere (Lall 1992; Evenson and Westphal 1995). Microeconometric studies using panel data sets of enterprises have come to more ambiguous conclusions. Some analysts have found a statistically significant negative relationship between the size of foreign investment in an industry or economy and the productivity performance of domestic firms (see, for example, Harrison 1996; Haddad and Harrison 1993).

This chapter investigates the impact of foreign investment on productivity performance of firms in the Czech Republic during the initial postreform period (1992–96).[1] It distinguishes between firms that established partnerships with foreign firms—through either a joint venture or the direct sale of a majority equity stake—and those that did not and asks whether total factor productivity (TFP) growth rates of these groups of firms differ. TFP is used as an indirect measure of technology transfer. Data constraints prohibit using more direct measures, such as R&D effort or the turnover of managers and highly skilled labor. Our results suggest that TFP growth is higher in firms with foreign partnerships and that there is a clear hierarchy: firms that have been acquired by foreign owners have the highest TFP growth, followed by firms involved in joint ventures. Firms without foreign partnerships have the lowest TFP growth as a group. This result continues to obtain if an adjustment is made for the higher initial level of productivity observed in firms that attract foreign participation.

We find a statistically significant negative spillover effect of foreign participation in an industry (through joint ventures and foreign direct investment [FDI]) on firms without such links. This finding is consistent with the results of Aitken and Harrison (1999) for Colombia and Haddad and Harrison (1993) for Morocco. It suggests that although foreign ownership or collaboration had a beneficial impact on the performance of the domestic partners, this benefit did not spill over to the rest of the industry in the time period studied. In part, this result may simply reflect the fact that in a transition economy like the Czech Republic, the amount of time required to independently adapt and learn to apply more efficient techniques is longer than the four years for which we have data. The result should also be interpreted in light of the fact that on average firms with foreign partnerships account for almost 50 percent of total assets and more than 40 percent of total employment in our sample. If the analysis of spillovers is restricted to the impact of FDI (foreign majority ownership) on the rest of the industry, the spillover effect remains negative, but it becomes much smaller and is no longer significant. This is likely to reflect not only the fact that joint venture firms have higher TFP growth than firms without any foreign partnerships but also the fact that know-how spillovers from FDI require a minimum level of technological capacity and effort to be absorbed. Survey data on training and investment in technologies suggests that many domestic firms may have relatively weak capacities in this regard compared with firms with joint ventures. This illustrates the importance of taking into account differences in the characteristics of firms in each industry, in particular their endowment of technological abilities and investments in upgrading that ability (as proxied, for example, by the level of R&D spending).

Channels of Technology Transfer

While there is little doubt that technologies make their way across international boundaries, the mechanisms through which this occurs are not well understood. With the exception of case studies, most of the empirical evidence is based on aggregate data or cross-sectional surveys and is subject to multiple interpretations. Various transmission channels may play a role in the technology transfer process. New technologies may be embodied in goods and transferred through imports of new varieties of differentiated products or capital goods and equipment or through arm's length trade in intellectual property (for example, licensing contracts). Firms may learn about technologies by exporting to knowledgeable buyers who share product designs and production techniques with them. Technology transfer will also occur in the context of formal cooperative arrangements between foreign and local firms, through, for example, FDI (acquisition) or project-specific joint ventures.[2] In all these cases, technology acquisition will require the availability of workers with appropriate training and expertise to allow technology absorption and adaptation. The absence of such a capacity is often held to explain why TFP is frequently lower in developing countries firms than in industrial nations even if identical equipment is utilized (Pack 1987).

It is helpful to differentiate between technology transfers that are realized in the context of formal cooperative arrangements between a foreign and a domestic firm and those that occur at arms length. Arm's length transactions, which include arm's length trade in machinery and components and direct purchases of knowledge (payment for patents, blueprints, and so forth), can be a major avenue of technology transfer. However, not all technologies are available at arms length. Many may be obtainable only through formal cooperation—either majority ownership (acquisition) or project-specific joint ventures.[3] In theory, firms will be adverse to unbundling and selling know-how or products if there are important internalization incentives. In this case, FDI may be the preferred route to exploit knowledge advantages (Markusen 1995).

Foreign investment is likely to be associated with transfer of both hard technologies (machinery, blueprints) and soft ones (management, information). It will have two dimensions: "generic" know-how, such as management skills and quality systems, and specific know-how, such as enforcement of intellectual property rights, that cannot be obtained at arms length because of weaknesses in the existing policy environment or internalization incentives.[4] Regarding weaknesses in the policy environment, foreign partners may reduce the cost of upgrading and learning by helping identify and implement systems that ensure that production meets technical specifications, is delivered on time, and meets other customer needs. Interviews by the authors with managers of enterprises with foreign partnerships suggest that all of these dimensions are prevalent in the Czech Republic. But more important is presumably the access to unique parent firm–specific information, as well as production and distribution networks.

A question is whether and to what extent the knowledge "transferred" by multinationals to affiliates diffuses to other firms in the industry.[5] Theoretical models of foreign investment suggest there should be a positive relationship between FDI and diffusion. Know-how will diffuse from firm to firm through demonstration effects, labor turnover, or reverse engineering.

Das (1987) models a foreign subsidiary as a price leader and domestic firms as a competitive fringe. If learning by domestic firms is proportional to the output of the multinational firm—that is, the larger the multinational is relative to the domestic industry, the easier the learning—this creates incentives for a firm to transfer technology to its subsidiary, as profits are higher if more advanced technology is used. The greater output of the subsidiary then induces domestic firms to learn and adopt the foreign technology at a higher rate.

Wang and Blomström (1992) use a similar set-up, but they endogenize both the level of technology transfer from the parent company to the subsidiary and the investment in learning activities by the domestic firm. Foreign firms transfer technology at a higher rate if domestic firms invest more in the learning activities. Blomström, Kokko, and Zejan (1994) find some empirical support for this prediction.

The empirical evidence on spillovers from foreign-owned affiliates to indigenous firms is mixed (Blomström and Kokko 1997). An extensive case study literature seeks to determine whether spillovers from R&D exist and how large they are. Much of this literature focuses on industrial countries.[6] The literature on developing countries has documented that the magnitude of potential knowledge spillovers depends on the existence of technological capabilities allowing the assimilation of know-how by indigenous firms (Pack and Westphal 1986). A unique feature of many transition economies is that technological ability is substantially greater than in most developing countries. In principle, this should facilitate adoption of new technologies and allow rapid convergence toward best practice.

Much of the econometric literature has focused on productivity measures as a proxy measure of technology diffusion. Early studies, such as Blomström and Persson (1983), using industry-level data, found that domestic labor productivity is positively influenced by foreign presence in an industry, measured by the foreign share of industry employment. More recent studies using firm-level data are less supportive of the existence of spillovers. Aitken and Harrison (1999) and Haddad and Harrison (1993) find that foreign investment has a negative effect on the performance of domestically owned firms. Harrison (1996) suggests that in imperfectly competitive markets, entry by foreign investors implies that domestic incumbents lose market share, which impedes their ability to attain scale economies. The negative spillover results contrast with the findings of the case study–based literature and may to some extent reflect the omission of important variables, such as the level of R&D spending, expenditures on training, and the magnitude of employment of personnel with technical degrees (engineers, scientists).[7]

The analysis in this chapter relies on the estimation of production functions and the use of TFP as a proxy measure for technology transfer. Our reliance on TFP as the dependent variable assumes that the adoption of new technologies will, with some lag, lead to an improvement in productivity. A serious problem with this assumption is that the case study literature has documented that such productivity improvements depend on the technological abilities of domestic firms. Nelson and Pack (1998) have demonstrated that the production function methodology can underestimate or ignore the role of technological effort at the level of the firm and thus affect TFP growth estimates. Differences in technological capacity across firms in an industry may be an important determinant of TFP performance, but there is no information on this at the level of the firm, as firm-level data on technology-relevant variables such as R&D expenditure or the composition of the workforce are not available. However, the Czech Republic is not a developing country—it has a long-standing industrial base and is well-endowed with engineering and scientific human capital. For the economy as whole, therefore, the capacity to rapidly upgrade productive efficiency through the adoption of best-practice techniques (both hard and soft) should be considerable. This makes the Czech case less relevant as a comparator for developing countries that do not have equivalent endowments.

Profile of Czech Firms

Information on Czech enterprises was compiled for the 1992–97 period from surveys, using a questionnaire prepared by the authors and a database containing financial and ownership information. Financial variables were defined using international accounting standards from the onset of the survey in 1992. The database includes 513 firms that are quoted on the Prague Stock Exchange, whose shares traded at least four times in a given year, and that report the financial information required. Of the sample firms, 340 did not establish joint ventures or attract FDI, 91 concluded joint ventures with foreign companies, and 82 attracted majority foreign equity investment.

TABLE 8.1 Descriptive Statistics of the Sample

Sector	Sample size	Number of foreign partners	Type of foreign partner (FDI or joint venture)
Mining	11	8	3
Construction	82	55	27
Food and beverages	54	36	18
Textiles and apparel	39	28	11
Furniture and other wood products	11	5	6
Pulp and paper	14	10	4
Printing and publishing	13	6	7
Chemicals	30	18	12
Shoes and leather products	6	5	1
Nonmetallic mineral products	21	16	5
Basic metals	13	9	4
Fabricated metal products	24	12	12
Electric and electronics	82	54	28
Transport equipment	12	5	7
Other manufacturing	10	6	4
Retail services	15	11	4
Financial services	76	56	20
Number of observations	513	340	173
Share of total (percent)	100.0	66.3	33.7

Source: Authors.

Thus 34 percent of the sample (173 firms) had a foreign link(either a joint venture or FDI, with relatively uniform distribution across sectors (table 8.1).

The criterion used in the sample to determine the existence of a foreign partnership or ownership relationship was that at least 20 percent of the equity was owned by a single foreign entity or that the firm had established one or more joint ventures with a foreign partner. Because minority shareholders have little protection under Czech law, equity investors have an incentive to take a majority stake. Most firms with foreign equity ownership in the sample are majority foreign owned. While the share of firms with foreign linkages appears high, it is representative of Czech industry. Aggregate statistics using a 5 percent or more foreign equity ownership share as a criterion reveal that during the 1994–97 period, 42 percent of all manufacturing

FIGURE 8.1 Labor Productivity of Czech Firms with and without Foreign Partners, 1991

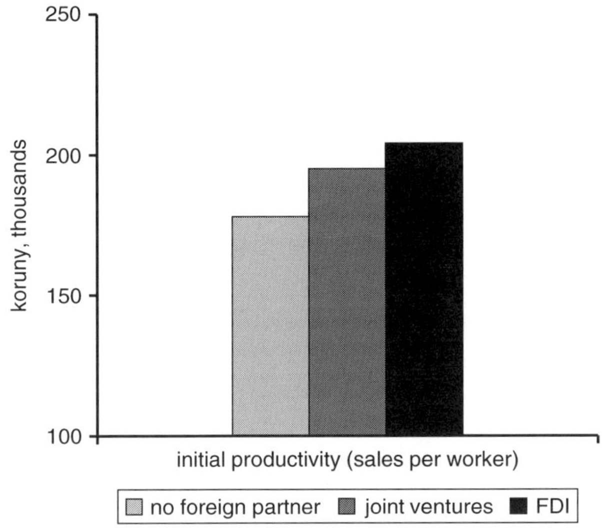

firms with more than 10 employees were involved in some kind of foreign partnership (Czech Statistical Office 1997).

Firms with foreign partnerships tend to be significantly larger than firms that remain independent: the median level of total employment is 689 in firms with FDI, 578 in firms with joint ventures, and 352 in firms without foreign links. Foreign affiliates or joint ventures also have higher levels of initial labor productivity, measured as sales per worker in 1991 (figure 8.1). This suggests that foreign investors are attracted to firms with above-average performance and size.

Average TFP growth performance is also highest in firms with FDI, followed by firms with joint ventures and domestic enterprises (figure 8.2). This may reflect the better than average initial level of productivity, suggesting that foreign investors choose the "best" firms as partners. In the statistical analysis, we therefore correct for the possibility of selection bias. The magnitude of TFP growth rates is highest in the earlier years and tapers off toward the end of the sample period. This reflects a marked deterioration in macroeconomic conditions in 1996, a common effect for all firms. TFP growth rates initially diverge substantially, with firms with foreign investment increasing growth while other firms experience a reduction in TFP growth rates. Thereafter some convergence occurs, suggesting that spillover effects may be occurring toward the end of the period.

Questionnaires administered in early 1997 suggest that both joint ventures and FDI are associated with technology transfers (figure 8.3). Two questions concerned training

FIGURE 8.2 Total Factor Productivity Growth in Czech Firms with and without Foreign Partners, 1992–96

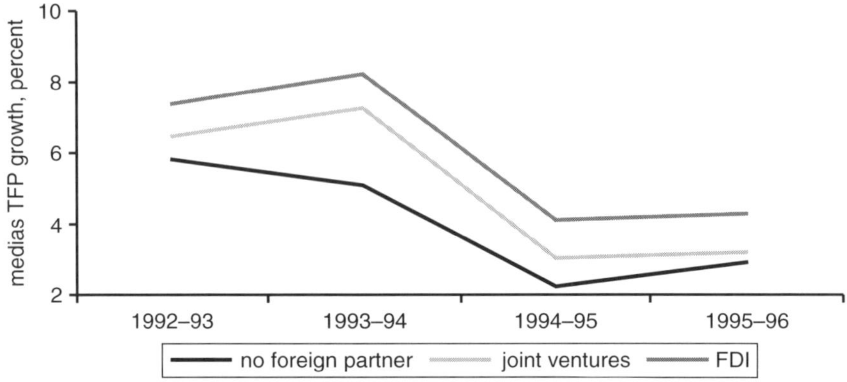

and acquisition of new technologies. Managerial responses clearly reveal what appears to be a significant difference between firms with and without foreign partnerships. The first question was, "Have your workers undergone any training in the last two years?" Managers were given discrete choices of yes or no. In firms without foreign partners, only 18 percent replied in the affirmative, while 47 percent of joint ventures and 60 percent of firms with FDI indicated that their workers answered positively.

FIGURE 8.3 Training and New Technology in Czech Firms with and without Foreign Partners

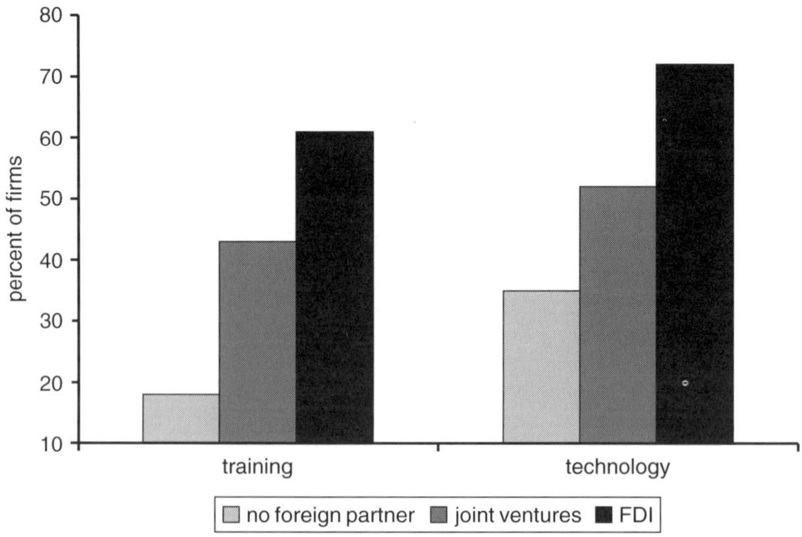

The second question was whether new technology (machinery, equipment) or related know-how had been obtained in the previous two years. Foreign investment was again associated with substantially greater positive responses—in more than 70 percent of the firms with FDI and 50 percent of the joint ventures, the partner had acquired some kind of new technology; for firms with no foreign linkages, the figure was just 35 percent. The difference between the two sets of firms is greater on the training variable ("software") than on the technology ("hardware") variable. These figures illustrate that although firm-level data on technological effort are not available, foreign partnerships are associated with greater investments in training and knowledge acquisition.

Estimation Procedure

We estimate production functions for the firms included in the sample. Each firm i has a production function for gross output:

$$Y_i = F^i(L_i, M_i, K_i, T_i), \tag{8.1}$$

where Y is gross output; K, L, and M are inputs of capital, labor, and materials; and T indexes technology. The firm's production function, F, is homogeneous of degree g ($g \neq 1$) in K, L, and M. Firms are assumed to be price takers on factor markets but may have market power in output markets. The price taker assumption is reasonable, since wages were largely set centrally during the sample period and most materials were bought abroad at world prices.

The production function in equation (8.1) implies the following relation between marginal physical products and outputs:[8]

$$F_L^i L_i + F_M^i M_i + F_K^i K_i = g_i Y_i, \tag{8.2}$$

where F_J^i is the marginal product of input J. The optimal choice of inputs by a firm with some monopoly power implies

$$P_i F_J^i = \mu_i P_{Ji}, \tag{8.3}$$

where P_{Ji} is the price of factor J, P_i is the price of the firm's output, and μ_i is the mark-up of price over marginal cost: $\mu_i = P_i / MC_i$, where MC_i is marginal cost. Combining equations (8.2) and (8.3) yields

$$s_{Li} + s_{Mi} + s_{Ki} = g_i / \mu_i, \tag{8.4}$$

where $s_{Ji} = P_{ji} J_i / P_i Y_i$ are the expenditures on each factor J_i relative to total enterprise revenues. Since firms do not necessarily produce under constant returns to scale, the sum of these shares is not always unity. Using equation 8.4, the revenue share of capital can be defined as

$$s_{Ki} = 1 - s_{Li} - s_{Mi} = s_{Ki} + (1 - g_i / \mu_i). \tag{8.5}$$

The productivity equation can then be derived from equation (8.1) as

$$dy_i = \mu_i[s_{Li}dl_i + \hat{s}_{Ki}dk_i + s_{Mi}dm_i] + \mu_i[s_{Ki} - \hat{s}_{Ki}]\, dk_i + \frac{F_T^iT_i}{F^i}dt_i, \tag{8.6}$$

where dy_i is output growth and $(F_T^iT/F^i)dt_i$ measures the technology change or TFP growth. The second term on the right-hand side can be simplified to $(g_i - \mu_i)dk_i$ using equation (8.5). Equation (8.6) is estimated in log-differences, using actual enterprise-level data to construct the first right-hand-side term. There are two terms to estimate for each industry (g_i and μ_i, the scale and mark-up parameters), as well as the TFP parameter for each enterprise. We use the reported book value of fixed assets to construct the capital revenue share.

To take into account the likelihood that foreign investment choices are not randomly distributed—the descriptive statistics reported in table 8.1 suggest that firms attracting FDI and joint ventures have above-average initial performance—we correct for the possible endogeneity of foreign investment choices by using the generalized Heckman two-step procedure for correcting sample selection bias, as developed by Amemiya (1984). The procedure involves separately estimating the foreign investment decision and the subsequent firm productivity growth performance. The first step uses a probit model to determine the probability of foreign investment based on initial efficiency (proxied by the share of variable costs in total revenues), firm size, and type of industry. The second step involves an estimation using only observations on firms with foreign linkages. This results in an omitted variable sample selection bias. The Amemiya procedure provides for a specification of the omitted variable that can be used in the full sample to alleviate sample selection. An additional variable estimated in the first step is included in the second-step regression.

Since the primary focus here is to test the association between productivity growth and foreign investment, we augment equation (8.6) by including a dummy for firms with foreign partners as an additional "factor of production." The dummy (*FOREIGN*) is 1 if a firm had either FDI or a joint venture in the preceding year and 0 if otherwise. This approach is similar to the empirical design used in Harrison (1994). In addition, the effects of other changes in the economic environment have to be controlled for. We do not have good proxies for these changes, nor can we account individually for each of them. Instead, we include annual dummies in the estimating equation that pick up the net effects of changes in the economic environment at the aggregate level.

We are also interested in investigating whether there have been any spillover effects from foreign investment on other firms that operate in the same sectors but do not have foreign partners. To analyze this, we run equation (8.6) on local firms only and include as an additional independent variable (*SPILLOVER*), the share of assets of firms with FDI or joint ventures in total assets of all firms in each sector. If foreign participation has beneficial spillover effects on other firms, we would expect the coefficient to be positive.

Because of the probable correlation between productivity effects and the independent variables, ordinary least squares (OLS) may give biased and inconsistent estimates. This simultaneity problem is endemic to the empirical literature on productivity measurement. We address the issue by using *F*-tests to reveal whether OLS is appropriate and relying on the Hausman specification test to choose between random or fixed effects frameworks in cases where OLS should not be used. These tests suggest that a random effects model is most appropriate.[9] Coefficient estimates for the major coefficients or variables of interest as well as information on the share of assets of firms with either joint ventures of FDI in total assets (48 percent) and the share of firms with majority foreign ownership (19 percent) are reported in table 8.2.

TABLE 8.2 Revenue Shares of Inputs, Mark-Up, and Scale Estimates by Sector

Sector	S_m	S_l	S_k	μ_i	g_i	Share of foreign assets	Share of FDI assets
Mining	0.538	0.215	0.246	1.246	1.200	0.398	0.124
Construction	0.720	0.169	0.111	1.137	1.088	0.432	0.325
Food and beverage	0.629	0.206	0.165	1.388	1.264	0.635	0.311
Textiles and apparel	0.677	0.180	0.142	1.284	1.132	0.294	0.182
Furniture and other wood products	0.743	0.145	0.110	1.152	1.001	0.542	0.261
Pulp and paper	0.791	0.129	0.079	1.211	1.113	0.715	0.521
Printing and publishing	0.730	0.136	0.133	0.889	0.992	0.885	0.605
Chemicals	0.757	0.151	0.091	1.201	1.163	0.547	0.281
Shoes and leather products	0.612	0.224	0.162	1.182	1.119	0.128	0.000
Nonmetallic mineral products	0.615	0.191	0.193	0.958	0.996	0.408	0.241
Basic metals	0.702	0.155	0.142	1.211	0.880	0.367	0.134
Fabricated metal products	0.733	0.121	0.145	1.192	1.100	0.785	0.191
Electric and electronics	0.657	0.191	0.151	1.201	1.039	0.356	0.110
Transport equipment	0.687	0.117	0.195	1.272	1.070	0.428	0.127
Other manufacturing	0.594	0.171	0.233	—	—	0.524	0.229
Retail services	0.257	0.453	0.289	1.352	1.198	0.402	0.221
Financial services	0.190	0.609	0.200	1.079	1.324	0.368	0.141
Average	0.625	0.209	0.164	1.184	1.104	0.483	0.191

Source: Authors.

Note: — = not available.

Results

The results of estimating equation (8.6) are reported in table 8.3, using both OLS and a random effects specification. The estimated coefficient on the dummy for FDI is positive and statistically significant for both specifications. This suggests that, as predicted, foreign investment involves an additional "transfer of technology." The dummy for joint ventures also has a positive sign, but it is slightly smaller in magnitude and is not statistically significant.

The possibility of a positive spillover impact of foreign investment is considered by including the share in total assets of firms with foreign partners (lagged one year) as a separate regressor (table 8.4). This is a continuous, not a categorical, variable. As noted previously, this approach assumes that spillovers are sector specific and therefore ignores possible interindustry spillovers. Contrary to what is predicted, spillovers are negative: greater foreign participation in an industry has a statistically significant negative effect on the performance of other firms. Each 10 percent increase in the foreign asset share is associated with a 1.7 percent fall in sales growth of domestic firms.

TABLE 8.3 Panel Regression Estimates (full sample)

Dependent variable: Growth in sales	OLS	Random effects
Amemiya selection bias correction variable	Yes	Yes
Sector-specific returns to scale and mark-ups	Yes	Yes
FDI dummy	0.015**	0.015*
	(2.011)	(1.937)
JV dummy	0.011	0.010
	(1.372)	(1.286)
Dummy for 1994	–0.012*	–0.011
	(–1.873)	(–1.672)
Dummy for 1995	–0.052**	–0.052**
	(–7.034)	(–6.942)
Dummy for 1996	–0.054**	–0.053**
	(–7.062)	(–7.534)
F-test ($A, B = A_i, B$)	0.89	
Hausman test (random versus fixed effects)		25.66 (30.19)
Adjusted R^2	0.894	0.861

Source: Authors.

Note: $N = 513$. Heteroskedasticity consistent (White correction); t-statistics are in parentheses, except for the Hausman test, for which the parenthetical figure indicates the cut-off point. A constant term is included in both regressions.

*Significant at the 10 percent level.

**Significant at the 5 percent level.

TABLE 8.4 Spillover Effects on Firms without Foreign Linkages

Dependent variable: Growth in sales	OLS	Random effects
Amemiya selection bias correction variable	Yes	Yes
Sector-specific returns to scale and mark-ups	Yes	Yes
Spillovers (share of assets of firms with joint ventures and FDI)	−0.178* (3.125)	−0.172* (2.054)
Dummy for 1994	0.002 (0.215)	0.002 (0.178)
Dummy for 1995	−0.038* (−4.201)	−0.037* (−3.934)
Dummy for 1996	−0.036* (−3.534)	−0.035* (−3.642)
F-test	0.92	
Hausman test (random versus fixed effects)		4.57 (14.45)
Adjusted R^2	0.887	0.843

Source: Authors.

Note: $N = 340$. Heteroskedasticity consistent (White correction); t-statistics are in parentheses, except for the Hausman test, for which the parenthetical figure indicates the cut-off point. A constant term is included in both regressions.

*Significant at the 5 percent level.

It has been argued that spillovers from joint ventures should be higher than those from FDI (establishment of majority-owned affiliates), as the foreign partner has less ability to control the behavior of the domestic partner and the domestic partner has a greater incentive to pursue R&D itself (see, for example, Pack and Saggi 1997). Internalization through FDI, in contrast, should offer greater opportunities to limit "technology leakage." If this is indeed the case, it implies that excluding joint ventures from the *SPILLOVER* measure of foreign "ownership" share and re-estimating the equation should increase the magnitude of the negative spillovers.

The evidence does not support this argument (table 8.5). The magnitude of the spillover effect becomes smaller and statistically insignificant, although it remains negative. Thus excluding joint ventures has an offsetting effect. In part, this reflects the fact that joint venture firms have higher TFP growth than firms without any foreign partnerships, which raises the average of the non–FDI group. This result illustrates that the initial negative spillover result may not be robust and that tests for spillovers with the methodology used here (and in the literature more generally) require some assurance that in distinguishing between two subsets of firms in an industry on the basis of whether or not there is majority foreign ownership (or, more generally, foreign linkages of some kind) one is not ignoring other important determinants of the performance of firms.

TABLE 8.5 Spillover Effects on Firms without FDI

Dependent variable: Growth in Sales	OLS	Random effects
Amemiya selection bias correction variable	Yes	Yes
Sector-specific returns to scale and mark-ups	Yes	Yes
Spillovers (share of assets of foreign affiliates in total assets of the sector)	−0.077 (1.425)	−0.074 (1.218)
Dummy for 1994	0.003 (0.897)	0.002 (0.178)
Dummy for 1995	−0.032** (−2.985)	−0.031** (−2.257)
Dummy for 1996	−0.027* (−1.847)	−0.025 (−1.514)
F-test	0.91	
Hausman test (random versus fixed effects)		4.13 (14.45)
Adjusted R^2	0.894	0.857

Source: Authors.

Note: $N = 431$. Heteroskedasticity is consistent (White correction); *t*-statistics are in parentheses, except for the Hausman test, for which the parenthetical figure indicates the cut-off point. A constant term is included in both regressions.

*Significant at the 10 percent level.

**Significant at the 5 percent level.

One such determinant likely to be important is the technological effort of firms. The survey questionnaire revealed that joint venture firms invested significantly more in training and new technologies than pure "domestic" firms. It may be that the technological ability and effort expended by many of the firms without foreign partners is too low to be able to absorb spillovers when they occur or that the firms with foreign linkages have absorbed a significant share of the available stock of labor with requisite skills. Given that FDI and joint ventures together account for significant shares of total assets, sales, and employment in the Czech Republic, the potential for positive spillovers among firms with foreign partnerships may be significant. It is suggestive that if "domestic" firms are excluded from the sample, FDI has a positive effect on firms with joint ventures, although this is not statistically significant (the *t*-statistic is 1.42, possibly reflecting the small sample size).

Account should also be taken of the short time frame on which the study focuses. Spillovers may require more time before they show up in TFP growth rates. And the absorption of new techniques requires significant in-house technological effort, which may not be captured adequately by the production function methodology used. Clearly, further research is required.

Concluding Remarks

Firm-level data for the Czech Republic during the 1992–96 period suggest that foreign investment has the predicted positive impact on the TFP growth of recipient firms. This result is robust to corrections for the sample selection bias that prevails because foreign investment tends to go to firms with above-average initial productivity performance. This result is not surprising, given that there is a presumption that foreign investors should be transferring new technologies and knowledge to partner firms. With some lag, this is likely to be reflected in greater TFP growth. FDI appears to have a greater impact on TFP growth than joint ventures, suggesting that parent firms transfer more know-how (soft or hard) to affiliates than joint venture firms obtain from their partners.

Taken together, joint ventures and FDI appear to have a negative spillover effect on firms in each industry that do not have foreign partnerships. This effect is relatively large and statistically significant. However, if the focus of attention is restricted to the impact of foreign-owned affiliates (firms with FDI) on all other firms in an industry, the magnitude of the negative effect becomes much smaller and loses statistical significance. In conjunction with the fact that, taken together, joint ventures and FDI account for significant shares of total output in many industries, this result suggests that further research is required to determine the extent to which knowledge diffuses from firms having strong linkages to foreign firms to those that do not have such relationships. Of particular importance in this connection is exploring the extent of spillovers among joint venture firms and between foreign affiliates and firms with joint ventures. Insofar as joint venture firms invest more in technology capacity (as suggested by their training efforts), one expects these firms to be better able to absorb and benefit from know-how diffusion. The absence of such capacity may be a factor underlying the observed negative spillover effect. Longer times series and collection of data on variables that measure in-house technological effort by firms would help identify the magnitude and determinants of technological spillovers.

Notes

1. A separate but related literature on technology diffusion focuses largely on analysis of the determinants of the number of firms or the proportion of industry output produced by a new technology (aggregate diffusion) and on analysis of the determinants of the time at which a firm adopts a new technology relative to other firms (so-called duration models) (see, for example, Ray [1964] and Karshenas and Stoneman [1994]). Data constraints prohibit analysis of the types of questions asked in the diffusion literature, as it is not possible to identify specific technologies in our data set.

2. See, for example, Helleiner (1973) and Keesing and Lall (1992) on subcontracting; Feenstra, Markusen, and Zeile (1992) on imports of inputs; Blomström and Kokko (1997) for a recent survey of the literature on FDI; and Pack and Saggi (1997) for a general survey of the literature on technology transfer.

3. Notions of arm's length exchange used in the literature vary. For example, Pack and Saggi (1997) distinguish between intrafirm exchange (FDI) and contractual exchanges (licensing, joint ventures, turnkey projects, and so forth). They call contractual exchanges arm's length arrangements.

4. See Javorcik (2004) for a recent analysis of the relationship between intellectual property protection and FDI in transition economies.

5. Equally important may be spillovers across industries. This issue is not explored in this chapter, although it may be important in the transition context.

6. See Nelson and Wolff (1997).

7. The literature on technology acquisition and adoption in developing countries is substantial. See, for example, Evenson and Westphal (1995); Lall (1987, 1992); and Pack and Westphal (1986). Westphal, Rhee, and Pursell (1981) discuss the case of the Republic of Korea in some depth.

8. The authors are grateful to a referee for suggesting the formulation used below.

9. A fixed-effects estimation assumes firm productivity growth to be constant over time. This assumption is objectionable, since changes in productivity due to increased competition is the phenomenon we seek to explore. The random effects model avoids the imposition of constant productivity growth over time, but it has the drawback that productivity shocks at the firm level are assumed to be uncorrelated over time. This may not be a reasonable restriction if there is convergence or divergence in corporate performance.

References

Aitken, Brian J., and Ann E. Harrison. 1999. "Do Domestic Firms Benefit from Direct Foreign Investment? Evidence from Venezuela." *American Economic Review* 89 (3): 605–18.

Amemiya, T. 1984. "Tobit Models: A Survey." *Journal of Econometrics* 24: 3–61.

Blomström, Magnus, and Ari Kokko. 1997. "How Foreign Investment Affects Host Countries." Policy Research Working Paper 1745, World Bank, Washington, DC.

Blomström, Magnus, and Hakan Persson. 1983. "Foreign Investment and Spillover Efficiency in an Underdeveloped Economy: Evidence from the Mexican Manufacturing Industry." *World Development* 11: 493–501.

Blomström, Magnus, Ari Kokko, and M. Zejan. 1994. "Host Country Competition and Technology Transfer by Multinationals." *Weltwirtschaftliches Archiv* 130: 521–33.

Burnside, Craig, Michael Eichenbaum, and Sergio Rebelo. 1996. "Sector Solow Residuals." *European Economic Review* 40 (3–5): 861–69.

Clerides, Sofronis, Saul Lach, and James Tybout. 1998. "Is Learning by Exporting Important? Micro-Dynamic Evidence from Colombia, Mexico and Morocco." *Quarterly Journal of Economics* 113 (3): 903–47.

Coe, David, and Elhanan Helpman. 1995. "International R&D Spillovers." *European Economic Review* 39: 214–42.

Coe, David, Elhanan Helpman, and Alexander Hoffmaister. 1997. "North-South R&D Spillovers." *Economic Journal* 107: 134–49.

Czech Statistical Office. 1997. *Statistical Yearbook 1996*. Prague: Czech Statistical Office.

Das, S. 1987. "Externalities, and Technology Transfer through Multinational Corporations: A Theoretical Analysis." *Journal of International Economics* 22: 171–82.

Evenson, R.E., and L. Westphal. 1995. "Technological Change and Technology Strategy." In *Handbook of Development Economics,* vol. III, ed. J. Behrman and T.N. Srinivasan. Amsterdam: North Holland.

Feenstra, Robert, James Markusen, and William Zeile. 1992. "Accounting for Growth with New Inputs: Theory and Evidence." *American Economic Review* 82: 415–21.

Grossman, Gene, and Elhanan Helpman. 1995. "Technology and Trade." In *Handbook of International Economics*, vol. III, ed. G. Grossman and K. Rogoff. Amsterdam: North Holland.

Haddad, M., and A. Harrison. 1993. "Are There Positive Spillovers from Direct Foreign Investment? Evidence from Panel Data for Morocco." *Journal of Development Economics* 42: 51–74.

Harrison, Ann. 1994. "Productivity, Imperfect Competition and Trade Reform." *Journal of International Economics* 36: 53–73.

———. 1996. "Determinants and Consequences of Foreign Investment in Three Developing Countries." In *Industrial Evolution in Developing Countries: Micro Patterns of Turnover, Productivity and Market Structure*, ed. Mark Roberts and James Tybout. Oxford: Oxford University Press.

Helleiner, G. 1973. "Manufactured Exports from Less Developed Countries and Multinational Firms." *Economic Journal* 83: 21–47.

Javorcik, Beata S. 2004. "The Composition of FDI and Protection of Intellectual Property Rights in Transition Economies." *European Economic Review* 48: 39–62.

Karshenas, M., and P. Stoneman. 1994. "Technological Diffusion." In *Handbook of Economics of Innovation and Technological Change*, ed. P. Stoneman. Oxford: Blackwell.

Keesing, Don, and S. Lall. 1992. "Marketing Manufactured Exports from Developing Countries: Learning Sequences and Public Support." In *Trade Policy, Industrialization and Development: New Perspectives*, ed. G. Helleiner. Oxford: Clarendon Press.

Lall, Sanjaya. 1987. *Learning to Industrialize: The Acquisition of Technological Capabilities in India*. London: MacMillan.

———. 1992. "Technological Capabilities and Industrialization." *World Development* 20: 165–86.

Markusen, James. 1995. "The Boundaries of the Multinational Enterprise and the Theory of International Trade." *Journal of Economic Perspectives* 9: 169–89.

Nelson, Richard, and Howard Pack. 1998. "The Asian Miracle and Modern Growth Theory." Policy Research Working Paper 1881, World Bank, Washington, DC.

Nelson, Richard, and Edward Wolff. 1997. "Factors behind Cross-Industry Differences in Technical Progress." *Structural Change and Economic Dynamics* 8: 205–20.

Pack, Howard. 1987. *Productivity Technology and Industrial Development*. New York: Oxford University Press.

Pack, Howard, and Kamal Saggi. 1997. "Inflows of Foreign Technology and Indigenous Technological Development." *Review of Development Economics* 1: 81–98.

Pack, Howard, and Larry Westphal. 1986. "Industrial Strategy and Technological Change." *Journal of Development Economics* 22: 87–128.

Ray, G.F. 1964. *The Diffusion of Mature Technologies*. Cambridge: Cambridge University Press.

Wang, Jian-Ye, and Magnus Blomström. 1992. "Foreign Investment and Technology Transfer: A Simple Model." *European Economic Review* 36: 137–55.

Westphal, Larry, Yung Rhee, and Garry Pursell. 1981. "Korean Industrial Competence: Where It Came from." Staff Working Paper 469, World Bank, Washington, DC.

9

TECHNOLOGICAL LEADERSHIP AND THE CHOICE OF ENTRY MODE BY FOREIGN INVESTORS

Beata Smarzynska Javorcik

During the past several decades, there has been a significant change in the attitudes of governments, especially those in developing countries, toward foreign direct investment (FDI). Rather than viewed as evil exploiters, foreign investors are now welcomed as a source of new technologies, better management and marketing techniques, and skilled jobs. Not all types of foreign investment, however, are perceived as equally beneficial to host countries. Governments tend to favor joint ventures over other forms of FDI, since they believe that active participation of local firms facilitates the absorption of new technologies and marketing skills.[1] This chapter leaves aside the issue of whether this perception is true and instead attempts to compare the potential of joint ventures and wholly owned foreign subsidiaries for such transfers.

In contrast to the existing literature, which demonstrates that differences in R&D and advertising intensities *between* industries influence a foreign investor's choice of entry mode, this chapter focuses on patterns present *within* industries. It provides empirical evidence indicating that industry structure affects the choice of entry mode and that treating industries as homogenous, as done in earlier studies, is thus not appropriate. It finds that joint ventures in manufacturing sectors tend to be undertaken by foreign investors possessing fewer intangible assets than their counterparts involved in wholly owned projects. These effects are present in high-technology but

The author wishes to thank Andy Bernard, Bruce Blonigen, Philip Levy, Robert Lipsey, T.N. Srinivasan, and Louis Wells for their helpful suggestions and Hans Peter Lankes for making the results of a foreign invesment survey by the European Bank for Reconstruction and Development available to her. The author benefited from comments received at seminars at Yale University, the World Bank, and the U.S. International Trade Commission.

not in low-technology sectors. The findings therefore suggest that joint ventures in more R&D-intensive sectors may present less potential for transfers of technology and marketing skills than wholly owned foreign subsidiaries.

The theoretical literature suggests that costs involved in drafting and enforcing contracts guiding the transfers of proprietary know-how to joint ventures as well as the threat of knowledge dissipation may make shared ownership less attractive and encourage the establishment of wholly owned subsidiaries. Firms differentiating their products through advertising may also seek full ownership to ensure the quality of their products and prevent debasing of their trademarks. Empirical studies confirm these predictions by finding a negative relationship between industry- or firm-level R&D and advertising intensities and the probability of shared ownership (Stopford and Wells 1972; Gatignon and Anderson 1988; Gomes-Casseres 1989; Asiedu and Esfahani 2001).[2]

In contrast to existing research, this chapter proposes that the choice of entry mode depends not only on the R&D or advertising intensity of a sector but also on an investor's endowment of intangible assets relative to the industry average (hereafter referred to as relative R&D and advertising intensity). Relative endowment of intangible assets may affect a firm's choice of entry mode in two opposing ways.

On the one hand, technological and marketing leaders in an industry may have greater bargaining power in negotiations with local firms and authorities and may be able to secure more favorable terms under joint venture agreements. Thus they are likely to gain more from such arrangements than industry laggards.[3] In R&D-intensive sectors, foreign technological leaders may also be so advanced compared with domestic firms that technology leakage may not pose a severe threat, since local firms may be unable to use independently the knowledge acquired through a joint venture. Thus one should expect to observe that industry leaders are more likely to share ownership than industry laggards.

On the other hand, if the gap between foreign and domestic firms does not guarantee protection against dissipation of intangible assets, industry leaders may be more averse to shared ownership than industry laggards. Moreover, they may use their bargaining power to negotiate sole ownership in countries in which foreign ownership is restricted by the government. In that case, the relative endowment of intangible assets would be positively correlated with the probability of full ownership. Thus the theoretical predictions are ambiguous, and an empirical investigation is necessary to shed light on this issue.

This chapter tests whether intraindustry differences in intangible assets affect a foreign investor's choice of entry mode. It also examines whether these effects are the same in industries with different levels of R&D spending. The analysis focuses on manufacturing industries and uses a unique data set on foreign investment projects in Eastern Europe and the former Soviet Union in the early 1990s. The data set is based on a worldwide survey of companies conducted by the European Bank for Reconstruction and Development (EBRD) that contains information on foreign investments in 22 transition economies.

The analysis begins by estimating a model similar to those found in earlier studies. As expected, firm- or industry-level R&D spending and firm advertising expenditure are negatively related to the probability of a joint venture and positively correlated with the likelihood of a wholly owned project. Factors governing the choice of entry mode in transition economies are thus similar to those present in other regions.

To test the first hypothesis, I include in the model industry-level R&D intensity as well as the ratio of a firm's R&D spending to the industry mean, which serves as a proxy for a firm's technological sophistication. R&D intensity is not a perfect measure of a firm's intangible assets; firms may have other intangible assets they are concerned about dissipating, such as distribution and marketing techniques. To take this into account, I add an analogous variable capturing sophistication of marketing techniques. I also control for other firm-specific characteristics and progress of reform in a host country. I improve over the existing studies by taking into account possible sample selection bias arising from the fact that the same firm characteristics affect both a firm's decision to invest and choice of entry mode.

The empirical analysis supports the hypothesis that an investor's endowment of intangible assets relative to other firms in the industry influences its choice of entry mode. The relationship between investors' technological and marketing sophistication and the probability of shared ownership is negative. Thus the results indicate that industry structure affects investors' behavior and industrial sectors should not be treated as homogenous in an examination of entry mode.

Next I test whether these effects vary across industries by allowing for different coefficients in high-/medium- and low-technology sectors. I show that the impact of technological and marketing sophistication on the choice of entry mode is statistically significant in the high- and medium-technology sectors but not in low-technology industries. This finding is robust to different classifications of sectors.

Related Literature

Theory Theory suggests that in order to compete successfully in a foreign market, a firm must possess so-called ownership advantages, which can take the form of superior technology, proprietary knowledge, managerial and marketing skills, and so on. A firm can earn rents on these assets through arm's length transactions (for example, licensing, franchising, turnkey contracts) or by creating a subsidiary or forming a joint venture in a foreign country. Assessing the value of intangible assets is a difficult task and is associated with information asymmetry. The seller may not receive adequate payment, since he may not be able to disclose the full potential of future profits generated by a given technology to a prospective buyer without giving away private information on the technology itself (Dunning 1988). Thus firms possessing more sophisticated technologies may face more uncertainty in pricing and may prefer wholly owned projects to joint ventures or arm's length transactions.

A wholly owned venture might also be preferred to shared ownership in order to guard against leakage of sensitive information. For example, a foreign investor may be concerned that in the case of the dissolution of a joint venture, the local partner will remain in possession of the technology acquired from the multinational and will become a competitor in third markets. The joint venture agreement may not offer full protection against this possibility, since it may be difficult to specify all contingencies in the contract.[4] Additionally, the local partner may use proprietary information obtained from the multinational in its own wholly owned operations, thus hurting the joint venture and the foreign partner (Gomes-Casseres 1989).[5] The multinational may also fear that if the local partner controls the employment policy, it may not put enough effort into keeping key employees, who may leave and reveal their knowledge of the production process to the competition.[6]

Firms investing heavily in advertising also have reasons to seek full ownership. A joint venture partner may have a strong incentive to free ride on the reputation of a foreign partner by debasing the quality of the product carrying the foreign trademark. In such a case, the local partner appropriates the full benefits of debasement while bearing only a small fraction of the costs (Caves 1982; Horstmann and Markusen 1987). Full ownership also allows foreign investors to retain control over the marketing strategy and eliminates the need to persuade the local partner about the optimal level and mix of marketing expenditure. In summary, theory predicts that firms with greater intangible assets should prefer full ownership to joint ventures.

The preferences of a host country government regarding the extent of foreign ownership may differ from those of a multinational corporation. Therefore, the entry mode used may be a result of a bargaining process between the two parties (Gomes-Casseres 1990). Studies of the bargaining approach predict that multinational corporations in R&D-intensive industries enjoy greater bargaining power in negotiations with local authorities (Gomes-Casseres 1990; see also UNCTAD 1992).

Empirical Evidence Earlier empirical research on the relationship between intangible assets and the choice of entry mode found a negative correlation between firm or industry R&D intensity and the probability of shared ownership (Stopford and Wells 1972; Gatignon and Anderson 1988; Asiedu and Esfahani 2001). Some studies, however, did not produce statistically significant results (Gomes-Casseres 1989; Blomström and Zejan 1991). Advertising intensity was shown to be negatively related to the probability of a joint venture (Gomes-Casseres 1989, 1990), but in some studies this relationship was not statistically significant (Hennart 1991).

The only examination of entry modes used by foreign firms investing in Eastern Europe was undertaken by Meyer (1998). He analyzed characteristics of British and German companies engaged in minority and majority joint ventures and wholly owned projects in the region. For each of these categories, he estimated a logit model with the dependent variable equal to one if the particular entry mode was chosen and zero otherwise. He found a significant positive correlation between a firm's R&D intensity and the probability of a wholly owned subsidiary and a negative relationship between a firm's R&D intensity and the likelihood of a joint venture. The negative relationship was present in the case of minority, majority, and all types of joint ventures combined.

Gomes-Casseres (1990) analyzed the choice between full and shared ownership in the context of bargaining theory. Among other variables, his model included industry-level R&D spending as well as its interaction with a dummy variable for countries restricting the extent of foreign ownership. A positive correlation between the interaction term and the probability of a wholly owned subsidiary would indicate that investors in more R&D-intensive sectors can more easily negotiate full ownership. Neither of the two variables, however, turned out to be statistically significant.[7]

Hypotheses

In contrast to the earlier literature, I postulate that the choice of entry mode is influenced not only by the R&D intensity of a given industry but also by the R&D spending of a foreign investor relative to other firms within the sector. Thus there exist both interindustry and intraindustry effects. However, the sign of the intraindustry effect is unclear.[8]

On the one hand, investors enjoying a technological lead in their sectors are perceived as more attractive joint venture partners by local firms and governments. Thus they are able to negotiate more favorable terms in a joint venture agreement. Additionally, industry leaders may be able to form joint ventures with more successful local companies than foreign technological laggards.[9] Moreover, the technology gap between foreign leaders and domestic producers may be so large that even in the presence of knowledge transfer to a joint venture, the threat of losing intangible assets may be minimal. Thus industry leaders may be more willing to form joint ventures than industry laggards.

On the other hand, the technology gap may not be enough to prevent knowledge dissipation. Investors possessing a technological advantage over other firms in

Figure 9.1 R&D Intensity and Probability of a Joint Venture: Case 1

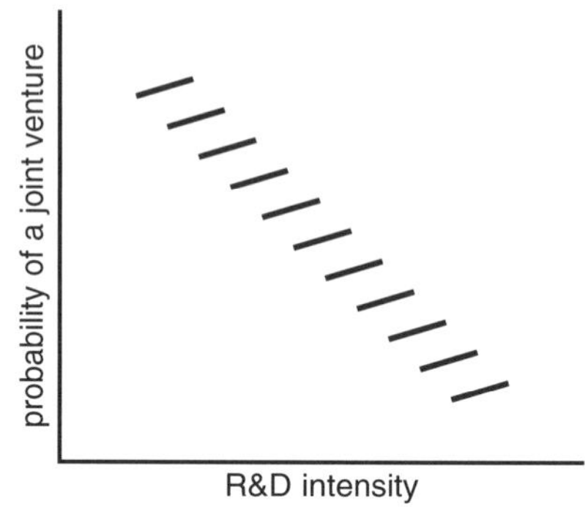

Source: Author.
Note: Each segment represents an industry.

their sectors may potentially incur greater losses from knowledge dissipation than investors with less sophisticated technologies. It may also be more difficult to price cutting-edge technologies than mature ones. Thus shared ownership may be less appealing to industry leaders, who may use their bargaining power to negotiate with the host country government an exemption from the restrictions on the extent of foreign ownership, should such restrictions be present.[10] In summary, the theoretical predictions about the relationship between relative R&D intensity and the likelihood of a joint venture are ambiguous. Figures 9.1 and 9.2 illustrate the two possibilities described above.

Figure 9.2 R&D Intensity and Probability of a Joint Venture: Case 2

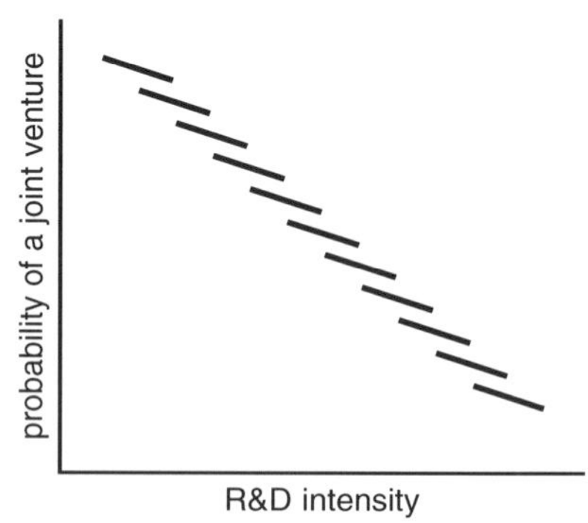

Source: Author.
Note: Each segment represents an industry.

An alternative assumption, often employed in the literature (see, for example, Ethier and Markusen 1996), is that knowledge dissipation takes place regardless of the form a multinational corporation uses to service a local market. Whether the multinational enters through a wholly owned subsidiary or a joint venture with a local partner, it will lose its intangible assets after a certain time because of employee turnover. This loss is, however, likely to be more costly to a firm with more sophisticated technologies. Under this assumption, the question is whether it is cheaper for the multinational to buy loyalty from employees of its wholly owned subsidiary or from a local partner. Ethier and Markusen (1996) show that this depends on the parameters of the model.[11] Thus theory again provides ambiguous predictions.

The second hypothesis is that the impact of relative R&D intensity may not be the same in all industries. It is plausible that relative R&D intensity plays an important role in high- and medium- but not in low-technology sectors. In low R&D sectors, technological sophistication may be less important to potential local partners. At the same time, foreign investors may be less concerned about losing their technological superiority. Thus there may exist no correlation between relative R&D intensity and the choice of entry mode in these industries. If this is the case, one would observe the relationship portrayed in figure 9.3.

Figures 9.1–9.3 suggest that studies treating industries as homogenous, and thus ignoring the impact of relative R&D intensity, would produce a negative relationship between industry R&D efforts and the likelihood of a joint venture. A negative correlation would also be found if only a firm's R&D expenditure were included in the model. Hence while the results of the earlier studies are consistent with the proposed framework, they capture only part of the true relationship. If the data do not

Figure 9.3 R&D Intensity and Probability of a Joint Venture: Case 3

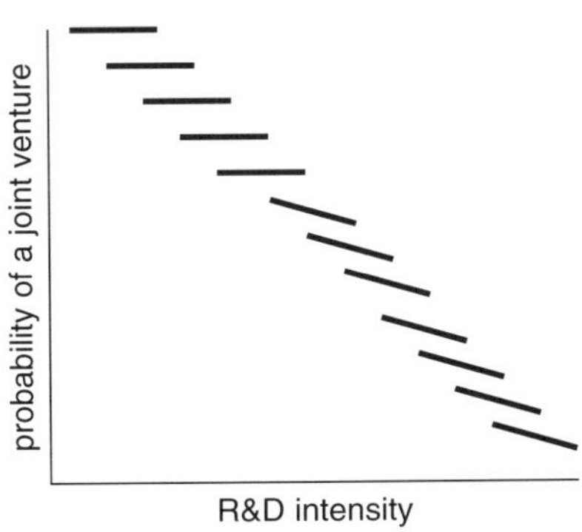

Source: Author.

Note: Each segment represents an industry.

support the hypothesis, one should observe a negative correlation between the probability of a joint venture and firm- or industry-level R&D spending. However, the effect of relative R&D intensity should be insignificant.

One caveat of using relative R&D expenditure as a proxy for technological leadership is that R&D intensity is not a perfect measure of a firm's success in innovative activities. Furthermore, in low-technology sectors, differences in R&D activities, which are generally modest, may not have strong effects. Finally, technology is not the only intangible asset firms may be concerned about losing. Leadership in terms of managerial techniques, marketing strategies, and distribution skills may be far more important in some industries. To capture this effect, I control for the advertising intensity of an industry and the relative advertising intensity of a firm.

I assume that all foreign investors have the option of engaging in a joint venture with a local partner, should they want to do so. Thus the supply of local joint venture partners is not constrained, and the observed entry patterns are determined entirely by foreign investors' demand. Considering that the aggregate FDI inflows into transition economies were quite small during the period covered by the sample, this assumption is quite realistic. Additionally, since most firms in the sample are relatively large, they most likely enjoyed an advantage in their search for local partners.[12]

I also assume that the available local partners fulfill some minimum standards set by foreign investors. Note that a difference in technological sophistication between foreign and local firms may be not be a serious obstacle to a successful joint venture agreement, since, as surveys indicate, local partners are expected to contribute their knowledge of market conditions, distribution networks, and ability to deal with government officials rather than provide technological expertise (OECD 1994).

In addition to proxies for intangible assets, the estimated model includes other variables. Blomström and Zejan (1991) mention that larger firms are more willing to take higher risks and are therefore more likely than smaller companies to engage in wholly owned projects. Thus I control for a firm's size and expect to find that it is negatively correlated with the probability of a joint venture. Note, however, that empirical investigations have often produced the opposite result (Blomström and Zejan 1991; Meyer 1998).[13] Stopford and Wells (1972) point out that more diversified firms tend to be more tolerant of minority ownership and are more likely to engage in joint ventures. This finding has been confirmed by Meyer (1998). Therefore, I also control for production diversification and anticipate a positive sign on its coefficient.

Usually, a major contribution of a local partner to a joint venture is the knowledge of the business environment in a host country. As foreign investors learn more about local conditions, their need for a local partner declines (Kogut and Singh 1988). Thus I control for investors' regional experience and expect to obtain a negative sign.[14] Similarly, I take into account firms' international experience. As

Blomström and Zejan (1991) show, firms with greater experience in foreign operations are less likely to share ownership.

The choice between full and shared ownership is also likely to be influenced by the investment climate in a host country. On the one hand, having a local partner that is well connected with local authorities may be more useful in countries with less friendly an attitude toward FDI. On the other hand, in economies with a more developed legal system and better corporate governance, foreign investors may be confident that potential disputes with local partners can be resolved fairly through the court system. They may thus be less averse to shared ownership. In summary, the relationship between business environment in a host country and the form of investment is ambiguous.

Data and Statistical Models

The data set used in this study is based on the EBRD survey of foreign investors, supplemented with the information obtained from the Worldscope database.[15] In January 1995 a brief questionnaire was sent to all companies (about 9,500) listed in Worldscope. Responses were obtained from 1,405 firms, which answered questions regarding their investments in Eastern European transition economies and the former Soviet republics, a total of 22 countries. The data set contains information on both investors and noninvestors.

In the case of investors, the type of entry mode for existing and planned projects in the region is known. Survey respondents were asked to classify each of their projects as a joint venture, acquisition, or greenfield project. For the purpose of this study, I consider all greenfield and acquisition projects not associated with joint ventures to be wholly owned. Unfortunately, no information is available on the size of investments.

The data set does not include any information on the timing of each investment. Since the magnitude of FDI inflows was marginal before 1989, the information collected pertains mostly to the period 1989–94.[16] Joint ventures outnumber wholly owned projects in all but one host country in the data set and constitute 62 percent of all projects (table 9.1).

During the past decade, transition economies undertook dramatic liberalization of their FDI regimes. In the Soviet Union a presidential decree issued in October 1990 allowed foreign wholly owned companies to be established in the form of branches or subsidiaries. The decree also created the legal basis for foreign investors to buy existing Soviet enterprises as they were privatized (McMillan 1996). Act XXIV of 1988 on the Investment of Foreigners in Hungary allowed non-Hungarian companies to own equity up to 100 percent (WTO 1998). In Poland the 1988 Law on Economic Activity with the Participation of Foreign Parties permitted 100 percent foreign equity participation (GATT 1992).

TABLE 9.1 Entry Modes Chosen by Investors in the Sample, by Country

Country	Number of joint venture projects in the sample	Number of wholly owned projects in the sample	Total number of projects in the sample	Population (millions) 1993	Average of transition indicators 1994
Albania	3	1	4	3.2	2.50
Azerbaijan	1	1	2	7.4	1.33
Belarus	5	3	8	10.4	1.67
Bulgaria	16	13	29	8.5	2.50
Croatia	7	4	11	4.8	3.17
Czech Republic	55	53	108	10.3	3.50
Estonia	16	8	24	1.5	3.33
FYR Macedonia	2	1	3	2.2	2.83
Georgia	4	2	6	5.4	1.33
Hungary	50	48	98	10.3	3.33
Kazakhstan	10	6	16	16.9	1.67
Latvia	13	6	19	2.6	2.83
Lithuania	8	5	13	3.7	3.00
Moldova	2	0	2	4.3	2.17
Poland	84	51	135	38.5	3.33
Romania	21	12	33	22.7	2.67
Russian Fed.	83	31	114	148.3	2.67
Slovak Republic	26	19	45	5.3	3.33
Slovenia	13	5	18	2.0	3.17
Turkmenistan	1	0	1	4.1	1.17
Ukraine	20	5	25	52.1	1.33
Uzbekistan	5	1	6	22.0	2.00
Mean or total	445	275	720	16.0	2.45

Source: Author.

Note: Sample includes manufacturing sectors only.

To the best of my knowledge, in none of the countries in the sample does legislation specifically forbid full ownership by foreign investors. It is possible, however, that in practice permissions for fully owned projects are denied in some economies. To take this possibility into account, I include host country dummies in the model and show that this change does not strongly affect the results. In many transition economies, however, FDI in sectors such as production of military equipment and extraction of natural resources is subject to restrictions on the extent of foreign ownership.[17] I therefore exclude firms in the coal, gas, and oil industries from the sample.

TABLE 9.2 Entry Modes Chosen by Investors in the Sample, by Industry

Industry	Joint ventures as percentage of all projects in the industry	Wholly owned projects as percentage of all projects in the industry
Drugs, cosmetics, and health care products	12.1	87.9
Electronics	67.0	33.0
Aerospace	80.0	20.0
Chemicals	78.0	22.0
Machinery and equipment	68.8	31.3
Electrical equipment	32.3	67.7
Automotive	60.0	40.0
Diversified manufacturing	88.5	11.5
Metal	72.2	27.8
Metal products	65.4	34.6
Paper	67.6	32.4
Beverages	20.0	80.0
Food	61.7	38.3
Apparel	50.0	50.0
Textiles	55.6	44.4
Tobacco	100.0	0.0
Printing and publishing	60.0	40.0

Source: Author.

The sample includes only manufacturing sectors, since in many service industries (such as banking, insurance, and telecommunications), restrictions on the extent of foreign ownership exist. In other sectors, such as accounting and public relations services, it may be extremely difficult to measure the endowment of intangible assets. Joint venture is the dominant form of investment in the majority of industries in the sample (table 9.2). However, in the drugs, cosmetics, and health care products sectors, only 12 percent of all projects are joint ventures, while wholly owned subsidiaries account for 88 percent of investments.[18] Wholly owned projects constitute 80 percent of all investments in the beverage sector. The drugs, cosmetics, and health care products sector is the most R&D-intensive industry, and in the beverage sector marketing activities play a very important role. Fear of losing intangible assets is thus a likely explanation of the underrepresentation of joint ventures in these industries.

Following earlier studies, I begin the analysis by estimating a probit model with the dependent variable taking on the value of one if investor i has engaged in a joint venture with a local partner in country *c* and zero if the project was wholly owned. Thus the number of observations will be equal to the number of projects undertaken in the region by all firms in the sample.[19]

I improve over the existing literature by controlling for possible sample selection bias. It is likely that the coefficient estimates from a model describing the choice between full and shared ownership are inconsistent because the model does not take into account which firms undertake FDI in the first place. To address this issue, I estimate a bivariate probit model accounting for sample selection. The first equation in the model describes firm *i*'s decision to undertake FDI in country *c*; the second models the choice between full and shared ownership.[20]

With the exception of regional experience, which comes from the survey, all explanatory variables employed in the estimation are taken from a commercial database, Worldscope, and refer to 1993 or the closest year for which information is available. Industry R&D intensity is measured by R&D expenditure as a percentage of net sales. To find the industry averages, I use figures for all firms listed in Worldscope in a given industry. The industry averages were calculated at the three-digit SIC industry classification.[21] To proxy for investor's technological sophistication, I use relative R&D intensity, defined as the ratio of a firm's R&D intensity to the industry average.[22] Industry advertising intensity is measured by sales, general expenditure, and administrative expenditure divided by net sales. This variable is a standard proxy for advertising intensity used in the literature. The industry average is calculated at the three-digit level. I define the relative advertising intensity as the ratio of a firm's advertising intensity to the industry average.

I use firm sales in millions of U.S. dollars as a measure of firm size. Diversification is proxied by the number of four-digit SIC codes describing a firm's activities. To control for regional experience, I include a dummy variable taking on the value of one if a firm had a trading relationship with the region before transition and zero otherwise. International experience is measured by the share of foreign sales in a firm's total sales. Ideally, I would like to have used the share of foreign assets in a firm's total assets. This would, however, have severely reduced the size of the sample. Since the share of foreign sales is highly correlated with the share of foreign assets (0.82), I believe that it can serve as a proxy for international experience. To capture the investment conditions in a host country, I use an index of transition progress, defined as the average of EBRD transition indicators. Transition indicators rate the progress of a country's reforms in price liberalization and competition, trade and exchange system, large-scale privatization, small-scale privatization, enterprise restructuring, and banking reform.[23] (The values for 1994 for all host countries in the sample are presented in table 9.1.) In the selection equation of bivariate probit, I also control for the size of the local market as captured by the host's GDP, measured in millions of U.S. dollars; the figures come from EBRD (1994).

To test whether the same effects are present in high- and low-technology sectors, I allow for different coefficients in the two groups of industries. The classification of industries is based on the average R&D intensity of each three-digit SIC sector, with two different cut-off values. I also employ the classification suggested by Blomström, Lipsey, and Ohlsson (1991).

Empirical Findings

Comparison with Earlier Studies

I begin the investigation by estimating a model comparable to those found in the earlier literature. This exercise allows me to establish whether the same factors affect foreign investors' choice of entry mode in transition economies as in other parts of the world. I employ a probit model in which the dependent variable takes on the value of one if the project takes the form of a joint venture with a local partner and zero if it is wholly owned. Explanatory variables include investor size, R&D intensity, advertising intensity, a measure of production diversification, and proxies for international and regional experience. The results are presented in terms of marginal effects evaluated at the sample mean.

The estimates shown in the first column of table 9.3 are consistent with those obtained in earlier studies. They indicate that smaller firms and firms with higher R&D and advertising intensities are more likely to engage in wholly owned projects than in joint ventures. Regional experience, international experience, and production diversification, however, do not appear to be statistically significant. Next I add the transition index to the model and observe that everything else equal, investors are more likely to engage in joint ventures in countries that are less advanced in terms of reform. Thus I find support for the hypothesis that local partners are more useful in countries with a less friendly investment climate. Including the transition index has little effect on the other variables in the regression.

Since many of the earlier studies used industry- rather than firm-level proxies for intangible assets, I also estimate a model that includes average R&D and advertising intensity for three-digit SIC industries. The results, presented in the third and fourth columns of table 9.3, indicate that firms in sectors characterized by high R&D spending are less likely to share ownership. Advertising intensity, however, does not appear to have a significant effect on the form of investment. I also find that investors with less international experience and those entering host countries that are less advanced in terms of reform are more likely to form joint ventures. None of the other variables is statistically significant.

In summary, the results are broadly consistent with the findings of the earlier literature. They suggest that forces determining the entry mode of foreign investors in transition economies are similar to those operating in other parts of the world.

TABLE 9.3 Probit Model with Firm- and Industry-Level R&D and Advertising Intensities

Variable	Marginal effects			
Firm R&D	−0.0233*** (0.0067)	−0.0257*** (0.0068)		
Industry R&D			−0.0370*** (0.0074)	−0.0381*** (0.0075)
Firm ADV	−0.0181*** (0.0035)	−0.0185*** (0.0035)		
Industry ADV			0.0017 (0.0014)	0.0010 (0.0015)
Diversification	0.0106 (0.0151)	0.0116 (0.0153)	0.0132 (0.0100)	0.0136 (0.0101)
Regional experience	0.0507 (0.0681)	0.0314 (0.0685)	0.0382 (0.0427)	0.0201 (0.0431)
International experience	−0.0007 (0.0012)	−0.0003 (0.0012)	−0.0020*** (0.0007)	−0.0018** (0.0007)
Firm size	<.0001** (0.0000)	<.0001* (0.0000)	<.0001 (0.0000)	<.0001 (0.0000)
Transition index		−0.1318** (0.0516)		−0.1478*** (0.0387)
Number of observations	346	346	603	603
Chi^2	84.01	90.74	49.84	64.99
Degrees of freedom	6	7	6	7
Prob > chi^2	0.00	0.00	0.00	0.00
Log L	−194.76	−191.39	−378.26	−370.68

Source: Author.

Note: Standard errors, pertaining to coefficients underlying the marginal effects, are reported in parentheses. <0.0001 denotes coefficients with an absolute value below 0.0001. The dependent variable is equal to one in the case of a joint venture and to zero in the case of a wholly owned subsidiary.

*The coefficient underlying the marginal effect is significant at the 10 percent level.

**The coefficient underlying the marginal effect is significant at the 5 percent level.

***The coefficient underlying the marginal effect is significant at the 1 percent level.

Summary Statistics

The next step is to compare the average R&D intensity of investors engaged in wholly owned projects with that of investors sharing ownership in each three-digit SIC sector. I group sectors into high, medium, and low technology, following the classification used by Blomström, Lipsey, and Ohlsson (1991) (table 9.4).

In all but one high-technology industry, investors engaged in wholly owned projects are on average more R&D intensive than those sharing ownership (table 9.5). This group consists of industries such as drugs, medical instruments and equipment, and communications equipment. In medium-technology industries, which include industrial chemicals, motor vehicles, and household appliances, in half of the sectors in which both modes are present, investors engaged in wholly owned projects are characterized by higher-level R&D efforts. In low-technology sectors, this is true in 10 of 16 cases. In each of the three groupings, the average R&D intensity of firms with wholly owned projects is higher than that of firms engaged in joint ventures.

TABLE 9.4 Classification of Industries by Technology Level

Low technology	Medium technology	High technology
Grain mill, bakery products	Soap and other cleansers	Drugs
Other food products	Industrial chemicals	Office and computing machinery
Beverages	Agricultural chemicals	Electronic components
Primary ferrous metals	Farm and garden machinery	Communication equipment, except radio and TV equipment
Primary nonferrous metals	Construction machinery	Other electrical machinery
Fabricated metal products	Other nonelectrical machinery, except office and computing machinery	Aircraft
Lumber, wood, furniture	Household appliances	Instruments
Paper and pulp	Radio, TV equipment	
Printing and publishing	Motor vehicles and equipment	
Textiles and apparel	Transportation equipment other than aircraft and motor vehicles and equipment	
Glass products	Rubber products	
Stone and clay products	Miscellaneous plastic products	
Tobacco	Other manufactures	

Source: Blomström, Lipsey, and Ohlsson (1991).

TABLE 9.5 R&D Intensity of FDI Projects in Three-Digit SIC Industries

Sector	SIC code	Joint ventures	Wholly owned projects	All
High-technology				
Drugs	283	10.62	**15.71**	15.23
Measuring and controlling devices	382	9.94	9.08	9.61
Aircraft and parts	372	7.48	**9.44**	8.08
Communications equipment	366	5.60	**13.31**	7.06
Medical instruments and supplies	384	4.58	**5.07**	4.99
Electronic components and accessories	367	3.39	**5.63**	4.14
Computer and office equipment	357	4.09		4.09
Search and navigation equipment	381	3.20		3.20
Average		6.36	12.67	9.54
Medium-technology				
Refrigeration and service machinery	358		7.26	7.26
Electric distribution equipment	361	7.26		7.26
Hose, belting, gaskets, and packing	305	6.00	6.00	6.00
Plastics materials and synthetics	282	4.65	**4.86**	4.71
Special industry machinery	355	4.22	**5.68**	4.70
Industrial inorganic chemicals	281	4.09	**6.23**	4.46
Motor vehicles and equipment	371	3.91	**4.49**	4.17
Railroad equipment	374	1.49	**4.60**	3.05
Household audio and video equipment	365	5.79	1.03	2.93
Metalworking machinery	354	2.68	2.56	2.66
Soap, cleaners, and toilet goods	284	2.60		2.60
General industrial machinery	356	2.30		2.30
Ship and boat building and repair	373	2.14		2.14
Engines and turbines	351	2.11	2.11	2.11
Construction and related machinery	353	1.83	**2.49**	2.03
Industrial machinery not elsewhere classified	359		1.75	1.75
Miscellaneous manufactures	399	1.59	1.59	1.59
Miscellaneous chemical products	289	1.31		1.31
Miscellaneous plastic products not elsewhere classified	308	1.22	0.11	1.11
Farm and garden machinery	352	0.00	**3.68**	0.74
Electric lightning, wiring equipment	364	0.67		0.67
Rubber and plastics footwear	302	0.00	0.00	0.00
Average		3.21	3.76	3.35

TABLE 9.5 (*Continued*)

Sector	SIC code	Joint ventures	Wholly owned projects	All
Low-technology				
Printing trade services	279		5.25	5.25
Preserved fruits and vegetables	203	4.24		4.24
Broadwoven fabric mills, wool	223	4.00		4.00
Nonferrous rolling and drawing	335	1.54	**5.11**	3.16
Heavy construction, excluding highways	162		2.70	2.70
Electrical work	173	2.67		2.67
Copper ores	102	1.75	**2.84**	2.29
Cutlery, handtools, and hardware	342	2.22	**2.28**	2.27
Nonresident building construction	154	1.25	**2.94**	1.93
Miscellaneous food and kindred products	209		1.86	1.86
Sugar and confectionery products	206		1.83	1.83
Miscellaneous metal ores	109	1.73		1.73
Manifold business forms	276	1.43		1.43
Miscellaneous textile goods	229	1.40		1.40
Clay, ceramic, and refractory minerals	145	1.35		1.35
Secondary nonferrous metals	334		1.34	1.34
Primary nonferrous metals	333	1.23	1.23	1.23
Iron ores	101		1.21	1.21
Miscellaneous converted paper products	267	0.21	**1.34**	1.15
Miscellaneous nonmetallic mineral products	329	0.76	**2.43**	1.13
Metal cans and shipping containers	341	1.20	0.79	0.99
Blast furnace and basic steel products	331	0.93		0.93
Meat products	201	0.79	**0.91**	0.85
Grain mill products	204	0.68	**1.10**	0.72
Glass and glassware pressed or blown	322		0.65	0.65
Miscellaneous wood products	249	0.63	0.63	0.63
Paper mills	262	0.60	**0.67**	0.61
Dairy products	202	0.57		0.57
Highway and street construction	161	0.55		0.55
Fabricated structural metal products	344	0.00	**0.82**	0.55

(*continued*)

TABLE 9.5 R&D Intensity of FDI Projects in Three-Digit SIC Industries (*Continued*)

Sector	SIC code	Joint ventures	Wholly owned projects	All
Paperboard containers and boxes	265	0.44	0.33	0.40
Carpets and rugs	227		0.36	0.36
Cement, hydraulic	324	0.28		0.28
Fats and oils	207	0.15	0.15	0.15
Beverages	208	0.35	0.13	0.15
Gold and silver ores	104	0.00		0.00
Commercial printing	275		0.00	0.00
Average		0.87	1.76	1.2

Source: Author.

Note: Figures in bold denote cases in which R&D intensity of wholly owned projects is greater than R&D intensity of joint ventures.

Testing the Hypotheses

To test whether intraindustry differences in intangible assets affect the choice of entry mode, I include in the model the average R&D intensity of each industry as well as the ratio of a firm's R&D spending to the industry mean. I also add the corresponding variables for advertising intensity. The results of the probit model lend support to the hypothesis (table 9.6). Firms with large R&D and advertising efforts relative to the industry average are more likely to undertake wholly owned projects than to engage in joint ventures. This finding holds when transition progress, host country dummies, and dummies for European and U.S. firms are included.

As before, joint ventures tend to be undertaken by firms in less R&D-intensive industries, by smaller and more diversified firms, and by investors with less international experience. As before, transition progress has a negative marginal effect on the probability of a joint venture relative to a wholly owned subsidiary. Industry advertising intensity, regional experience, and dummies for European and U.S. investors do not appear to be statistically significant. With the exception of international experience, which ceases to be significant at the conventional levels, the inclusion of host country dummies does not strongly affect the results.

A major drawback of employing the probit specification is that it takes into account firms that invested in the region but not those that decided against FDI. This may be a source of sample selection bias and may lead to inconsistent estimates. To the best of my knowledge, none of the previous studies addressed this issue. I correct for the sample selection bias by estimating a bivariate probit model in which the first

TABLE 9.6 Probit Model with Relative R&D and Advertising Intensities

Variable	Marginal effects			
Relative R&D	−0.0611** (0.0262)	−0.0638** (0.0263)	−0.0672** (0.0272)	−0.0738*** (0.0279)
Industry R&D	−0.0462*** (0.0106)	−0.0472*** (0.0106)	−0.0542*** (0.0112)	−0.0550*** (0.0113)
Relative ADV	−0.1782** (0.0718)	−0.1934*** (0.0724)	−0.2191*** (0.0755)	−0.1800** (0.0770)
Industry ADV	<0.0001 (0.0023)	−0.0008 (0.0024)	<0.0001 (0.0025)	−0.0003 (0.0025)
Diversification	0.0293* (0.0152)	0.0314** (0.0153)	0.0311* (0.0161)	0.0218 (0.0166)
Regional experience	0.1011 (0.0669)	0.0824 (0.0673)	0.0709 (0.0696)	0.0866 (0.0714)
International experience	−0.0025** (0.0012)	−0.0022* (0.0012)	−0.0018 (0.0012)	−0.0033** (0.0013)
Firm size	<.0001* (0.0000)	<.0001* (0.0000)	<.0001* (0.0000)	<.0001 (0.0000)
Transition index		−0.1217** (0.0508)		
U.S. parent				−0.1146 (0.1917)
European parent				0.1641 (0.1861)
Host dummies	No	No	Yes	Yes
Number of observations	346	346	345	345
Chi^2	66.31	72.18	86.51	97.34
Degrees of freedom	8	9	27	29
Prob > chi^2	0.00	0.00	0.00	0.00
Log L	−203.61	−200.67	−192.94	−187.52

Source: Author.

Note: Standard errors, pertaining to coefficients underlying marginal effects, are reported in parentheses. Marginal effects whose absolute value is less than 0.0001 are indicated by <.0001. The dependent variable is equal to one in the case of a joint venture and to zero in the case of a wholly owned subsidiary.

* The coefficient underlying the marginal effect is significant at the 10 percent level.

** The coefficient underlying the marginal effect is significant at the 5 percent level.

*** The coefficient underlying the marginal effect is significant at the 1 percent level.

equation describes a firm's decision to invest in a host country and the second focuses on the choice between full and shared ownership. The selection equation includes all variables used in the second stage as well as the market size of the host country, proxied by the GDP. Market size is an important factor in the decision to undertake FDI, but it is unlikely to affect the choice of entry mode.

The estimation results lead to conclusions similar to those suggested by the probit model (table 9.7). Firms that are leaders in terms of technology or marketing skills are more likely to undertake wholly owned projects than to share ownership. Thus the data support the hypothesis that intraindustry differences in intangible assets are important and that treating industries as homogenous is not appropriate. The signs and significance levels of the other coefficients are very similar to those found in the probit model. The only exception is production diversification, which ceases to be statistically significant. In summary, taking sample selection into account does not have a large impact on the estimation results.

The results from the selection equation indicate that FDI is more likely to take place in countries with larger market size and more advanced reforms. More diversified and larger firms as well as those with greater regional and international experience are more likely to invest in the region. This is also true of firms in more advertising-intensive industries. Relative endowments of intangible assets and industry-level R&D spending do not have a statistically significant effect on the probability of investing. With the exception of industry R&D intensity, all these results are consistent with the stylized facts.[24]

Relative endowments of intangible assets affect an investor's choice of entry mode when all manufacturing sectors are taken into account. It is likely, however, that these effects differ across industries. Technological leadership may play an important role in high-technology industries but may be of little significance in low R&D sectors. To test this hypothesis, I reestimate the model allowing for different coefficients for high-/medium- and low-technology sectors. I group sectors into the two categories based on the classification proposed by Blomström, Lipsey, and Ohlsson (1991). I also use an alternative grouping method based on the average R&D intensity of the sector, with R&D spending of 1 and 2 percent of net sales serving as the cut-off values.[25]

The results indicate that relative R&D intensity and industry R&D spending affect the choice of entry mode in high- and medium-technology sectors but not in low-technology industries (table 9.8). The difference between the coefficients underlying the marginal effects of relative R&D spending in the two groups of industries is significant at the 5 percent level. Relative advertising expenditure influences the entry mode in both types of industries, though in the case of low-technology sectors it is significant in only one regression. Firm size is negatively related to the probability

TABLE 9.7 Bivariate Probit with Sample Selection

Variable	Marginal effects	
	Entry mode equation	Investment equation
Relative R&D	−0.0698** (0.0742)	0.0003 (0.0330)
Industry R&D	−0.0496*** (0.0298)	−0.0002 (0.0114)
Relative ADV	−0.2297*** (0.1899)	0.0098 (0.0788)
Industry ADV	−0.0022 (0.0072)	0.0011*** (0.0025)
Diversification	0.0247 (0.0442)	0.0025* (0.0169)
Regional experience	0.0457 (0.2049)	0.0433*** (0.0702)
International experience	−0.0026* (0.0038)	0.0007*** (0.0013)
Firm size	<.0001** (0.0000)	<.0001*** (0.0000)
Transition index	−0.1766*** (0.1585)	0.0491*** (0.0443)
Market size		<0.0001*** (0.0000)
Number of observations		7,152
Log L		−1225.16

Source: Author.

Note: Standard errors, pertaining to coefficients underlying the marginal effects, are reported in parentheses. Marginal effects whose absolute value is less than 0.0001 are indicated by <.0001. In the investment equation, the dependent variable is equal to one if FDI takes place and to zero otherwise. In the entry mode equation, the dependent variable is equal to one in the case of a joint venture and to zero in the case of a wholly owned subsidiary.

*The coefficient underlying the marginal effect is significant at the 10 percent level.

**The coefficient underlying the marginal effect is significant at the 5 percent level.

***The coefficient underlying the marginal effect is significant at the 1 percent level.

TABLE 9.8 Marginal Effects of Bivariate Probit with Sample Selection: Entry Mode Equation

Variable	Blomström and others' classification	R&D = 1 percent cut-off	R&D = 2 percent cut-off
High* relative R&D	−0.2275*** (0.1601)	−0.1271*** (0.1162)	−0.2828*** (0.1898)
Low* relative R&D	−0.0305 (0.1621)	0.0390 (0.2423)	−0.0597 (0.1290)
High* Industry R&D	−0.1049*** (0.0400)	−0.0535*** (0.0318)	−0.1140*** (0.0484)
Low* industry R&D	−0.0997 (0.3960)	1.1569 (3.0105)	−0.0997 (0.3792)
High* relative ADV	−0.4345*** (0.3127)	−0.2837*** (0.2321)	−0.5281*** (0.3674)
Low* relative ADV	−0.2873 (0.4282)	−0.8189* (1.1316)	−0.0984 (0.3005)
High* industry ADV	−0.0014 (0.0089)	−0.0028 (0.0077)	−0.0026 (0.0096)
Low* industry ADV	−0.0122 (0.0203)	−0.0071 (0.0664)	−0.0091 (0.0182)
High* diversification	−0.0163 (0.0567)	0.0028 (0.0519)	−0.0241 (0.0638)
Low* diversification	0.0794* (0.1013)	0.1870** (0.2066)	0.0552* (0.0800)
High* regional experience	0.1760 (0.3044)	0.1104 (0.2391)	0.1382 (0.3627)
Low* regional experience	−0.3374* (0.4880)	−0.9271 (1.4678)	−0.1039 (0.3454)
High* international experience	−0.0007 (0.0049)	−0.0022 (0.0046)	0.0002 (0.0059)
Low* international experience	−0.0040 (0.0076)	−0.0034 (0.0130)	−0.0025 (0.0070)
High* firm size	<.0001 (0.0000)	<.0001* (0.0000)	<.0001 (0.0000)

TABLE 9.8 (*Continued*)

Variable	Blomström and others' classification	R&D = 1 percent cut-off	R&D = 2 percent cut-off
Low* firm size	−0.0001** (0.0001)	−0.0001* (0.0001)	−0.0001** (0.0001)
High* transition	−0.2308*** (0.2102)	−0.1896*** (0.1736)	−0.2163** (0.2150)
Low* transition	−0.1705 (0.3524)	−0.0973 (0.9443)	−0.1305 (0.3305)
Number of observations	7,152	7,152	7,152
LogL	−1159.79	−1188.00	−1156.67

Source: Author.

Note: Standard errors, pertaining to coefficients underlying marginal effects, are reported in parentheses. Marginal effects whose absolute value is less than 0.0001 are indicated by < .0001. In the investment equation (not reported in the table), the dependent variable is equal to one if FDI takes place and to zero otherwise. In the entry mode equation, the dependent variable is equal to one in the case of a joint venture and to zero in the case of a wholly owned subsidiary.

*The coefficient underlying the marginal effect is significant at the 10 percent level.

**The coefficient underlying the marginal effect is significant at the 5 percent level.

***The coefficient underlying the marginal effect is significant at the 1 percent level.

of shared ownership in all industries. Transition progress is significant at the conventional levels only in high– and medium–R&D sectors, though the underlying coefficients for the two groups of industries are not significantly different from each other. In low-technology sectors, more diversified firms and investors without regional experience are more tolerant toward shared ownership.[26]

The empirical evidence presented in this chapter indicates that differences in relative endowments of intangible assets across firms within an industry influence choice of entry mode. Technological and marketing leaders tend to avoid joint ventures and prefer to engage in wholly owned projects. The effect is present mainly in high- and medium-technology sectors.

Concluding Remarks

The choice of entry mode by foreign investors has been of interest to both policymakers and researchers in the field of international business. Changing attitudes toward FDI, greater openness to foreign investment, and the loosening of restrictions on foreign ownership have increased the need to understand the impact of foreign investors on host economies. Developing country governments are especially interested in the question of technology and know-how transfer resulting from FDI. To be able to assess the potential magnitude of such benefits, it is important to understand the preferences of different types of investors with respect to the entry mode. This study sheds some light on this issue by analyzing entry modes chosen by foreign firms entering transition economies in Eastern Europe and the former Soviet Union in the early 1990s.

This chapter contributes to the literature by formulating and providing empirical support for new hypotheses relating the intraindustry differences in R&D and marketing efforts to the choice of entry mode. It improves on other studies by taking into account the potential sample selection bias. The results indicate that industry structure has a significant impact on foreign investment decisions. Treating industries as homogenous in investigations of forces governing FDI flows may therefore be inappropriate.

The findings show that foreign investors that are technological and marketing leaders in their sectors are less likely to form joint ventures than firms lagging behind. This effect is most prominent in high- and medium-technology industries. Thus while it is widely believed that joint ventures with local firms are more conducive to transferring knowledge and know-how than wholly owned FDI projects, the potential magnitude of transfers from joint ventures in high R&D sectors may be smaller than that from fully owned subsidiaries.

Notes

1. See Beamish (1988), Blomström and Zejan (1991), and Blomström and Sjöholm (1999). Such views have led some host countries to restrict the extent of foreign ownership or to offer special incentives to foreign investors undertaking joint ventures with local partners. For instance, in the 1980s China, India, Indonesia, the Republic of Korea, Malaysia, Mexico, Nigeria, Pakistan, and Sri Lanka placed restrictions on foreign ownership (UNCTC 1987).

2. R&D intensity is defined as expenditure on R&D expressed as a percentage of net sales. In the text the terms *R&D expenditure, R&D spending*, and *R&D efforts* refer to R&D intensity. Sectors with a high average R&D intensity are described as high-technology or R&D-intensive sectors.

3. Some evidence indicates that firms with higher R&D spending enjoy greater bargaining power in negotiations with host country governments (Stopford and Wells 1972; UNCTAD 1992).

4. A significant number of joint ventures terminate during the first few years of their existence. For instance, 35 of 92 joint ventures examined by Kogut (1989) failed within seven years. Twenty-seven percent of joint ventures surveyed by Miller and others (1996) were not expected to survive by their partners. Killing (1982) reported that 36 percent of partners rated the performance of their joint ventures as unsatisfactory.

5. Unilever's joint venture in Shanghai may serve as an example. The Chinese partner began to manufacture a washing detergent that had a similar formula and was packaged in a strikingly similar box as the Omo brand produced by the joint venture (*The Economist* 1997).

6. In Bulgaria the Commission for the Protection of Competition has investigated cases of violation of business secrets by former employees. Some of these cases have been brought by foreign companies operating in the country (Hoekman and Djankov 1997).

7. See Gomes-Casseres (1990) for a brief review of the earlier tests of bargaining theory and a description of their limitations.

8. Cohen and Klepper (1992) show that the distributions of R&D intensity within different industries "display a strikingly regular pattern" (p. 773).

9. In general, foreign investors tend to choose more successful local companies for acquisitions or joint venture arrangements. Analysis of firm-level data shows that foreign investment flows to local firms of above-average size, initial profitability, and initial labor productivity (for evidence from the Czech Republic, see chapter 8 of this volume; for evidence from Indonesia, see Arnold and Javorcik 2005).

10. Foreign investors often take precautions against losses of their proprietary knowledge. Warhurst (1991) provides some examples of such efforts on the part of multinational corporations engaged in joint ventures in China. For instance, foreign experts training employees of the Chinese partner were reported to withhold certain technical knowledge, including blueprints. In one case the Chinese company was not allowed to have access to the "'know-why' which would enable it to absorb fully and alter the technology for future needs" (p. 1063). In another case the training of a Chinese team in the investor's home country took place in a rented section of a university rather than at the company research center, possibly out of concern about protecting proprietary technology.

11. Ethier and Markusen (1996) focus on the choice of exporting, licensing, and acquiring a subsidiary.

12. Brouthers and Bamossy (1997) report that some state-owned enterprises in the region had only limited knowledge of Western firms, which led them to restrict their search for foreign joint venture partners to major, well-known Western multinationals.

13. I also experimented with including firm size relative to the industry average, but this variable did not appear to be statistically significant and adding it to the regressions had little impact on other coefficients.

14. However, a study by Meyer (1998) produced a positive coefficient.

15. Worldscope is a commercial database that provides detailed financial statements, business descriptions, and historical pricing information on thousands of public companies located in more than 50 countries.

16. Eastern European countries and the Soviet Union were virtually closed to foreign investment before 1989 (see Dunning and Rojec 1993; Meyer 1995; Hunya 1997).

17. See Dunning and Rojec (1993) for a description of these restrictions.

18. In the drugs sector (SIC code 283), 10 percent of projects are joint ventures and 90 percent are wholly owned subsidiaries.

19. The number of observations is smaller than the number of projects in table 9.1 because of missing firm-specific information.

20. This model was first proposed by Wynand and van Praag (1981). See Greene (1993) for a brief description of the model.

21. In calculating industry averages, I removed two outliers from the drug sector and one from the communications equipment industry. These firms reported R&D intensities equal to 16,598, 1,815, and 2,560, respectively. All three firms reported annual sales of less than $500,000. They are thus likely to be start-up companies. The conclusions of this chapter remain unchanged even if this correction is not performed.

22. If firm- and industry-level figures were both equal to zero, relative R&D intensity took on the value of one.

23. See EBRD (1994) for a detailed description.

24. The evidence in Javorcik (2004) suggests that low R&D intensity of foreign investors in transition economies may be linked to weak protection of intellectual property rights.

25. The average R&D intensity of all three-digit SIC sectors is 1.29 percent of sales.

26. The equation I estimated includes the following R&D terms:

$$Y_{ic}^* = \beta_0 + \beta_1 \frac{\text{R\&D}_{firm}}{\text{R\&D}_{ind}} + \beta_2 \text{R\&D}_{ind} + \cdots$$

$$Y_{ic} = 1 \text{ if } Y_{ic}^* > 0$$

$$Y_{ic} = 0 \text{ otherwise.}$$

It is possible that β_1 is a function of R&D intensity of an industry. For instance, it is conceivable that

$$\beta_1 = \alpha_0 + \alpha_1 \text{R\&D}_{ind}.$$

Then

$$Y_{ic}^* = \beta_0 + \alpha_0 \frac{\text{R\&D}_{firm}}{\text{R\&D}_{ind}} + \alpha_1 R\&D_{firm} + \beta_2 \text{R\&D}_{ind} + \cdots$$

$$Y_{ic} = 1 \quad \text{if } Y_{ic}^* > 0$$

$$Y_{ic} = 0 \quad \text{otherwise.}$$

Estimation of the above equation, however, produced insignificant coefficients α_0 and α_1.

References

Arnold, Jens Matthias, and Beata Smarzynska Javorcik. 2005. "Gifted Kids or Pushy Parents? Foreign Acquisitions and Plant Performance in Indonesia." Policy Research Working Paper 3597, World Bank, Washington, DC.

Asiedu, Elizabeth, and Hadi Salehi Esfahani. 2001. "Ownership Structure in Foreign Direct Investment Projects." *Review of Economics and Statistics* 83 (4): 647–62.

Beamish, Paul W. 1988. *Multinational Joint Ventures in Developing Countries.* Routledge: London and New York.

Blomström, Magnus, and Fredrik Sjöholm. 1999. "Technology Transfer and Spillovers: Does Local Participation with Multinationals Matter?" *European Economic Review* 43 (4–6): 915–23.

Blomström, Magnus, and Mario Zejan. 1991. "Why Do Multinational Firms Seek Out Joint Ventures?" *Journal of International Development* 3 (1): 53–63.

Blomström, Magnus, Robert E. Lipsey, and Lennart Ohlsson. 1991. "What Do Rich Countries Trade with Each Other? R&D and the Composition of U.S. and Swedish Trade." NBER Reprint No. 1551 (from *Banca Nazionale del Lavoro Quarterly Review* 173 [June 1990] 215–35), National Bureau of Economic Research, Cambridge, MA.

Brouthers, Keith D., and Gary J. Bamossy. 1997. "The Role of Key Stakeholders in International Joint Venture Negotiations: Case Studies from Eastern Europe." *Journal of International Business Studies* 28 (2): 285–308.

Caves, R. 1982. *Multinational Enterprise and Economic Analysis.* New York: Cambridge University Press.

Cohen, Wesley M., and Steven Klepper. 1992. "The Anatomy of Industry R&D Intensity Distributions." *American Economic Review* 82 (4): 773–99.

Dunning, John H. 1988. *Explaining International Production*. London: Unwin Hyman.

Dunning, John H., and Matija Rojec. 1993. *Foreign Privatization in Central and Eastern Europe.* Central and Eastern European Privatization Network, Ljubljana, Slovenia.

EBRD (European Bank for Reconstruction and Development). 1994. *Transition Report.* London.

The Economist. 1997. "Multinationals in China: Going It Alone." April 19.

Ethier, Wilfred, and James Markusen. 1996. "Multinational Firms, Technological Diffusion and Trade." *Journal of International Economics* 41 (1–2): 1–28.

Gatignon, Hubert, and Erin Anderson. 1988. "The Multinational Corporation's Degree of Control over Foreign Subsidiaries: An Empirical Test of a Transaction Cost Explanation." *Journal of Law, Economics, and Organization* 4 (2): 305–36.

GATT (General Agreement on Trade and Tariffs). 1992. *Trade Policy Review: Poland.* Vol. 1. Geneva.

Gomes-Casseres, Benjamin. 1989. "Ownership Structures of Foreign Subsidiaries: Theory and Evidence." *Journal of Economic Behavior and Organization* 11 (1): 1–25.

———. 1990. "Firm Ownership Preferences and Host Government Restrictions: An Integrated Approach." *Journal of International Business Studies* 21 (1): 1–22.

Greene, William H. 1993. *Econometric Analysis,* 2nd. ed. New York: Macmillan Publishing Company.

Hennart, Jean–François. 1991. "The Transaction Cost Theory of Joint Ventures: An Empirical Study of Japanese Subsidiaries in the United States." *Management Science* 37 (4): 483–97.

Hoekman, Bernard, and Simeon Djankov. 1997. "Competition Law in Post-Central Planning Bulgaria." Discussion Paper 1723, Centre for Economic Policy Research, London.

Horstmann, Ignatius, and James R. Markusen. 1987. "Licensing versus Direct Investment: A Model of Internalization by the Multinational Enterprise." *Canadian Journal of Economics* 20 (3): 464–81.

Hunya, Gabor. 1997. "Large Privatisation, Restructuring and Foreign Direct Investment." In *Lessons from the Economic Transition: Central and Eastern Europe in the 1990s,* ed. Salvatore Zecchini, 275–300. Dordrecht: Kluwer Academic Publishers.

Javorcik, Beata Smarzynska. 2004. "The Composition of Foreign Direct Investment and Protection of Intellectual Property Rights: Evidence from Transition Economies." *European Economic Review* 48 (1): 39–62.

Killing, J.P. 1982. "How to Make a Global Joint Venture Work." *Harvard Business Review* 60 (May): 120–27.

Kogut, Bruce. 1989. "The Stability of Joint Ventures: Reciprocity and Competitive Rivalry." *Journal of Industrial Economics* 37: 183–98.

Kogut, Bruce, and Harbir Singh. 1988. "The Effect of National Culture on the Choice of Entry Mode." *Journal of International Business Studies* 19 (3): 411–32.

McMillan, Carl H. 1996. "Foreign Investment in Russia: Soviet Legacies and Post-Soviet Prospects." In *Foreign Investment in Russia and Other Soviet Successor States*, ed. Patrick Artisien-Maksimenko and Yuri Adjubei, 41–72. New York: St. Martin's Press.

Meyer, Klaus. 1995. "Direct Foreign Investment in Eastern Europe. The Role of Labor Costs." *Comparative Economic Studies* 37 (1): 69–88.

———. 1998. *Direct Investment in Economies in Transition.* Cheltenham, UK, and Northampton, MA: Edward Elgar.

Miller, Robert, Jack Glen, Frederick Jaspersen, and Yannis Karmokolies. 1996. "International Joint Ventures in Developing Countries: Happy Marriages?" Discussion Paper 29, International Finance Corporation, Washington, DC.

OECD (Organisation for Economic Co-operation and Development). 1994. *Assessing Investment Opportunities in Economies in Transition.* Paris.

Stopford, John M., and Louis T. Wells, Jr. 1972. *Managing the Multinational Enterprise.* New York: Basic Books.

UNCTAD (United Nations Conference on Trade and Development). 1992. *World Investment Report 1992: Transnational Corporations as Engines of Growth.* New York: United Nations.

UNCTC (United Nations Centre on Transnational Corporations). 1987. *Arrangements between Joint Venture Partners in Developing Countries.* Advisory Study No. 2. New York.

Warhurst, Alyson. 1991. "Technology Transfer and the Development of China's Offshore Oil Industry." *World Development* 19 (8): 1055–73.

WTO (World Trade Organization). 1998. *Trade Policy Review: Hungary.* Geneva.

Wynand, P., and B. van Praag. 1981. "The Demand for Deductibles in Private Health Insurance: A Probit Model with Sample Selection." *Journal of Econometrics* 17 (2): 229–52.

10

DOES FOREIGN DIRECT INVESTMENT INCREASE THE PRODUCTIVITY OF DOMESTIC FIRMS? IN SEARCH OF SPILLOVERS THROUGH BACKWARD LINKAGES

Beata Smarzynska Javorcik

Policymakers in many developing countries and transition economies place attracting foreign direct investment (FDI) high on their agenda, expecting FDI inflows to bring much-needed capital, new technologies, marketing techniques, and management skills. While all of these potential benefits of FDI are viewed as important, emphasis is placed on the contribution of FDI to increasing the productivity and competitiveness of the domestic industry. It is often hoped that technology transfer resulting from FDI will go beyond actual projects undertaken by foreign investors and, through knowledge spillovers, will benefit domestic firms.

Yet there is no evidence that positive externalities generated by foreign presence actually exist. As Rodrik (1999) notes, "Today's policy literature is filled with

This chapter originally appeared in the *American Economic Review* 94 (3): 605–27. The author wishes to thank Mary Amiti, Enrique Aldaz-Carroll, Andrew Bernard, Simon Evenett, Ana Fernandes, Michael Ferrantino, Caroline Freund, Holger Görg, Mary Hallward-Driemeier, Wolfgang Keller, Pravin Krishna, Hiau Looi Kee, Maryla Maliszewska, Jacques Morisset, Marcelo Olarreaga, Nina Pavcnik, Maurice Schiff, Matt Slaughter, Mariana Spatareanu, Jim Tybout, two anonymous referees, and the participants of the Tuck International Trade Conference for valuable comments and suggestions. Financial support from the Foreign Investment Advisory Service (FIAS), a joint facility of the International Finance Corporation and the World Bank, is gratefully acknowledged. This chapter is part of a larger FIAS effort to improve the understanding of spillovers from multinational corporations to local firms.

extravagant claims about positive spillovers from FDI, but the evidence is sobering." Indeed, data limitations and the difficulties associated with disentangling different effects at play have prevented researchers from providing conclusive evidence of positive externalities resulting from FDI. While recent firm-level studies have overcome many of the difficulties faced by earlier literature, the emerging message is not very positive.

The literature on this subject is of three kinds. First, there are case studies, which are often very informative and include a wealth of valuable information (see, for instance, Moran 2001). Because they pertain to particular FDI projects or specific countries, however, these studies cannot easily be generalized.

Second, there is a plethora of industry-level studies, most of which show a positive correlation between foreign presence and the average value-added per worker in the sector. Because most of these studies rely on cross-sectional data, it is difficult to establish the direction of causality. It is possible that the positive association is caused by the fact that multinationals tend to locate in high-productivity industries rather than by genuine productivity spillovers. The positive correlation may also be a result of FDI inflows forcing less productive domestic firms to exit or of multinationals increasing their share of the host country market, both of which would raise the average productivity in the industry.[1]

Third, there is research based on firm-level panel data, which examines whether the productivity of domestic firms is correlated with the extent of foreign presence in their sector. Most of these studies—such as the careful analyses done by Haddad and Harrison (1993) on Morocco, Aitken and Harrison (1999) on República Bolivariana de Venezuela, Djankov and Hoekman (2000) on the Czech Republic, and Konings (2001) on Bulgaria, Romania, and Poland—cast doubt on the existence of spillovers from FDI in developing countries. The researchers either fail to find a significant effect or produce evidence of negative horizontal spillovers (the effect of the presence of multinational corporations on domestic firms in the same sector). The picture is more positive in the case of industrial countries: recent work by Haskel, Pereira, and Slaughter (2002) and Keller and Yeaple (2003) provides convincing evidence of positive FDI spillovers in the United Kingdom and the United States, respectively.[2]

It is possible, though, that researchers have been looking for FDI spillovers in the wrong place. Since multinationals have an incentive to prevent information leakage that would enhance the performance of their local competitors but at the same time may benefit from transferring knowledge to their local suppliers, spillovers from FDI are more likely to be vertical than horizontal in nature. In other words, spillovers are most likely to take place through backward linkages (contacts between domestic suppliers of intermediate inputs and their multinational clients) and thus would not have been captured by earlier studies.[3] It is also plausible that spillovers from multinational presence in upstream sectors exist thanks to the provision of inputs that either were previously unavailable in the country or are technologically more advanced, less expensive, or accompanied by provision of complementary services. As Blomström, Kokko, and Zejan (2000) point out, very few empirical studies analyze vertical spillovers. The notable exceptions are Blalock (2001), who uses firm-level panel data from Indonesia, and Schoors and van der Tol (2001), who rely on cross-sectional enterprise-level data from Hungary. Both provide evidence of positive FDI spillovers through backward linkages.[4]

The purpose of this chapter is twofold. First, it examines whether the productivity of domestic firms is correlated with the presence of multinationals in downstream sectors (potential customers) or upstream industries (potential suppliers of intermediate inputs). Detecting such effects would be consistent with the existence of vertical spillovers. The analysis improves over the recent literature by taking into account econometric problems that may have biased the results of earlier work. The semiparametric estimation method suggested by Olley and Pakes (1996) is employed to account for endogeneity of input demand. Moreover, standard errors are corrected to take into account the fact that the measures of potential spillovers are industry specific while the observations in the data set are at the firm level. As Moulton (1990) notes, failing to make such a correction will lead to a serious downward bias in the estimated errors, resulting in a spurious finding of statistical significance for the aggregate variable of interest.

Second, this study goes beyond the existing literature by shedding light on determinants of vertical spillovers. It examines whether benefits stemming from vertical linkages are related to the extent of foreign ownership in affiliates. Based on case studies and investor surveys, these factors have been conjectured to influence the reliance on local sourcing on the part of multinationals and thus the potential benefits of backward linkages, but their impact does not appear to have been systematically examined.[5]

The analysis is based on data from the annual enterprise survey conducted by the Lithuanian Statistical Office. The survey coverage is extensive, as firms accounting for about 85 percent of output in each sector are included. The data constitute an unbalanced panel covering the 1996–2000 period. Focusing on a transition economy such as Lithuania is very suitable for this project, as the endowment of skilled labor enjoyed by transition economies makes them a particularly likely place for productivity spillovers to manifest themselves.[6]

The findings can be summarized as follows. The empirical results are consistent with the existence of positive spillovers from FDI taking place through backward linkages, but there is no robust evidence of spillovers occurring through either the horizontal or the forward linkage channel. In other words, the productivity of Lithuanian firms is positively correlated with the extent of potential contacts with multinational customers but not with the presence of multinationals in the same industry or the existence of multinational suppliers of intermediate inputs. The magnitude of the effect is economically meaningful. A one standard deviation increase in the foreign presence in the sourcing sectors is associated with a 15 percent rise in output of each domestic firm in the supplying industry. The productivity effect is found to originate from investments with joint foreign and domestic ownership but not from fully owned foreign affiliates, which is consistent with the evidence of a larger amount of local sourcing undertaken by jointly owned projects.

Overview of Spillover Channels

Spillovers from FDI take place when the entry or presence of multinational corporations increases the productivity of domestic firms in a host country and the multinationals do not fully internalize the value of these benefits. Spillovers may take place when local firms improve their efficiency by copying technologies of foreign affiliates operating in the local market either through observation or by hiring workers trained by the affiliates. Another kind of spillover occurs if multinational entry leads to more severe competition in the host country market and forces local firms to use their existing resources more efficiently or to search for new technologies (Blomström and Kokko 1998).

To the extent that domestic firms and multinationals operating in the same sector compete with one another, multinationals have an incentive to prevent technology leakage and spillovers from taking place. This can be achieved by formally protecting their intellectual property, engaging in trade secrecy, paying higher wages to prevent labor turnover, or locating in countries or industries in which domestic firms have limited imitative capacities to begin with.[7] This observation is consistent with the results of recent studies that failed to produce evidence of positive horizontal spillovers from FDI.

Multinationals have no incentive to prevent technology diffusion to upstream sectors, as they may benefit from improved performance of intermediate input suppliers. Thus backward linkages—contacts between multinational firms and their local suppliers—should be the most likely channel through which spillovers manifest themselves. These spillovers may take place through direct knowledge transfer from foreign customers to local suppliers;[8] higher requirements for product quality and on-time delivery introduced by multinationals, which provide incentives to domestic suppliers to upgrade their production management or technology;[9] and multinational entry increasing demand for intermediate products, which allows local suppliers to reap the benefits of scale economies.

Similarly, domestic firms may become more productive as a result of gaining access to new, improved, or less costly intermediate inputs produced by multinationals in upstream sectors (forward linkage channel). Sales of these inputs by multinationals may be accompanied by provision of complementary services that may not be available in connection with imports.

Anecdotal evidence confirms that spillovers take place through backward linkages in transition economies. For instance, after a Czech producer of aluminum alloy castings for the automotive industry signed its first contract with a multinational customer, the staff from the multinational would visit the Czech firm's premises two days each month over an extended period to work on improving the quality control system. Subsequently, the Czech firm applied these improvements to its other production lines (not serving this particular customer) and reduced the number of defective items produced (interview with company management in the Czech Republic, May 2003).

The results of a recent enterprise survey conducted in Latvia by the World Bank (2003a) are consistent with the expectation that positive spillovers take place through backward linkages but are ambiguous with respect to the intraindustry effect. The evidence from Latvia is particularly relevant as, in addition to being neighboring countries, Lithuania and Latvia share many similarities in terms of their history and economic conditions. The survey demonstrates that a majority of multinationals are engaged in local sourcing: 82 percent of those interviewed had at least one Latvian supplier of intermediate inputs, and on average 47 percent of intermediate inputs purchased by foreign firms came from Latvian producers. Thirty-six percent of Latvian firms supplying multinationals reported receiving assistance from their customers. As far as horizontal spillovers are concerned, one-third of Latvian firms stated that they had benefited from the presence of foreign firms in their sector (15 percent through sourcing inputs from multinationals, 14 percent by learning about new technologies, and 9 percent by learning about new marketing strategies). At the same time, 45 percent of survey respondents reported that foreign entry increased competition in their industry, with 6.5 percent of firms admitting to having lost market share to foreign firms. As Aitken and Harrison (1999) point out, knowledge spillovers within an industry may be counterbalanced by the competition effect; that is, as domestic firms lose market share to foreign entrants, their productivity falls, since their fixed costs are spread over a smaller market. Thus the reported increase in competition levels due to foreign entry is consistent with the lack of intraindustry spillovers found in the current analysis.

Different types of FDI projects may have different implications for vertical spillovers. For instance, it has been argued that affiliates established through mergers and acquisitions or joint ventures are likely to source more locally than those taking the form of greenfield projects (UNCTC 2001). While greenfield projects have to put time and effort into developing local linkages, affiliates established through mergers and acquisitions can take advantage of the supplier relationships of the acquired firm or its local partners. Empirical evidence to support this view has been found for Japanese investors (Belderbos, Capannelli, and Fukao 2001) and for Swedish affiliates in Eastern Europe (UNCTC 2000). Unfortunately, in the data set used here, it is impossible to distinguish among the three types of foreign investment. However, to the extent that full foreign ownership is a proxy for greenfield projects, it is expected that fully owned foreign affiliates will tend to rely more on imported inputs, while investment projects with shared domestic and foreign ownership will tend to source more locally.[10] This hypothesis is supported by the World Bank (2003a) survey, which shows that while more than half of partially owned foreign affiliates operating in Latvia purchased their intermediate inputs locally, only 9 percent of fully owned foreign subsidiaries did so. Similarly, the results of a study of the largest exporters in Hungary (Toth and Semjen 1999) indicate that foreign affiliates with larger shares of foreign equity tend to purchase fewer inputs from Hungarian companies. In sum, it is expected that larger spillovers are associated with partially rather than fully owned foreign projects.

Data and Estimation Strategy

This section reviews recent developments in FDI inflows to Lithuania and describes the data and the estimation strategy.

Foreign Direct Investment in Lithuania

Like other former Soviet republics, Lithuania had been virtually closed to foreign investment until 1990, when it regained its independence and began the process of transition to a market economy. The first stage of the privatization process, which began in 1991, offered limited opportunities for foreign investors. It was not until 1997 that FDI inflows into Lithuania increased significantly, as a result of the second stage of the privatization program. FDI inflows peaked in 1998, when 60 percent of the shares of Lietuvas Telekomas (Lithuanian Telecom), the fixed-line monopoly operator, were sold to Amber Teleholdings, a consortium of Swedish Telia and Finish Sonera (EBRD 2001) (figure 10.1).[11]

Due to its late start, Lithuania has attracted less FDI than other Central and Eastern Europe countries. Cumulative FDI inflows between 1993 and 2000 reached $694 per capita, placing Lithuania seventh among Central and Eastern European countries, above Slovenia, Bulgaria, and Romania. In terms of the value of cumulative FDI inflows, Lithuania ranks eighth, above Estonia and Slovenia (table 10.1).

In terms of sectoral distribution, 44 percent of the FDI stock in 1996 was in manufacturing. Following large inflows into the telecommunications and financial sectors, this figure decreased to 32 percent in 2000. Within manufacturing, food products, beverages, and tobacco attracted the largest share of investment (12 percent of the total FDI stock), followed by textiles and leather products (4 percent) and

FIGURE 10.1 Net FDI Inflows to Lithuania, 1993–2000

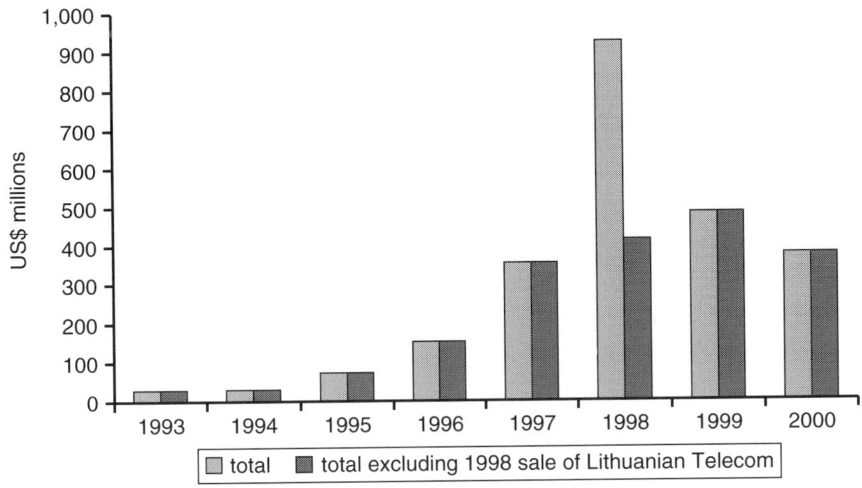

Source: IMF (2003).

refined petroleum and chemicals (4 percent). Electrical machinery, optical instruments, and wood products also received significant foreign investments (OECD 2000) (table 10.2).

Lithuania's population, GDP, and FDI inflows are concentrated in three cities: Vilnius, Kaunas, and Klaipeda. At the beginning of 2000, Vilnius accounted for 60.5 percent of Lithuania's FDI, Klaipeda accounted for 11.6 percent, and Kauna accounted for 10.5 percent. Direct investment in manufacturing sectors is concentrated around Klaipeda, while the bulk of FDI inflows into wholesale and retail trading are found in Vilnius (OECD 2000, 2001).

Data Description

The data used in this study come from the annual enterprise survey conducted by the Lithuanian Statistical Office. The survey coverage is extensive, as firms accounting for about 85 percent of output in each sector are included in the sample.

The Lithuanian enterprise data have been praised for their high quality and reliability.[12] The data constitute an unbalanced panel covering the 1996–2000 period. The number of firms per year varies from a low of 12,000 in 1996 to a high of 21,000 in 1999. Due to financial constraints in some years, the Statistical Office was forced to reduce the scope of the exercise. In each year, however, the same sampling technique was used.

TABLE 10.1 FDI Inflows to Central and Eastern European Countries, 1993–2000

Country	FDI inflow (millions of dollars)								FDI inflows, 2000		FDI inflows, 1993–2000	
	1993	1994	1995	1996	1997	1998	1999	2000	As percent of GDP	Per capita ($)	Value (millions of $)	Per capita ($)
Bulgaria	40	105	90	109	505	537	806	1,002	7.9	123	3,194	393
Czech Republic	654	878	2,568	1,435	1,286	3,700	806	1,002	7.9	123	3,194	393
Estonia	162	214	201	150	266	581	305	387	7.5	283	2,268	1,656
Hungary	2,350	1,144	4,519	2,274	2,167	2,037	1,977	1,646	3.5	163	18,113	1,790
Latvia	45	214	180	382	521	357	348	410	5.7	173	2,456	1,036
Lithuania	30	31	73	152	355	926	486	379	3.4	108	2,432	694
Poland	1,715	1,875	3,659	4,498	4,908	6,365	7,270	9,341	5.9	242	39,631	1,025
Romania	94	341	419	263	1,215	2,031	1,041	1,037	2.7	46	6,441	287
Slovak Republic	199	270	236	351	174	562	354	2,052	10.4	380	4,198	777
Slovenia	113	117	150	173	334	216	107	136	0.7	68	1,345	676

Source: IMF (2003) and World Bank (2003b).

TABLE 10.2 Distribution of Firms with Foreign Capital, by Industry, 2000

NACE code	Sector	Distribution of FDI across sectors (percent)[1]	Domestic firms (1)	Firms with foreign capital[2] (2)	All firms (3)	(2)/(3) * 100	Horizontal	Backward	Forward
15	Food products and beverages	19.6	396	50	446	11.2	26.6	1.5	4.8
17	Textiles	12.5	74	30	104	28.8	39.7	13.7	1.7
18	Wearing apparel	1.9	172	43	215	20.0	33.5	2.7	25.6
19	Leather and leather products	0.1	19	3	22	13.6	6.6	6.8	15.3
20	Wood and wood products, except furniture	4.2	382	43	425	10.1	34.3	12.5	8.4
21	Pulp, paper, and paper products	2.3	17	6	23	26.1	39.4	17.2	10.9
22	Publishing, printing, and recorded media	0.2	204	12	216	5.6	7.0	3.5	18.3
24	Chemicals and chemical products	10.7	44	17	61	27.9	20.9	7.4	3.7
25	Rubber and plastic products	3.8	111	25	136	18.4	31.4	11.0	10.9
26	Other nonmetallic mineral products	7.4	141	17	158	10.8	35.3	3.1	6.8
27	Basic metals	0.6	6	3	9	33.3	50.3	16.7	4.3
28	Fabricated metal products	0.7	156	24	180	13.3	10.7	8.4	22.3
29	Machinery and equipment	1.1	94	12	106	11.3	23.2	6.9	15.3
30	Office machinery and computers	0.0	8	2	10	20.0	8.0	6.3	22.5
31	Electrical equipment and apparatus	1.2	37	4	41	9.8	65.3	7.1	15.0
32	Radio, television, and communication equip.	4.3	24	5	29	17.2	32.2	14.4	17.0
33	Medical, precision, and optical instruments	0.8	42	7	49	14.3	23.8	11.9	18.7
34	Motor vehicles	0.8	9	1	10	10.0	59.8	4.4	12.6
35	Other transport equipment	7.6	39	8	47	17.0	71.5	0.2	13.1
36	Furniture	0.6	154	20	174	11.5	9.7	6.9	14.5
	Total	80.4	2,129	332	2,461	13.5	31.5	8.1	13.1

Source: Author.

1. Shares do not add up to 100 percent because the table excludes tobacco (NACE 16), which accounts for 0.9 percent of the FDI stock, and manufacturing of refined petroleum products (NACE 23), which account for 18.7 percent.

2. Foreign share of at least 10 percent of total capital.

This chapter focuses on manufacturing firms (sectors 15–36 in *Nomenclature générale des activités économiques dans les Communautés européennes* [NACE]), which lowers the sample size to 2,500–4,000 firms a year. The number of observations is further reduced by deleting those with missing values, no sales, and no employment, as well as observations failing to satisfy other basic error checks. Two sectors—tobacco (NACE 16) and manufacturing of refined petroleum products (NACE 23)—are excluded, because the small number of firms makes it impossible to apply the Olley-Pakes technique (discussed below) to these industries. Thus the final sample size varies between 1,918 and 2,711 firms.

The data set contains information on foreign ownership, sales, inventories, employment, fixed assets, input costs, investment, location, and share of exports in total sales. Firms with foreign capital participation are defined as firms in which the share of subscribed capital (equity) owned by foreign investors is equal to at least 10 percent. More than 12 percent (1,414 of the 11,630 observations) meet this definition.

Lithuania and other transition economies in Central and Eastern Europe are suitable objects for an analysis of FDI spillovers because of their high endowment of skilled labor, which makes them particularly likely locations for productivity spillovers. On the downside, the brief duration of the panel makes it more difficult to detect the presence of spillovers. Extending the panel to earlier years would not mitigate this problem, because FDI during the early 1990s was limited. Furthermore, a high level of aggregation in the industry classification (NACE two-digit) and the fact that the data set pertains to firms rather than plants also works against finding a significant spillover effect.

Estimation Strategy

To examine the correlation between firm productivity and FDI in the same industry or other sectors, an approach similar to that taken by earlier literature is followed and several variations of the following equation are estimated:

$$\begin{aligned}\ln Y_{ijrt} = \alpha + \beta_1 \ln K_{ijrt} + \beta_2 \ln L_{ijrt} + \beta_3 \ln M_{ijrt} + \beta_4\, Foreign\ Share_{ijrt} \\ + \beta_5\, Horizontal_{jt} + \beta_6\, Backward_{jt} + \beta_7\, Forward_{jt} \\ + \alpha_t + \alpha_r + \alpha_j + \varepsilon_{ijrt}. \end{aligned} \tag{10.1}$$

Y_{ijrt} stands for the real output of firm i operating in sector j and region r at time t, which is calculated by adjusting the reported sales for changes in inventories of finished goods and deflating the resulting value by the producer price index for the appropriate two-digit NACE sector. K_{ijrt}, capital, is defined as the value of fixed assets at the beginning of the year, deflated by the simple average of the deflators for five

NACE sectors: machinery and equipment; office, accounting, and computing machinery; electrical machinery and apparatus; motor vehicles, trailers, and semi-trailers; and other transport equipment. Since in the data set it is impossible to distinguish between skilled and unskilled workers, labor is expressed in terms of efficiency units, which are computed by dividing the wage bill by the minimum wage (L_{ijrt}).[13] M_{ijrt}, materials, is equal to the value of material inputs adjusted for changes in material inventories, deflated by an intermediate inputs deflator calculated for each sector based on the input-output matrix and deflators for the relevant industries. $Foreign\ Share_{ijrt}$ measures the share of firm's total equity owned by foreign investors.

Turning to proxies for spillovers, $Horizontal_{jt}$ captures the extent of foreign presence in sector *j* at time *t* and is defined as foreign equity participation averaged over all firms in the sector, weighted by each firm's share in sectoral output.[14] In other words,

$$Horizontal_{jt} = (\Sigma_{i \text{ for all } i \in j}\ Foreign\ Share_{it} * Y_{it}) / \Sigma_{i \text{ for all } i \in j} Y_i. \qquad (10.2)$$

Thus the value of the variable increases with the output of foreign investment enterprises and the share of foreign equity in these firms.

$Backward_{jt}$ is a proxy for the foreign presence in the industries supplied by sector *j*. It is intended to capture the extent of potential contacts between domestic suppliers and multinational customers. It is defined following Blalock (2001), and Schoors and van der Tol (2001) as

$$Backward_{jt} = \Sigma_{k \text{ if } k \neq j}\ \alpha_{jk}\ Horizontal_{kt} \qquad (10.3)$$

where α_{jk} is the proportion of sector *j*'s output supplied to sector *k* taken from the 1996 input-output matrix at the two-digit NACE level.[15] The proportion is calculated excluding products supplied for final consumption but including imports of intermediate products.[16] As the formula indicates, inputs supplied within the sector are not included, since this effect is already captured by the *Horizontal* variable.[17] The greater the foreign presence in sectors supplied by industry *j* and the larger the share of intermediates supplied to industries with a multinational presence, the higher the value of the variable.

The *Forward* variable is defined as the weighted share of output in upstream (or supplying) sectors produced by firms with foreign capital participation. As only intermediate goods sold in the domestic market are relevant to this study, goods produced by foreign affiliates for exports (X_{it}) are excluded. The following formula is used:

$$Forward_{jt} = \Sigma_{m \text{ if } m \neq j}\ \sigma_{jm}\ (\Sigma_{i \text{ for all } i \in m}\ Foreign\ Share_{it} * (Y_{it} - X_{it})) / (\Sigma_{i \text{ for all } i \in m} (Y_{it} - X_{it})), \qquad (10.4)$$

where σ_{jm} is the share of inputs purchased by industry *j* from industry *m* in total inputs sourced by sector *j*. For the same reason as before, inputs purchased within the sector are excluded. The value of the variable increases with the share of foreign affiliates in the (domestically sold) output of upstream sectors.

The proxies for horizontal and vertical linkages are time-varying sector-specific variables. While the coefficients taken from the input-output table remain fixed, changes in the level of foreign investment and firm output are observed during the period in question.

There is significant variation across sectors and time in all variables (tables 10.2, 10.3, and 10.4). For instance, the value of *Horizontal* ranges from 6.6 percent in leather and leather products to 65.3 percent in electrical equipment and apparatus and 71.5 percent in other transport equipment. The average value increases from almost 12 percent in 1996 to more than 31 percent in 2000. Similarly, the value of the *Backward* variable rises from 3.6 percent in 1996 to 6.0 percent in 1998 and 8.1 percent in 2000. The highest values are registered in pulp, paper, and paper products (17.2 percent); basic metals (16.7 percent); and radio, television, and communications equipment (14.4 percent). The lowest value (0.2 percent) is in manufacturing other transport equipment. The *Forward* proxy ranges from 1.7 percent in manufacturing textiles to 25.6 percent in manufacturing wearing apparel. The *Forward* variable increases over time, from 3.3 percent in the first year to 13.1 percent in the last year.

Changes between 1996 and 2000 in the value of all spillover variables are calculated for each sector (figures 10.2–10.4). Seven industries registered a rise in the *Backward* measure of more than 5 percentage points, and 10 other sectors experienced increases of more than 2 percentage points. The largest changes were observed in textiles, pulp and paper, wood, rubber, and plastics. Changes in the *Horizontal* variable were even more pronounced: 13 industries experienced an increase of more than 10 percentage points, with motor vehicles leading the ranking. In contrast, apparel, metal products, and office machinery saw the greatest changes in the *Forward* measure.

In an exploratory regression, the model described above is estimated using ordinary least squares (OLS) with White's correction for heteroskedasticity. A firm's output is the dependent variable; explanatory variables include capital, labor, materials, foreign equity share, and proxies for FDI spillovers operating through horizontal, backward, and forward channels. Since knowledge externalities from the foreign presence may take time to manifest themselves, two specifications are employed, one with contemporaneous and one with lagged spillover variables. The estimation is performed on the full sample and on the sample of domestic firms only.[18] The model includes fixed effects for years (4), industries (19), and regions (9). The results indicate that firms with foreign capital tend to be more productive than purely Lithuanian firms (table 10.5). And, more important for the purpose of this study, there is a significant and positive coefficient on both the *Backward* and *Horizontal* variables in all four specifications. The coefficients on lagged values appear larger and (in the case of *Backward*) of higher statistical significance. The third spillover variable, *Forward*, does not appear to be

TABLE 10.3 Summary Statistics

Variable	Summary statistics for levels			Summary statistics for first differences		
	Number of observations	Mean	Standard deviation	Number of observations	Mean	Standard deviation
ln Y	11,630	13.5	2.0	6,853	0.01	0.6
ln L	11,630	6.1	1.8	6,853	–0.06	0.6
ln K	11,630	12.0	2.4	6,853	0.26	0.7
ln materials	11,630	12.3	2.5	6,853	–0.02	1.1
ln gross investment	8,262	10.6	2.4	3,765	–0.04	1.8
Foreign share	11,630	7.8	23.0	6,853	0.42	9.1
Exports share	9,762	21.1	34.0	5,757	–1.20	22.6
Horizontal	11,630	19.7	12.3	6,853	3.99	4.7
Backward	11,630	4.9	3.9	6,853	1.05	1.1
Backward (fully foreign owned)	11,630	1.9	2.0	6,853	0.41	0.6
Backward (partially foreign owned)	11,630	3.0	2.5	6,853	0.64	1.1
Backward (concentrated)	11,630	1.9	2.1	6,853	0.37	0.8
Backward (competitive)	11,630	2.9	3.0	6,853	0.68	1.1
Forward	11,630	6.9	5.5	6,853	2.38	2.6
ln demand	11,630	18.9	1.4	6,853	0.06	0.1
H4	11,630	576.9	844.4	6,853	–8.03	209.3

Source: Author.

TABLE 10.4 Additional Summary Statistics for Spillover Variables

	Horizontal spillovers			Backward spillovers			Forward spillovers		
Year	Mean	Standard deviation	Number of industries	Mean	Standard deviation	Number of industries	Mean	Standard deviation	Number of industries
1996	11.85	12.92	20	3.62	3.05	20	3.29	2.42	20
1997	17.32	15.70	20	5.17	4.03	20	4.27	2.83	20
1998	21.95	15.58	20	6.02	4.59	20	6.16	3.14	20
1999	28.93	19.93	20	7.72	4.93	20	8.81	4.27	20
2000	31.46	19.20	20	8.13	5.00	20	13.08	6.70	20

Source: Author.

FIGURE 10.2 Change in Horizontal Measure, 1996–2000

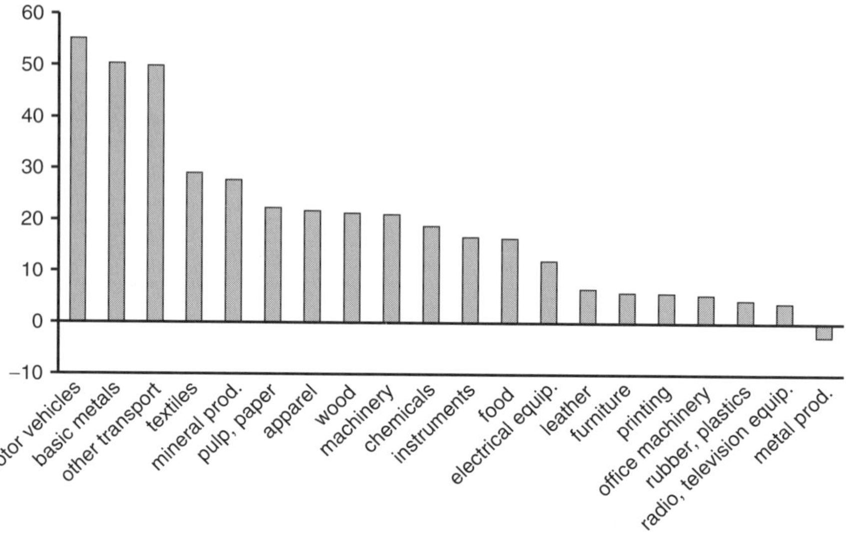

Source: Author.

FIGURE 10.3 Change in Backward Measure, 1996–2000

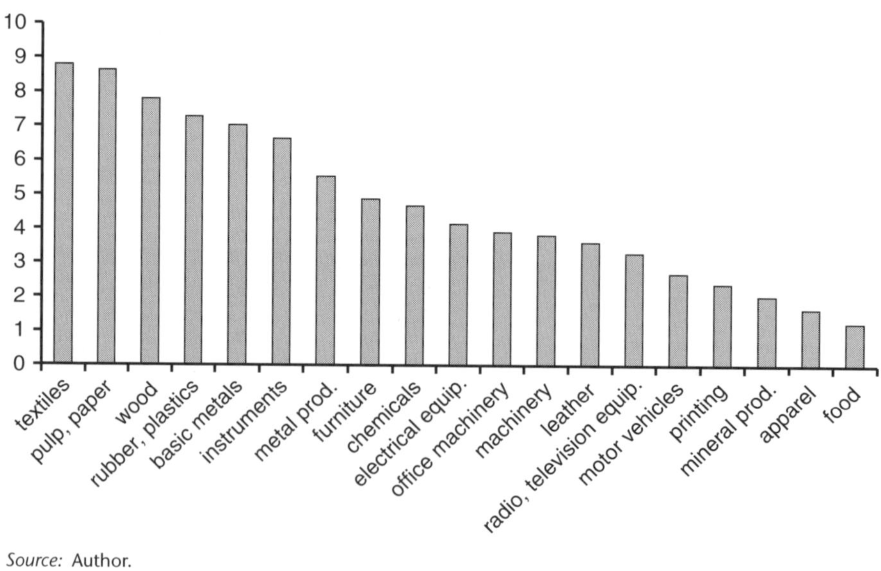

Source: Author.

FIGURE 10.4 Change in Forward Variable, 1996–2000

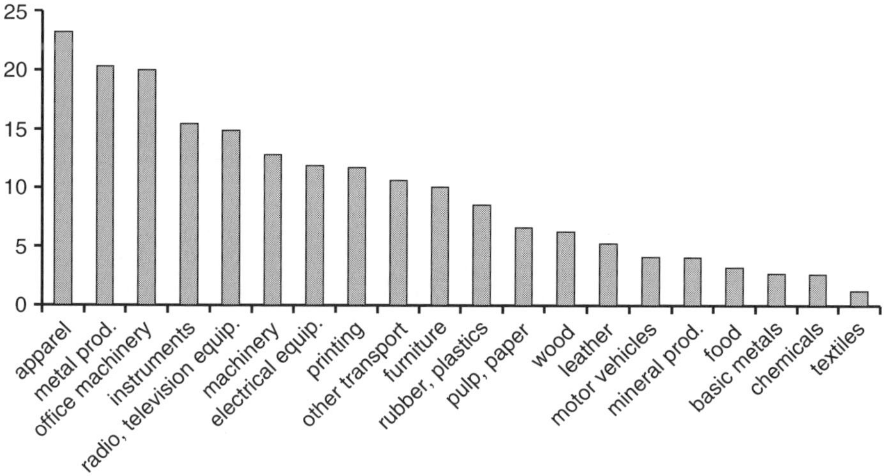

Source: Author.

statistically significant. In sum, the results are consistent with productivity spillovers from FDI taking place within industries and flowing from multinational customers to their domestic suppliers.

To be more confident about isolating the effects of productivity spillovers, one must control for other factors that may influence firm productivity. If multinational entry decreases industry concentration, leading to more competition and forcing domestic firms to improve their efficiency, this situation may be regarded as a broadly defined spillover effect. Since, however, our interest is primarily knowledge transfer, it would be useful to separate the two phenomena. Thus the Herfindahl index (H4) is included as a proxy for the level of industry concentration.[19] Foreign entry into downstream sectors may increase demand for intermediate products, which in turn will allow local suppliers to reap the benefits of scale economies. To separate this effect, the regression includes the demand for intermediate goods, calculated based on information on sourcing patterns from the input-output matrix and the value of production in using sectors.[20] A positive correlation between demand for intermediates (*Demand*) and firm productivity is anticipated.

Several econometric concerns need to be addressed in the analysis. The first is the omission of unobserved variables. Firm-, time-, and region-specific factors that are unknown to the econometrician but known to the firm may affect the correlation between firm productivity and foreign presence. Examples of these variables include

TABLE 10.5 OLS with Lagged and Contemporaneous Spillover Variables

Variable	All firms	Domestic	All firms	Domestic
Foreign share	0.0025*** (0.0002)		0.0025*** (0.0003)	
Backward	0.0105** (0.0048)	0.0086* (0.0051)		
Backward lagged			0.0173*** (0.0060)	0.0177*** (0.0066)
Forward	–0.0030 (0.0024)	0.0001 (0.0027)		
Forward lagged			–0.0029 (0.0040)	–0.0007 (0.0044)
Horizontal	0.0029** (0.0013)	0.0040*** (0.0014)		
Horizontal lagged			0.0038* (0.0021)	0.0046** (0.0023)
Intercept	5.2323*** (0.0805)	5.2082*** (0.0876)	5.1599*** (0.1007)	5.1582*** (0.1108)
Number of observations	11.630	10.216	8.214	7.118
R^2	0.93	0.92	0.93	0.92

Source: Author.

Note: Robust standard errors are presented in parentheses. The dependent variable is ln firm output. Each regression includes ln capital stock, ln effective employment, and ln materials, as well as industry, region, and year fixed effects.

*Significant at the 10 percent level.

**Significant at the 5 percent level.

***Significant at the 1 percent level.

high-quality management in a particular firm or better infrastructure in a given region. This problem is addressed by following Haskel, Pereira, and Slaughter (2002) and using time differencing as well as a full set of fixed effects for year, industry, and region. In addition to removing any fixed firm-specific unobservable variation, differencing will remove fixed regional and industrial effects, such as infrastructure and technological opportunity. Time, industry, and regional dummy variables will control for unobservables that may be driving changes in, for instance, the attractiveness of a particular region or industry.[21] Thus the specification becomes

$$\begin{aligned}\Delta \ln Y_{ijrt} = {} & \delta_1 \Delta \ln K_{ijrt} + \delta_2 \Delta \ln L_{ijrt} + \delta_3 \Delta \ln M_{ijrt} + \delta_4 \Delta Foreign\ Share_{ijrt} \\ & + \delta_5 \Delta Horizontal_{jt} + \delta_6 \Delta Backward_{jt} + \delta_7 \Delta Forward_{jt} \\ & + \delta_8 \Delta H4_{jt} + \delta_9 \Delta \ln Demand_{jt} + \alpha_t + \alpha_r + \alpha_j + \varepsilon_{ijrt} \qquad (10.5)\end{aligned}$$

This model is estimated in first, second, and fourth differences. The examination of longer differences gives relatively more weight to more persistent changes in the variables of interest and hence reduces the influence of noise. Its disadvantage is that longer time differences reduce the size of the sample. As a compromise, the specifications in differences are employed, but only the relationship between contemporaneous changes in FDI and firm-level total factor productivity is considered, because adding lags would seriously strain the time span of the data set.

Second, Moulton (1990) shows that in the case of regressions performed on micro units but including aggregated market (or in this case, industry) variables, the standard errors from OLS will be underestimated. As he demonstrates, failing to take this into account leads to a serious downward bias in the estimated errors, resulting in spurious findings of statistical significance for the aggregate variable of interest. To address this issue, the standard errors are corrected for a correlation between observations belonging to the same industry in a given year (in other words, standard errors are clustered for all observations in the same industry and year).

It has been argued that the use of OLS is inappropriate when estimating productivity, because this method treats labor and other inputs as exogenous variables. Griliches and Mairesse (1995) made the case that inputs should be considered endogenous, because they are chosen by the firm based on its productivity, which is observed by the producer but not by the econometrician. Not taking into account the endogeneity of input choices may bias the estimated coefficients. Since the focus of this chapter is on firm productivity, the consistency of the estimates is crucial for the analysis.

For this reason, the semiparametric estimation procedure suggested by Olley and Pakes (1996), which allows for firm-specific productivity differences exhibiting idiosyncratic changes over time, is used. Following Olley and Pakes, it is assumed that at the beginning of every period a firm chooses variable factors and a level of investment that, together with the current capital value, determine the capital stock at the beginning of the next period. The capital accumulation equation is given by

$$k_{it+1} = (1 - \delta)k_{it} + i_{it}, \qquad (10.6)$$

where k stands for capital, i for investment, and δ for the rate of depreciation.

Consider the following Cobb-Douglas production function model:

$$y_{it} = \alpha + \beta_l * l_{it} + \beta_k * k_{it} + \beta_m * m_{it} + \omega_{it} + \eta_{it}, \qquad (10.7)$$

where y_{it}, l_{it}, and m_{it} denote the logarithm of output, labor, and material inputs, respectively, and subscripts i and t stand for firm and time. ω_{it} denotes productivity, and η_{it} stands for either measurement error or a shock to productivity that is not forecastable during the period in which labor can be adjusted. Both ω_{it} and η_{it} are unobserved. The difference is that ω_{it} is a state variable in the firm's decision problem and thus affects the input demand, while η_{it} does not. Labor and materials are assumed to be freely variable inputs. Capital is a fixed factor and is affected only by the distribution of ω conditional on information at time $t-1$ and past values of ω. The fact that input choices are determined in part by the firm's beliefs about ω_{it} gives rise to simultaneity bias. The positive correlation between ω_{it} and inputs used in period t will cause an OLS estimation that does not take into account unobserved productivity differences to provide upwardly biased estimates of the coefficients on variable inputs.

The insight of the Olley-Pakes method is that the observable characteristics of the firm can be modeled as a monotonic function of the productivity of the firm. Since the investment decision depends on the capital stock and on firm productivity,

$$i_{it} = i_{it}\,(\omega_{it}, k_{it}). \qquad (10.8)$$

By inverting equation 10.8, one can express unobserved productivity ω_{it} as a function of observable investment and capital and thus control for ω_{it} in estimation:

$$\omega_{it} = h_{it}\,(i_{it}, k_{it}). \qquad (10.9)$$

By substituting equation (10.9) into equation (10.7), the equation to be estimated in the first stage of the procedure is obtained:

$$y_{it} = \alpha + \beta_l * l_{it} + \beta_k * k_{it} + \beta_m * m_{it} + h(i_{it},k_{it}) + \eta_{it}. \qquad (10.10)$$

The functional form of $h(\cdot)$ is not known. Therefore, the β_k coefficient cannot be estimated at this stage. A partially linear model, including a third-order polynomial expansion in capital and investment to approximate the form of the $h(\cdot)$, is estimated. From this stage, the consistent estimates of the coefficients on labor and material inputs as well as the estimate of the third-order polynomial in i_{it} and k_{it} (referred to as ψ_{it}) are obtained:

$$\psi_{it} = \alpha + \beta_k * k_{it} + h(i_{it},k_{it}) \qquad (10.11)$$

$$h(i_{it},k_{it}) = \psi_{it} - \beta_k * k_{it}. \qquad (10.12)$$

The second step of the estimation procedure considers the expectation of $y_{it+1} - \beta_m * m_{it+1} - \beta_l * l_{it+1}$:

$$\begin{aligned} &\mathrm{E}(y_{it+1} - \beta_m * m_{it+1} - \beta_l * l_{it+1} \mid k_{it+1}) \\ &= \alpha + \beta_k * k_{it+1} + \mathrm{E}(\omega_{it+1} \mid \omega_{it}) \\ &= \beta_k * k_{it+1} + g(\omega_{it}). \end{aligned} \qquad (10.13)$$

Assuming that ω_{it} follows a first-order Markov process, one can rewrite ω_{it+1} as a function of ω_{it}, letting ξ_{it+1} be the innovation in ω_{it+1}. Using equations (10.9) and (10.12), equation (10.13) becomes a function of i_{it} and k_{it}:

$$y_{it+1} - \beta_m * m_{it+1} - \beta_l * l_{it+1} = \beta_k * k_{it+1} + g(\psi_{it} - \beta_k * k_{it}) + \xi_{it+1} + \eta_{it+1}, \quad (10.14)$$

where g is a third-order polynomial of $\psi_{it} - \beta_k * k_{it}$. This is the equation to be estimated in the second stage of the procedure. Only in this stage is it possible to obtain consistent estimates of β_k. Since the capital in use in a given period is assumed to be known at the beginning of the period and ξ_{it+1} is mean independent of all variables known at the beginning of the period, ξ_{it+1} is mean independent of k_{it+1}. A nonlinear least squares method is used to estimate equation (10.14).

A production function with the Olley-Pakes correction is estimated for each industry separately. From the estimation, the measure of total factor productivity (the difference between the actual and predicted output) is recovered and used in the estimation of the basic model:[22]

$$tfp_{it} = y_{it} - \beta_l * l_{it} - \beta_k * k_{it} - \beta_m * m_{it}. \quad (10.15)$$

The Olley-Pakes correction appears to work quite well. If the procedure successfully corrects for biases, one would expect to find a decrease in coefficients on labor and material inputs and an increase in the capital coefficient relative to the OLS results. The material and labor coefficients move in the predicted direction in 17 cases each, while the magnitude of the capital coefficient increases in 16 of 20 cases (table 10.6).

Estimation Results from a Model in Differences

This section describes the results from a baseline specification, compares spillovers associated with fully and partially owned foreign affiliates, and discusses robustness checks.

Baseline Specification

A model estimated in first differences produces findings that are consistent with the hypothesis that domestic firms benefit from the foreign presence in sectors they supply. The first two columns of table 10.7 contain the results for the full sample and the subsample of domestic firms. (Because of lack of space, the coefficients on inputs are not reported.) In both regressions a positive and significant coefficient on the proxy for spillovers through backward linkages can be found. The third and fourth columns present the results from the regressions with the Olley-Pakes correction.[23] The estimations produce a positive and significant coefficient on the *Backward* variable

TABLE 10.6 Comparison of Coefficients from OLS and Olley-Pakes Regressions

Sector code	15	17	18	19
Coefficients from Olley-Pakes regressions				
Number of observations in Stage I	1150	271	498	68
ln (labor)	0.3395***	0.3823***	0.6211***	0.3201***
ln (materials)	0.5036***	0.4356***	0.2312***	0.5256***
ln (capital)	0.1002***	0.0176	0.0221	0.0547*
Sum of coefficients	0.94	0.84	0.87	0.90
Coefficients from OLS regressions				
ln (labor)	0.4114***	0.4500***	0.7357***	0.3318***
ln (materials)	0.5180***	0.4816***	0.2483***	0.5490***
ln (capital)	0.0396***	0.0028	–0.0003	0.0038
Sum of coefficients	0.97	0.93	0.98	0.88
Change in l coefficient	–	–	–	–
Change in m coefficient	–	–	–	–
Change in k coefficient	+	+	+	+
Sector code	**27**	**28**	**29**	**30**
Coefficients from Olley-Pakes regressions				
Number of observations in Stage I	22	465	256	23
ln (labor)	0.6059**	0.3917***	0.4885***	0.7412***
ln (materials)	0.5703**	0.4475***	0.3851***	0.3394***
ln (capital)	–0.0922***	0.0528***	0.0455**	–0.1521***
Sum of coefficients	1.08	0.89	0.92	0.93
Coefficients from OLS regressions				
ln (labor)	0.7614***	0.3970***	0.5261***	0.6215***
ln (materials)	0.2807***	0.4910***	0.4364***	0.2122
ln (capital)	–0.0606	0.0258**	–0.0617***	–0.1084
Sum of coefficients	0.98	0.91	0.90	0.73
Change in l coefficient	–	–	–	+
Change in m coefficient	+	–	–	+
Change in k coefficient	–	+	+	–

TABLE 10.6 (*Continued*)

20	21	22	24	25	26
828	66	610	174	311	364
0.3658***	0.1420*	0.4380***	0.2633***	0.3843***	0.4478***
0.4797***	0.5272***	0.3391***	0.4601***	0.4748***	0.4804***
0.0679***	0.1427***	0.0862***	0.1625***	0.0444***	0.0307***
0.91	0.81	0.86	0.89	0.90	0.96
0.4558***	0.2655***	0.5048***	0.3072***	0.4360***	0.5134***
0.4862***	0.6103***	0.3864***	0.5277***	0.5118***	0.4804***
0.0214***	0.0420	0.0512***	0.0730***	0.0357**	–0.0375***
0.96	0.92	0.94	0.91	0.98	0.96
–	–	–	–	–	–
–	–	–	–	–	0
+	+	+	+	+	+

31	32	33	34	35	36
84	68	117	23	100	400
0.4651***	0.2845***	0.3791***	0.2739*	0.5015***	0.4003***
0.4374***	0.3833***	0.5275***	0.3497***	0.2769***	0.4460***
–0.0167	0.0132	0.0510	0.0850	0.0945***	0.0737***
0.89	0.68	0.96	0.71	0.78	0.92
0.4296***	0.3429***	0.4704***	0.2542**	0.5233***	0.4567***
0.5189***	0.4611***	0.5311***	0.3531***	0.3158***	0.5030***
–0.0527	0.0862	–0.0339	0.1322**	0.0326	–0.0123
0.90	0.89	0.97	0.74	0.87	0.95
+	–	–	+	–	–
–	–	–	–	–	–
+	–	+	–	+	+

Source: Authors

*Significant at the 10 percent level.

**Significant at the 5 percent level.

***Significant at the 1 percent level.

TABLE 10.7 Results of OLS and Olley-Pakes Regressions in First, Second, and Fourth Differences

Variable	First differences				Second differences				Fourth differences			
	OLS		Olley-Pakes method		OLS		Olley-Pakes method		OLS		Olley-Pakes method	
	All	Domestic	All	Domestic	All	Domestic	All	Domestic	All	Domestic	All	Domestic
Foreign share	0.0006 (0.0007)		0.0009 (0.0007)		0.0008 (0.0007)		0.0003 (0.0006)		0.0015 (0.0010)		0.0006 (0.0008)	
Backward	0.0382*** (0.0101)	0.0360*** (0.0103)	0.0407** (0.0163)	0.0347* (0.0193)	0.0321*** (0.0120)	0.0301** (0.0125)	0.0539*** (0.0101)	0.0523*** (0.0113)	0.0232** (0.0105)	0.0256** (0.0111)	0.0590*** (0.0080)	0.0706*** (0.0087)
Forward	−0.0050 (0.0033)	−0.0073** (0.0034)	−0.0060 (0.0055)	−0.0118* (0.0063)	−0.0079* (0.0047)	−0.0088 (0.0055)	−0.0061 (0.0060)	−0.0039 (0.0073)	−0.0027 (0.0067)	0.0026 (0.0040)	0.0004 (0.0067)	0.0192*** (0.0041)
Horizontal	−0.0003 (0.0013)	−0.0006 (0.0013)	−0.0019 (0.0025)	−0.0022 (0.0024)	0.0015 (0.0021)	0.0013 (0.0022)	0.0024 (0.0026)	0.0012 (0.0027)	0.0103*** (0.0013)	0.0114*** (0.0015)	0.0108*** (0.0011)	0.0078*** (0.0012)
H4	0.0000 (0.0000)	0.0000 (0.0000)	0.0001*** (0.0000)	0.0001*** (0.0000)	0.0000 (0.0000)	0.0000 (0.0000)	0.0000 (0.0000)	0.0001 (0.0000)	0.0002** (0.0001)	0.0003*** (0.0000)	0.0000 (0.0001)	0.0003*** (0.0000
Demand	0.6103*** (0.1945)	0.6752*** (0.1929)	0.3699 (0.2934)	0.5341* (0.2806)	0.3527* (0.1869)	0.3911** (0.1872)	0.2464 (0.2970)	0.4137 (0.3003)	−0.2196* (0.1133)	−0.2344* (0.1305)	−0.3966*** (0.1007)	−0.3806*** (0.1132)
Number of observations	6,853	5,916	3,765	3,084	4,551	3,923	2,379	1,920	1,135	964	833	681
R^2	0.49	0.49	0.08	0.08	0.62	0.61	0.09	0.09	0.70	0.69	0.06	0.06

Source: Author.

Note: Standard errors in parentheses have been corrected for clustering for each industry in each year. In the regressions without the Olley-Pakes correction, the dependent variable is Δln firm output and the right-hand side includes Δln capital stock, Δln labor, and Δln materials. In models employing the Olley-Pakes procedure, the dependent variable is Δln total factor productivity. All regressions include industry, region, and year fixed effects.

*Significant at the 10 percent level.

**Significant at the 5 percent level.

***Significant at the 1 percent level.

in both the full sample and the subsample of domestic firms. The size of the coefficients is similar across columns and slightly larger in the case of the full sample. The magnitude of the effect is economically meaningful. A one standard deviation increase in the foreign presence in the sourcing sectors (that is, an increase of 4 percentage points in the backward variable) is associated with a 15 percent rise in output of each domestic firm in the supplying industry.[24]

There is little evidence that spillovers take place through the other channels. The coefficient on the *Horizontal* variable does not appear to be statistically significant, which is consistent with the literature that fails to find a positive intraindustry effect in developing countries (for example, Aitken and Harrison 1999; Djankov and Hoekman 2000; Konings 2001). The *Forward* variable, in contrast, bears a negative sign but appears to be statistically significant in only two regressions.

As for the other control variables, there is no indication of a positive association between changes in foreign equity share and productivity growth. As in Aitken and Harrison (1999), the *Foreign Share* is positively correlated with productivity *levels* (recall the results from table 10.5) but not with *growth rates,* suggesting that foreign firms may be investing in the most productive domestic enterprises.[25] A positive coefficient is found on demand in downstream sectors, indicating the existence of procyclical productivity effects. The data suggest a positive correlation between industry concentration and productivity growth, but the results are statistically significant in only two cases.[26]

To check the robustness of the results, a model in second and fourth differences is estimated. Since the sample covers only five years of data, the specification in fourth differences is the longest difference that can be employed. A positive and significant coefficient on the *Backward* variable is found in all specifications, evidence that is consistent with the hypothesis that productivity spillovers are taking place through contacts between domestic firms and their foreign customers in downstream sectors. There is no indication of the other type of vertical spillovers, as the *Forward* variable appears to be insignificant in the majority of cases. As for intrasectoral spillovers, only the results on the long differences suggest their existence. These results should be treated with caution, however, as they are based on a small number of observations. The reduction in the sample size may also be responsible for the change in the sign of demand from downstream sectors, which, in the fourth difference specification, appears to be negatively correlated with firm productivity.[27]

Full versus Partial Foreign Ownership

Next consider the hypothesis that backward linkages associated with partially owned foreign projects lead to greater spillovers than linkages associated with wholly owned foreign affiliates because of different propensities to engage in local sourcing. To examine this question, two measures of backward linkages are calculated for the two types of foreign investments. The proxy for fully owned foreign projects is defined as

$$Backward\ (Full\ Ownership)_{jt} = \Sigma_{k\ if\ k \neq j}\ \alpha_{jk} * (\Sigma_{i\ for\ all\ i \in k}\ WOS_{it} * Foreign\ Share_{it} * Y_{it})/\Sigma_{i\ for\ all\ i \in k}\ Y_{it}, \quad (10.16)$$

where *WOS* is a dummy for wholly owned subsidiaries. *WOS* is equal to 1 for firms whose share of foreign capital is at least 99 percent.[28] The measure for partially owned investments (those with foreign capital participation greater than 10 percent but less than 99 percent) is defined in an analogous manner.

The results support the hypothesis (table 10.8). A significant and positive correlation is found between changes in output of domestic firms and backward linkages associated with partially foreign owned projects but not wholly foreign owned affiliates. The difference between the magnitudes of the two coefficients is statistically significant in three out of four cases (in the case of the full sample, at the 1 percent level). These findings are consistent with the observation that projects owned jointly by domestic and foreign entities are more likely to source locally, thus creating greater scope for spillovers to firms operating in upstream sectors.

The other variables exhibit patterns similar to those observed in table 10.7. The only exception is the *Forward* measure, which appears to be negative and statistically significant in three out of four cases, suggesting that foreign presence in upstream sectors has a negative impact on the performance of local firms in using industries. This finding is similar to that obtained by Schoors and van der Tol (2001). A possible explanation is that after buying out domestic firms in supplying sectors, foreign owners upgrade production facilities and manufacture more sophisticated products, which are then sold at a higher price. Local firms in using sectors that purchase these inputs may have limited ability to benefit from their higher technological content but are forced to bear the higher cost.

Another reason why the extent of foreign ownership may matter for spillovers is the control over company operations. Foreign owners may be more inclined to import intermediate inputs (for example, due to their familiarity with foreign suppliers) and may be in better position to do so in enterprises in which they have majority ownership. Thus as a robustness check, a model comparing the effect of minority- versus majority-owned foreign investments on spillovers through backward linkages was estimated. Since no significant difference between vertical spillovers from the two types of projects was found, the results are not reported here. These findings are consistent with the observation that domestic capital participation in FDI projects lowers foreign investors' costs of using local suppliers and thus results in more local sourcing and greater productivity spillovers to domestic producers of intermediate inputs.

TABLE 10.8 Share of Foreign Ownership and Productivity Spillovers

Variable	OLS		Olley-Pakes method	
	All	Domestic	All	Domestic
Foreign share	0.0006 (0.0007)		0.0010 (0.0007)	
Backward (partial ownership)	0.0444*** (0.0085)	0.0394*** (0.0096)	0.0499*** (0.0146)	0.0401** (0.0190)
Backward (full ownership)	0.0040 (0.0110)	0.0154 (0.0133)	0.0020 (0.0171)	0.0090 (0.0223)
Forward	−0.0053* (0.0030)	−0.0074** (0.0032)	−0.0066 (0.0053)	−0.0121* (0.0062)
Horizontal	−0.0009 (0.0012)	−0.0009 (0.0012)	−0.0025 (0.0024)	−0.0026 (0.0023)
H4	0.0000 (0.0000)	0.0000 (0.0000)	0.0001*** (0.0000)	0.0001*** (0.0000)
Demand	0.6181*** (0.1778)	0.6817*** (0.1825)	0.3794 (0.2810)	0.5427** (0.2698)
R^2 *F*-stat (BKFO = BKPO)	0.49 12.01	0.49 2.91	0.08 6.41	0.08 1.68
Prob $F > 0$	0.00	0.09	0.01	0.20

Source: Author.

Note: Specification in first differences. Standard errors in parentheses have been corrected for clustering for each industry in each year. In the regressions without the Olley-Pakes correction, the dependent variable is Δln firm output and the right-hand side includes Δln capital stock, Δln labor, and Δln materials. In models employing the Olley-Pakes procedure, the dependent variable is Δln total factor productivity. All regressions include industry, region, and year fixed effects. BKFO = Backward (full ownership); BKPO = Backward (partial ownership).

*Significant at the 10 percent level.

**Significant at the 5 percent level.

***Significant at the 1 percent level.

Robustness Checks

This section describes three additional extensions and robustness checks. First, it is conceivable, though not very likely, that the results on the effect of backward linkages are driven by the level of concentration in purchasing industries (which may be correlated with foreign presence) rather than genuine knowledge spillovers from FDI. For instance, both domestic and foreign enterprises operating in concentrated sectors may have more resources to provide assistance to their suppliers, although at the same time they may be less inclined to do so. In contrast, firms in competitive industries may have fewer resources to support their suppliers but a greater incentive to transfer knowledge to downstream sectors in order to obtain higher-quality or less expensive inputs. Thus ex ante the effect of concentration is ambiguous.

To eliminate the alternative explanation driven by the above arguments, a model is estimated to test whether a differential effect of foreign presence in the two types of downstream industries exists. The U.S. Department of Justice definition of concentrated sectors (those with a Herfindahl index for the four largest firms exceeding 1,800) is employed to calculate separate measures of *Backward* for concentrated and competitive industries.[29] The results indicate that foreign presence in both types of upstream industries leads to positive spillovers to supplying sectors (table 10.9). The *Backward* variable is statistically significant seven out of eight times—the only exception being the case of spillovers from concentrated industries in the regression with the Olley-Pakes correction estimated on the subsample of domestic firms. In all four models, there is no statistically significant difference between the magnitude of the backward linkage effect for the two types of sectors, suggesting that the level of concentration in upstream sectors is not a concern in the model.

Second, the motivation for undertaking FDI is likely to affect the extent of local sourcing by foreign subsidiaries. It has been suggested that domestic market-oriented foreign affiliates tend to purchase more inputs locally than their export-oriented counterparts (Altenburg 2000; UNCTC 2000). Exporting affiliates that are part of international production networks are more likely to be dependent on the global sourcing policies of their parent company and thus may have less freedom to choose their suppliers. Moreover, quality and technical requirements associated with goods targeted for the domestic market may be lower, so local suppliers may find it easier to serve multinationals focused on the domestic market. In contrast, if multinationals serving global markets impose more stringent cost and quality requirements and thus necessitate greater adjustments and larger productivity improvements on the part of local suppliers, one may expect more spillovers to be associated with exporting multinationals. This effect would be reinforced by the fact that multinationals serving global markets may possess superior technologies, creating greater opportunities for learning by local suppliers. In summary, the theoretical predictions regarding the relationship between export orientation of multinationals and spillovers are ambiguous.

To examine whether the export orientation of foreign affiliates matters for spillovers, two separate measures of backward linkages are calculated: one for affiliates

TABLE 10.9 Concentration of Downstream Sectors and Productivity Spillovers

Variable	OLS		Olley-Pakes method	
	All	Domestic	All	Domestic
Foreign share	0.0006 (0.0007)		0.0009 (0.0007)	
Backward (concentrated)	0.0394*** (0.0131)	0.0360*** (0.0126)	0.0401** (0.0187)	0.0258 (0.0193)
Backward (competitive)	0.0379*** (0.0107)	0.0360*** (0.0108)	0.0409** (0.0187)	0.0383* (0.0214)
Forward	–0.0050 (0.0033)	–0.0073** (0.0033)	–0.0059 (0.0054)	–0.0115* (0.0061)
Horizontal	–0.0003 (0.0013)	–0.0006 (0.0013)	–0.0019 (0.0025)	–0.0020 (0.0024)
H4	0.0000 (0.0000)	0.0000 (0.0000)	0.0001*** (0.0000)	0.0001*** (0.0000)
Demand	0.6158*** (0.2210)	0.6754*** (0.2203)	0.3684 (0.3107)	0.5099* (0.3032)
Number of observations R^2	6,853 0.49	5,916 0.49	3,765 0.08	3,084 0.08
F-stat (BK concentrated = BK competitive) Prob F > 0	0.02 0.90	0.00 1.00	0.00 0.97	0.33 0.57

Source: Author.

Note: Specification in first differences. Standard errors in parentheses have been corrected for clustering for each industry in each year. In the regressions without the Olley-Pakes correction, the dependent variable is Δln firm output and the right-hand side includes Δln capital stock, Δln labor, and Δln materials. In models employing the Olley-Pakes procedure, the dependent variable is Δln total factor productivity. All regressions include industry, region, and year fixed effects. BK = Backward.

*Significant at the 10 percent level.

**Significant at the 5 percent level.

***Significant at the 1 percent level.

focused mostly on exporting and one for foreign firms targeting the domestic market. The measure for affiliates is defined as

$$Backward\ (Export\text{-}Oriented)_{jt} = \Sigma_{k\ if\ k \neq j}\ \alpha_{jk} * (\Sigma_{i\ for\ all\ i \in k}\ Export\text{-}Oriented_{it} * Foreign\ Share_{it} * Y_{it})/\Sigma_{i\ for\ all\ i \in k}\ Y_{it}, \quad (10.17)$$

where *Export-Oriented*$_{it}$ is equal to 1 if the share of output exported by firm *i* is above 50 percent and 0 otherwise. The measure for domestic market-oriented foreign affiliates is defined analogously. The results (not reported here) suggest that both types of foreign affiliates are associated with spillovers to upstream sectors. While the magnitude of the coefficient on domestic market-oriented affiliates is larger in three out of four cases, the difference between the two coefficients is not statistically significant.

The same exercise was performed for two additional cut-off points, 66 and 90 percent of output exported, but only in regressions estimated with the Olley-Pakes correction on the subsample of domestic firms was the coefficient on *Backward* (*Domestic market-oriented*) significantly larger than the coefficient on the measure of spillovers associated with exporting affiliates. Thus there is some indication that domestic market-oriented FDI projects are correlated with greater productivity spillovers to their local suppliers, although the evidence is not very robust.

To correct for potential biases in coefficients on variable factor inputs, the share of foreign capital as well as other sectoral variables (*Horizontal, Backward, Forward, H4*, and *Demand*) was included in the first stage of the Olley-Pakes procedure. For each of the exercises presented in tables 10.7–10.9, a separate Olley-Pakes procedure with the relevant spillover measures added to the first stage was estimated. The results from this estimation led to exactly the same conclusions as those presented here (and are therefore not shown). A likely reason why this modification did not produce significant changes in the results is that investment, which enters the first stage of the Olley-Pakes procedure in the polynomial form, picks up most of the effect that foreign entry and presence have on firm behavior.

Conclusions

In contrast to earlier literature, which focused on intraindustry spillovers from FDI, this chapter tests for productivity spillovers taking place through backward linkages (contacts between foreign affiliates and their domestic suppliers) and forward linkages (interactions between foreign suppliers of intermediate inputs and their domestic customers). The analysis, based on a firm-level panel data set from Lithuania, addresses econometric issues that may have biased the findings of earlier research, such as endogeneity of input demand and correction of standard errors to account for the fact that, while observations pertain to firms, the variables of interest are at the industry level.

The results are consistent with the presence of productivity spillovers taking place through backward linkages. They suggest that a one standard deviation increase in the foreign presence in downstream sectors is associated with a 15 percent rise in output of each domestic firm in supplying industries. Productivity benefits are found to be associated with partially but not fully owned foreign projects, a result that is in line with the evidence that suggests that partially owned foreign projects engage in more local sourcing than fully owned foreign projects. As was the case with the earlier firm-level studies of developing countries, no evidence of intrasectoral spillovers is found. Nor is there any indication of spillovers stemming from a multinational presence in sectors supplying intermediate inputs.

More research is needed to fully understand the effect of FDI on host countries. In particular, it would be useful to confirm the findings of this chapter using data that allow individual firms to be identified as suppliers to multinationals rather than relying on input-output matrices to measure interactions between sectors. It would also be interesting to learn more about host country and investor characteristics that determine the extent of spillovers operating through different channels. It is to be hoped that improved data availability will allow researchers to examine these questions in the future.

Notes

1. The pioneering work on this issue done by Caves (1974) focused on Australia. It was followed by studies looking at Mexico, where, due to the large technological gap between foreign and domestic firms, the scope for spillovers may have been higher (see Blomström and Persson 1983; Blomström and Wolff 1994; Blomström 1989). The criticism regarding reverse causality does not apply to all industry-level research, as some studies look at changes taking place between two points in time (Blomström 1986 on Mexico) or rely on panel data (Liu, Wang, and Wei 2000 on the United Kingdom) and still conclude that positive spillovers from FDI exist.

2. For surveys of the literature on spillovers from FDI, see Görg and Strobl (2001) and Lipsey (2002).

3. For a theoretical justification of spillovers through backward linkages, see Rodríguez-Clare (1996), Markusen and Venables (1999), and Lin and Saggi (2004). For case studies, see Moran (2001).

4. Kugler (2000) finds intersectoral technology spillovers from FDI in Colombia. However, he does not distinguish between the channels through which such spillovers may be occurring (backward versus forward linkages).

5. See UNCTC (2001, chapter 4) for a comprehensive review of this topic.

6. For instance, during the 1990–2000 period the number of scientists and engineers in research and development activities per 1 million people was 2,031 in Lithuania, 2,139 in the Republic of Korea, 711 in Argentina, 168 in Brazil, and 154 in Malaysia (World Bank 2003b).

7. Several studies (Aitken, Harrison, and Lipsey 1996; Girma, Greenaway, and Wakelin 2001) have shown that foreign firms pay higher wages than domestic firms. Multinationals have also been found to be sensitive to the strength of intellectual property rights protection in host countries (see Javorcik 2004).

8. As numerous case studies indicate, multinationals often provide technical assistance to their suppliers in order to raise the quality of their products or facilitate innovation (see Moran 2001). They help suppliers with management training and organization of the production process, quality control, purchase of raw materials, and even finding additional customers. The existence of linkages does not necessarily guarantee that spillovers take place, nor does the fact that multinationals may charge for services provided preclude the presence of spillovers. Spillovers take place when foreign affiliates are unable to extract the full value of the resulting productivity increase through direct payment or the lower prices they pay for intermediates sourced from the local firm.

9. For instance, many multinationals require their suppliers to obtain International Standards Organization (ISO) quality certifications.

10. Greenfield projects may be undertaken jointly by foreign and local entities, but in this case they should be lumped together with joint ventures, as the participation of a local company brings access to

domestic suppliers. This classification will, however, be problematic in the case of full acquisitions undertaken by foreigners.

11. The large jump in FDI inflows due to this transaction does not affect the results of this chapter, as only manufacturing sectors are included in the econometric analysis.

12. A recent study examining the quality of data collected by statistical offices ranked Lithuania second among 20 transition economies (see Belkindas, Dinc, and Ivanova 1999).

13. This approach was pioneered by Griliches and Ringstad (1971) and used by Tybout, de Melo, and Corbo (1991). Defining employment as the number of workers yields similar results.

14. This definition is analogous to that in Aitken and Harrison (1999), who use employment as weights. Schoors and van der Tol (2001) as well as Blalock (2001) employ output weights but do not take into account the share of foreign equity, treating as foreign total output of firms with at least 10 or 20 percent foreign equity, respectively.

15. To illustrate the meaning of the variable, suppose that the sugar industry sells half of its output to jam producers and half to chocolate producers. If no multinationals are producing jam but half of all chocolate production comes from foreign affiliates, the *Backward* variable will be calculated as follows: $(0.5 * 0) + (0.5 * 0.5) = 0.25$.

16. Since relationships between sectors may change over time (although a radical change is unlikely), using multiple input-output matrices would be ideal. Unfortunately, input-output matrices for later years are unavailable. Similarly, employing a matrix excluding imports would be preferable, but such a matrix does not exist. The results should be interpreted with these two caveats in mind.

17. Including the share of intermediates supplied within the sector in the *Backward* measure does not change the conclusions with respect to the correlation between firm productivity and foreign presence in the sourcing sectors.

18. Domestic firms are defined as those with less than 10 percent foreign equity.

19. The index is defined as the sum of the squared market shares of the four largest producers in a given sector. Its value may range from 0 to 10,000. As Nickell (1996) notes, predictions of the theoretical literature on the impact of competition on productivity are ambiguous. In his empirical analysis, however, Nickell finds evidence of a positive correlation between competition and productivity growth.

20. More precisely, $Demand_{jt} = \Sigma k\ a_{jk} * Y_{kt}$, where a_{jk} is the input-output matrix coefficient indicating that in order to produce one unit of good k, a_{jk} units of good j are needed. Y_{kt} stands for industry k output deflated by an industry-specific deflator.

21. In this case, the fixed effect for region r captures not just the fact that region r is an attractive business location but also that its attractiveness is changing over time.

22. While the Olley-Pakes method also allows for taking into account firm exit, it is not done here, because the data do not allow one to distinguish between firm exit from the sample due to liquidation and firm exit due to not being included in the group of enterprises surveyed in a given year.

23. The number of observations is smaller in these regressions, as the Olley-Pakes procedure can be applied only to firms reporting positive gross investment in a given year.

24. The calculation is based on the coefficient from the regression with the Olley-Pakes correction estimated on the subsample of domestic firms (table 10.7, column 4).

25. This conclusion is supported by the findings obtained by Djankov and Hoekman (2000). Aitken and Harrison (1999) report a similar pattern of results in their analysis of Indonesian data. As an additional check, an indicator variable for the cases in which the foreign share increases from less than 50 percent to more than 50 percent (thus giving the foreign investor majority ownership) was experimented with, but it did not appear to be statistically significant.

26. This finding would be consistent with a Schumpeterian-style argument that more monopolistic firms can more readily fund research and development expenditure because they face less market uncertainty and have a larger and more stable cash flow (see Levin, Cohen, and Mowery 1985).

27. This change is not due to multicolinearity with the *Backward* variable, as the correlation between the two is 0.3.

28. There are 342 observations pertaining to fully owned foreign affiliates and another 35 observations for firms whose foreign capital share is 99–100 percent. Together they constitute 27 percent of all observations of firms with foreign capital participation.

29. The following seven sectors fall into the concentrated category: chemicals and chemical products (NACE 24); basic metals (NACE 27); office machinery and computers (NACE 30); electrical equipment and apparatus (NACE 31); radio, TV, and communications equipment (NACE 32); motor vehicles (NACE 34); and other transport equipment (NACE 35).

References

Aitken, Brian J., and Ann E. Harrison. 1999. "Do Domestic Firms Benefit from Direct Foreign Investment? Evidence from Venezuela." *American Economic Review* 89 (3): 605–18.

Aitken, Brian, Ann E. Harrison, and Robert E. Lipsey. 1996. "Wages and Foreign Ownership: A Comparative Study of Mexico, Venezuela and the United States." *Journal of International Economics* 40 (3–4): 345–71.

Altenburg, Tilman. 2000. "Linkages and Spillovers between Transnational Corporations and Small and Medium-Sized Enterprises in Developing Countries: Opportunities and Best Policies." In *TNC-SME Linkages for Development: Issues-Experiences-Best Practices,* ed. United Nations Conference on Trade and Development, 3–61. New York and Geneva: United Nations.

Belderbos, Rene, Giovanni Capannelli, and Kyoji Fukao. 2001. "Backward Vertical Linkages of Foreign Manufacturing Affiliates: Evidence from Japanese Multinationals." *World Development* 29 (1): 189–208.

Belkindas, Misha, Mustafa Dinc, and Olga Ivanova. 1999. "Statistical Systems Need Overhaul in Transition Economies." *Transition* 10 (4): 22–24.

Blalock, Garrick. 2001. "Technology from Foreign Direct Investment: Strategic Transfer through Supply Chains." Paper presented at the "Empirical Investigations in International Trade" conference, Purdue University, West Lafayette, IN, November 9–11.

Blomström, Magnus. 1986. "Foreign Investment and Productive Efficiency: The Case of Mexico." *Journal of Industrial Economics* 35 (1): 97–110.

———. 1989. *Foreign Investment and Spillovers.* London: Routledge.

Blomström, Magnus, and Ari Kokko. 1998. "Multinational Corporations and Spillovers." *Journal of Economic Surveys* 12 (2): 1–31.

Blomström, Magnus, and Hakan Persson. 1983. "Foreign Investment and Spillover Efficiency in an Underdeveloped Economy: Evidence from the Mexican Manufacturing Industry." *World Development* 11 (6): 493–501.

Blomström, Magnus, and Edward N. Wolff. 1994. "Multinational Corporations and Productivity Convergence in Mexico." In *Convergence of Productivity: Cross-National Studies and Historical Evidence,* ed. William J. Baumol, Richard R. Nelson, and Edward N. Wolff, 263–84. Oxford: Oxford University Press.

Blomström, Magnus, Ari Kokko, and Mario Zejan. 2000. *Foreign Direct Investment: Firm and Host Country Strategies.* London: Macmillan Press.

Caves, Richard E. 1974. "Multinational Firms, Competition and Productivity in Host-Country Markets." *Economica* 41 (162): 176–93.

Djankov, Simeon, and Bernard Hoekman. 2000. "Foreign Investment and Productivity Growth in Czech Enterprises." *World Bank Economic Review* 14 (1): 49–64.

EBRD (European Bank for Reconstruction and Development). 2001. *Lithuania Investment Profile 2001.* London: EBRD.

Girma, Sourafel, David Greenaway, and Katharine Wakelin. 2001. "Who Benefits from Foreign Direct Investment in the UK?" *Scottish Journal of Political Economy* 48 (2): 119–33.

Görg, Holger, and Eric Strobl. 2001. "Multinational Companies and Productivity Spillovers: A Meta-Analysis." *Economic Journal* 111 (475): 723–39.

Griliches, Zvi, and Jacques Mairesse. 1995. "Production Functions: The Search for Identification." Working Paper 5067, National Bureau of Economic Research, Cambridge, MA.

Griliches, Zvi, and Vidar Ringstad. 1971. *Economies of Scale and the Form of the Production Function.* Amsterdam: North-Holland.

Haddad, Mona, and Ann E. Harrison. 1993. "Are There Positive Spillovers from Direct Foreign Investment? Evidence from Panel Data for Morocco." *Journal of Development Economics* 42 (1): 51–74.

Haskel, Jonathan E., Sonia C. Pereira, and Matthew J. Slaughter. 2002. "Does Inward Foreign Direct Investment Boost the Productivity of Domestic Firms?" Working Paper 8724, National Bureau of Economic Research, Cambridge, MA.

IMF (International Monetary Fund). 2003. *International Financial Statistics Database.* Washington, DC: IMF.

Javorcik, Beata Smarzynska. 2004. "The Composition of Foreign Direct Investment and Protection of Intellectual Property Rights: Evidence from Transition Economies." *European Economic Review* 48 (1): 39–62.

Keller, Wolfgang, and Stephen R. Yeaple. 2003. "Multinational Enterprises, International Trade and Productivity Growth: Firm Level Evidence from the United States." NBER Working Paper No. 9504, National Bureau of Economic Research, Cambridge, MA.

Konings, Jozef. 2001. "The Effects of Foreign Direct Investment on Domestic Firms." *Economics of Transition* 9 (3): 619–33.

Kugler, Maurice. 2000. "The Diffusion of Externalities from Foreign Direct Investment: Theory Ahead of Measurement." Discussion Papers in Economics and Econometrics, University of Southampton, United Kingdom.

Levin, Richard C., Wesley M. Cohen, and David C. Mowery. 1985. "R&D Appropriability, Opportunity, and Market Structure: New Evidence on Some Schumpeterian Hypotheses." *American Economic Review* 75 (2): 20–24.

Lin, Ping, and Kamal Saggi. 2004. "Multinational Firms and Backward Linkages: A Survey and a Simple Model." Department of Economics, Southern Methodist University, Dallas, TX.

Lipsey, Robert E. 2002. "Home and Host Country Effects of FDI." Working Paper 9293, National Bureau of Economic Research, Cambridge, MA.

Liu, Xiaming, Pamela Siler, Chengqi Wang, and Yingqi Wei. 2000. "Productivity Spillovers from Foreign Direct Investment: Evidence from UK Industry Level Panel Data." *Journal of International Business Studies* 31 (3): 407–25.

Markusen, James R., and Anthony J. Venables. 1999. "Foreign Direct Investment as a Catalyst for Industrial Development." *European Economic Review* 43 (2): 335–56.

Moran, Theodore H. 2001. *Parental Supervision: The New Paradigm for Foreign Direct Investment and Development.* Washington, DC: Institute for International Economics.

Moulton, Brent R. 1990. "An Illustration of a Pitfall in Estimating the Effects of Aggregate Variables on Micro Units." *Review of Economics and Statistics* 72 (2): 334–38.

Nickell, Stephen. 1996. "Competition and Corporate Performance." *Journal of Political Economy* 104 (4): 724–46.

Olley, Steven G., and Ariel Pakes. 1996. "The Dynamics of Productivity in the Telecommunications Equipment Industry." *Econometrica* 64 (6): 1263–97.

OECD (Organisation for Economic Co-operation and Development). 2000. *Lithuania: Foreign Direct Investment Impact and Policy Analysis.* Paris.

———. 2001. *Reviews of Foreign Direct Investment: Lithuania.* Paris.

Rodríguez-Clare, Andrés. 1996. "Multinationals, Linkages, and Economic Development." *American Economic Review* 86 (4): 852–73.

Rodrik, Dani. 1999. "The New Global Economy and Developing Countries: Making Openness Work." Overseas Development Council Policy Essay No. 24, Baltimore, MD.

Schoors, Koen, and Bartoldus van der Tol. 2001. "The Productivity Effect of Foreign Ownership on Domestic Firms in Hungary." Paper presented at the International Atlantic Economic Conference, Philadelphia, PA, October 11–14.

Toth, Istvan Janos, and Andras Semjen. 1999. "Market Links and Growth Capacity of Enterprises in a Transforming Economy: The Case of Hungary." In *Market Links, Tax Environment and Financial Discipline of Hungarian Enterprises,* ed. Istvan Janos Toth and Andras Semjen, 1–37. Budapest: Institute of Economics, Hungarian Academy of Sciences.

Tybout, James, Jaime de Melo, and Vittorio Corbo. 1991. "The Effects of Trade Reforms on Scale and Technical Efficiency." *Journal of International Economics* 31 (3–4): 231–50.

UNCTC (United Nations Centre on Transnational Corporations). 2000. *The Competitiveness Challenge: Transnational Corporations and Industrial Restructuring in Developing Countries.* New York and Geneva: United Nations.

———. 2001. *World Investment Report: Promoting Linkages.* New York and Geneva: United Nations.

World Bank. 2003a. "Developing Knowledge-Intensive Sectors, Technology Transfers, and the Role of FDI." Foreign Investment Advisory Service, Washington, DC.

———. 2003b. *World Development Indicators Database.* Washington, DC: World Bank.

11

PRODUCT QUALITY, PRODUCTIVE EFFICIENCY, AND INTERNATIONAL TECHNOLOGY DIFFUSION: EVIDENCE FROM PLANT-LEVEL PANEL DATA

Aart Kraay, Isidro Soloaga, and James R. Tybout

A casual tourist can confirm that technologies often make their way from the developed world to less developed countries. However, despite the critical importance of technology diffusion for economic development, the evidence on many aspects of this process remains sketchy. What mechanisms most frequently transmit foreign technologies to firms in less developed countries? Do these foreign technologies affect both productive efficiency and product quality in the recipient firms? Under what circumstances do firms pursue activities that give them access to foreign knowledge? This chapter develops a new methodology for addressing these issues and applies the framework to plant-level panel data from Colombia, Mexico, and Morocco.

The Literature

The limited understanding of international knowledge diffusion does not derive from neglect of the topic. In the empirical literature alone, at least three different methodological approaches have been deployed. At the very micro level, analysts have used case studies and qualitative surveys to generate descriptions of learning

The authors would like to thank Wolfgang Keller, Jim Levinsohn, Marc Melitz, and Mark Roberts for useful discussions and absolve them of blame for any methodological flaws that remain.

processes at individual firms (see, for example, Hobday 1995; Lall 1987; Katz 1987; Pack 1987; Rhee, Ross-Larson, and Pursell 1984). This literature provides invaluable details concerning firms' efforts to absorb technology, as Pack (chapter 2 of this volume) observes, but it has to little to say quantitatively about the results of these efforts in terms of productive efficiency or product quality. There are some exceptions (for example, Pack 1987), but they are too few and based on such small samples that they provide little basis for generalization.

At the other extreme, studies based on aggregated data have correlated cross-country patterns of productivity growth or productivity levels with various proxies for countries' exposure to foreign knowledge and their ability to absorb it. These proxies include capital goods imports (de Long and Summers 1991; Keller 2000); trade with countries possessing large R&D stocks (Coe and Helpman 1995; Keller 1998, 2000); foreign direct investment (FDI) inflows (Blomström, Lipsey, and Zejan 1994); and domestic patent stocks (Eaton and Kortum 1996).[1] Unlike case studies and descriptive surveys, cross-country regressions document broad patterns of association, providing a basis for generalization. However, most are subject to a variety of econometric criticisms, including aggregation bias, omitted variable bias, measurement error bias, and simultaneity bias.

Plant- or firm-level econometric studies correlate proxies for firms' exposure to foreign knowledge with their productivity levels or growth rates. To cite a few examples, Aitken and Harrison (1999), Haddad and Harrison (1993), and Djankov and Hoekman (chapter 8 of this volume) study FDI; Chen and Tang (1987); Aw and Hwang (1995); Clerides, Lach, and Tybout (1998); Kraay (chapter 7 of this volume); and Bigsten and others (1999) study exports.[2] By sacrificing the nuance and detail provided by case studies, these micro econometric studies gain the ability to treat large numbers of producers and make statistical inferences. They do better than the macro studies in terms of identifying specific correlates of productivity and avoiding aggregation bias.

Plant-level econometric studies have significant shortcomings, however. One common problem is failure to disentangle causality. For example, contemporaneous correlation between exports and efficiency tells us nothing about what caused what. Even studies that use lagged exports to predict current efficiency may miss the knowledge transmission mechanism, as Westphal (2001) emphasizes.

A second problem is that productivity is almost always poorly measured.[3] Manufactured products are quite heterogeneous, even within narrowly defined industries, so there is no single measure of output that can be compared across firms. Real revenue is typically used as a stand-in for physical product, but this variable responds to product-specific price adjustments as well as fluctuations in physical volume. Productivity measures consequently confound productive efficiency and market power. Furthermore, when technology diffusion leads to product innovation rather than process innovation, these productivity measures may miss the effect entirely. Finally, like cross-country regressions, studies in this literature usually focus on a single conduit for international technology diffusion and ignore the others, opening the door to omitted variable bias.

The Contribution of This Chapter

As an econometric study based on plant-level data, this chapter falls squarely in the third category mentioned before. It attempts to improve on existing methodologies in two respects. First, we abandon the standard approach to measuring productivity in favor of an alternative approach that treats the observed plant-level data on revenues, costs, and market shares as reflecting equilibrium in a differentiated product market. By using a new normalization and imposing sufficient demand-side structure, we avoid the problem of distinguishing real revenue from physical product. We are also able to separately measure process innovations, which are manifest in marginal cost reductions, and product innovations, which are manifest in heightened demand for a product at a given vector of goods prices. Accordingly, we can study the joint evolution of these processes and investigate whether improvements in one dimension are complemented or offset by changes in the other.

Second, we treat multiple channels for international technology diffusion in a single integrated framework. This would not be important if the various activities that transmit technology were unrelated to one another. But because of complementarities and indivisibilities, they tend to come in bundles, predictable sequences, or both.[4] Hence econometric models that treat any one of them as the unique source of foreign technology run a considerable risk of misattribution.

The main knowledge-transmitting activities identified by the case study literature include FDI in domestic enterprises, joint ventures, outsourcing, licensing arrangements, importation of intermediate and capital goods, learning from exporting to knowledgeable buyers, and learning from the importation of final goods and from reactions to changes in domestic market structures as these goods enter the country. It is not hard to identify reasons why firms' decisions regarding these activities will be related to one another. Exporters are more likely to use imported capital and intermediate goods, because they are granted preferential access to foreign exchange or because the product characteristics needed for exporting are best manufactured with these goods. Similar input and capital good requirements may accompany licensing agreements. Firms with FDI are more likely to use imported intermediates, because the parent company can internalize some costs by doing so. Multinationals sometimes locate plants abroad to exploit low wages while protecting their intangible assets, such as proprietary knowledge and product reputation, so FDI and exporting can also be complementary activities.

Methodology

To motivate our empirical model, we begin by sketching a dynamic model of industrial evolution in which firms make optimal decisions about their activity bundles. In turn, these bundles influence their future performance. Although the model is not very formal, the discussion allows us to be explicit about the causal relationships we assume generated the data.

Determinants of Performance

From the perspective of managers, the activities described above generate revenues by improving a firm's unit production costs (c_{jt}) or the appeal of its product (a_{jt}), which we combine to form the performance vector $\omega_{jt} = (a_{jt}, c_{jt})$ for the jth firm in period t. Then, presuming that the elements of ω_{jt} evolve according to autoregressive processes, conditioned on exogenous firm characteristics (hereafter x_{jt}) and the firm's history of activity bundles (hereafter B_{jt-1}), we write

$$\omega_{jt} = g_{\omega}(x_{jt}, B_{jt-1}, \Omega_{jt-1}, \varepsilon_{jt}), \qquad (11.1)$$

where $\Omega_{jt} = (\omega_{jt}, \omega_{jt-1}, \omega_{jt-2}, \ldots)$, $B_{jt} = (b_{jt}, b_{jt-1}, b_{jt-2}, \ldots)$, ε_{jt} is a serially uncorrelated vector of unobserved innovations in the ω_{jt} process, and the column vector of dummies b_{jt} indicates which activity bundle the jth firm is pursuing at time t, if any.[5]

Determinants of Activity

Of course, the activities themselves are endogenous. We envision that firms weigh four kinds of effects on their profit streams when choosing which combination to pursue. First, as described by equation (11.1), activities influence future realizations on the performance trajectories, ω_{jt}. Second, in so doing they may also affect the future activity choices and performance of competing firms. Third, given the realization of ω_{jt}, activity bundles can affect net operating profits by changing demand conditions—by, for example, providing access to foreign markets—or by affecting the share of operating profits retained by the firm's majority owners. Finally, the initiation of activities generally involves start-up or adjustment costs.

The specifics of these effects on profits depend on the activity in question. For example, joint ventures, subcontracting, and FDI may transmit knowledge, improve a firm's access to inputs, or both, thereby affecting the evolution of its performance vector, ω_{jt}. These activities may also affect the firm's operating profits by creating new markets for its products, branding its products, creating profit-sharing obligations, or diluting corporate control. Finally, joint ventures, subcontracting, and FDI involve up-front research costs and legal fees.

Similar observations apply to the other activities cited above. Firms that import intermediate or capital goods improve their performance trajectories by using higher-quality inputs and by extracting knowledge from these foreign goods. But they also incur higher material or capital costs, and before importing they must research foreign suppliers and learn about customs procedures. Firms that employ high-quality workers typically improve their processes and their products, but they also incur higher labor costs, and they bear the sunk costs of attracting and screening job applicants for these positions. Finally, firms that export improve their earnings by tapping new markets, and they may learn from knowledgeable buyers abroad. But to begin exporting, firms must establish distribution channels, research foreign markets, and repackage or even redesign their products.

When managers understand these linkages and correctly anticipate the behavior of their rivals, the activity choices of the jth firm in period t can be represented as determined by the decision rule

$$b_{jt} = g_b(j, Z_t, X_t, \Omega_{t-1}, B_{t-1}, \Gamma_{jt}). \tag{11.2}$$

Here the arguments of $g_b(\cdot)$ include everything that helps firms predict the future payoffs from each possible bundle: current and past exchange rates and demand levels, Z_t; the set of previous realizations on ω for all industry participants, $\Omega_{t-1} = (\Omega_{1t-1}, \ldots, \Omega_{jt-1}, \ldots, \Omega_{Nt-1})$; previous activity choices for all industry participants, $B_{t-1} = (B_{1t-1}, \ldots, B_{jt-1}, \ldots, B_{Nt-1})$; the exogenous characteristics of all industry participants, $X_t = (x_{1t-1}, \ldots, x_{jt-1}, \ldots, x_{Nt-1})$; and a set of beliefs, Γ_{jt}, about the decision rules that will be used by all of the other firms. (When the industry is in equilibrium, these beliefs must be consistent with observed behavior.) Finally, while the information set is common to all firms, $g_b(\cdot)$ depends upon j because the jth firm's own characteristics and history affect its payoffs asymmetrically from those of all other firms.

Inference

Our basic objective is to quantify the relationships described by equations (11.1) and (11.2). We view significant associations between B_{jt-1} and ω in equation (11.1) as evidence that the international activities Granger-cause performance. Similarly, when Ω_{jt-1} helps predict b_{jt} in equation (11.2), we view performance as Granger-causing activities. In both equations, the fact that we treat activities as bundles allows us to determine whether particular combinations of bundles are relatively potent performance determinants. Also, given that we examine the joint evolution of productive efficiency and product quality, we will be able to make inferences about the nature of the performance effects induced by international contacts.

Given sufficient variation in the data, this approach to inference should pick up most instances in which knowledge acquired through observable activities enhances future product quality or productive efficiency. It will fail to detect some types of linkages between activities and performance, however. For example, suppose a foreign corporation subcontracts with a plant to become a supplier of one of its products and it transmits the necessarily technical information to that plant. It may be years before the plant begins to export the product to the buyer (see Kim 1997), so when the associated exports show up in the data, the plant's performance trajectory will already have responded to the new knowledge and no association will be detected. This scenario would generate the misleading impression that the outsourcing activity responded to a productivity shock rather than vice versa.

Equation (11.1) will also miss technology diffusion that does not occur at the firm level. For example, if firms acquire imported intermediate or capital inputs through an intermediary rather than by purchasing directly from foreign suppliers, the associated improvement in performance will not be attributed to foreign sources. Similarly, if firms that learn by engaging in international business serve as valuable examples for others, the knowledge spillovers they generate will not be attributed to foreign sources. For all of these reasons, the results of the exercise that follows should be viewed as suggestive rather than definitive.

An Empirical Model

To render equations (11.1) and (11.2) empirically useful, we must surmount several obstacles. The first problem is measuring the performance vector, ω. Somehow we must measure unit production costs (c_{jt}) and product attractiveness (c_{jt}) concepts using plant-level data on revenues, intermediate input costs, market shares, labor costs, and crude capital stock proxies.

To render unit production costs observable, we use a novel normalization. Specifically, we define one unit of the *j*th plant's product to be whatever that plant can produce with a dollar's worth of intermediate inputs. Since products are differentiated, this does not imply that a plant using a relatively large amount of intermediate inputs is producing a relatively valuable output. However, if firms were to differ in the efficiency with which they convert intermediate goods into final output, this assumption would have the undesirable implication that efficiency gains reduce output when the physical volume of final production does not change.[6] Thus our normalization is accompanied by two strong assumptions: firms exhibit constant-returns homothetic technologies, and these technologies differ across producers only because of differences in primary factor efficiency.[7] That is, the only reason why marginal cost schedules differ across firms is that some firms use labor and capital more effectively than others.

With these assumptions we can calculate unit (variable) production costs and output prices at the *j*th plant in year *t* as $c_{jt} = (C_{jt}/I_{jt})$ and $p_{jt} = (R_{jt}/I_{jt})$, where C_{jt} is total variable costs (labor, intermediates, and energy); I_{jt} is intermediate input costs;

and R_{jt} is revenues. Our assumptions imply that unit variable production costs correspond to marginal costs.

Our approach to measuring product appeal or quality is more involved. It is based on the notion that, given the vector of prices for all available products in an industry (including a composite imported variety), large market shares imply high quality. Of course, market shares reflect more than product characteristics, so one should think of "quality" as a broad measure of product appeal that responds to reputational effects and advertising, as well as to physical characteristics of the products.

To impute this quality notion from prices and market shares, we need a demand system and a market equilibrium concept. For these we use Lu and Tybout's (2000) adaptation of Berry's (1994) representation of a differentiated product market, which is based on McFadden's (1974) nested logit demand system and the generalizations developed by Berry, Levinsohn, and Pakes (1995).

The Demand System

We begin by assigning each producer in the industry of interest to one of G geographic regions ("nests"). Producers in all regions compete with one another and with a composite imported good, but consumers view products within a region as closer substitutes than products from distinct regions. The price of the composite imported good is exogenously determined by the real exchange rate. There are N domestic establishments, indexed by $j \in \{1,\ldots N\}$, each supplying its own unique variety, so counting the composite imported good (identified by $j = 0$), there are $N + 1$ available varieties. The set of product varieties included in product j's nest (including product j itself) is given by Θ_j.

Domestic consumers have heterogeneous tastes, indexed by the real number $\ell \in (0, L_t]$. Each period, each consumer in the market chooses a single unit of the variety that yields him or her the largest net indirect utility, where variety j yields consumer net utility:

$$u_{\ell jt} = \overline{u}_{jt} + \varsigma_{\ell g\, jt} + \nu_{\ell jt}. \tag{11.3}$$

Here $\overline{u}_{jt} = \xi_{jt} - \alpha p_{jt}$ for the N domestic varieties and $\overline{u}_{0t} = \xi_{0t} - \gamma \cdot r_t$ for the imported variety, where ξ_{jt} indexes the "quality" of good j, p_{jt} is the price of the N domestic good, and r_t is the domestic currency price of the imported composite good.

The last two terms on the right-hand side of equation (11.3) are unobserved error components that capture individual taste differences. The first component, $\varsigma_{\ell g\, jt}$, varies across but not within nests, while $\nu_{\ell jt}$ exhibits within-nest variation. By assumption, both $[\varsigma + \nu]$ and ν are distributed type-I extreme values across consumers, with variances of and $(\pi\mu_1)^2/3$ and $(\pi\mu_2)^2/3 < (\pi\mu_1)^2/3$, respectively.

The indirect utility function parameters are identified only up to a scalar multiple, so we impose $\mu_1 = 1$.[8] We define $\sigma = 1 - \mu_2/\mu_1 (0 < \sigma < 1)$ to index the degree of substitutability among, versus within, nests.[9]

Integrating over domestic consumers yields the standard nested logit expression for the demand for the jth domestic variety as a fraction of total domestic demand for varieties in the jth product's nest:

$$s_{j|g,t} = \frac{\exp[(\bar{u}_{jt} - \bar{u}_{0t})/(1-\sigma)]}{\sum\limits_{k \in \Theta_j} \exp[(\bar{u}_{kt} - \bar{u}_{0t})/(1-\sigma)]}, \qquad j = 1,2,\ldots,N_t. \tag{11.4}$$

Similarly, total demand for group g varieties as a share of total domestic consumption is

$$s_{g,t} = \frac{D_{gt}^{1-\sigma}}{\sum\limits_{k=1}^{G} D_{kt}^{1-\sigma} + 1}, \qquad \text{where } D_{gt} = \sum_{k \in \Theta_g} \exp\left[\frac{(\bar{u}_{kt} - \bar{u}_{0t})}{(1-\sigma)}\right], \qquad g = 1,\ldots,G, \tag{11.5a}$$

and the demand for the imported variety as a share of total domestic consumption—that is, the import penetration rate—is

$$s_{0,t} = 1 - \sum_{g=1}^{G} s_{g,t} = \frac{1}{\sum\limits_{k=1}^{G} D_{kt}^{1-\sigma} + 1}. \tag{11.5b}$$

Hence demand for the jth variety as a fraction of total units sold is $s_{jt} = s_{j|g,t} \cdot s_{g,t}$.

Given our normalization rule, we can measure the market shares that appear in equations (11.4), (11.5a), and (11.5b) as

$$s_{j|g,t} = \frac{I_{jt}}{\sum\limits_{i \in \Theta_j} I_{it}}, \qquad s_{g,t} = \frac{\sum\limits_{i \in \Theta_g} I_{it}}{\sum\limits_{i=1}^{N_t} I_{it} + M_t}$$

and

$$s_{o,t} = 1 - \sum_{g=1}^{G} s_{g,t},$$

where M_t is the dollar value of imports converted to pesos at the same real exchange rate for all years.[10] Using these equations and the definition of $\bar{u}_{jt}$, it is then possible to solve for the quality of domestic good j by using the expression for mean utility:

$$\xi_{jt} = \alpha p_{jt} - \sigma \ln(s_{j|g,t}) + \ln\left(\frac{s_{jt}}{s_{0,t}}\right) \qquad j = 1,\ldots,N \qquad t = 1,\ldots,T.$$

Finally, we obtain our quality/appeal measure, a_{it}, by expressing ξ_{jt} relative to the quality of imports. Specifically, without loss of generality we set the mean utility from imports to zero ($\bar{u}_{0t}=0$) and measure the quality of domestic good j relative to the quality of imports as

$$a_{jt} \equiv \xi_{jt} - \xi_{0t} = \alpha p_{jt} - \gamma r_t - \sigma \ln(s_{j|g,t}) + \ln\left(\frac{s_{jt}}{s_{0,t}}\right) \quad j=1,\ldots,N \quad t=1,\ldots,T. \quad (11.6)$$

Note that expression (11.6) does not explain how a_{jt} is determined; rather it provides a way to solve for an unobserved matrix in terms of observed matrices and parameters that can be estimated.

Estimation

Expression (11.6) cannot be used to impute quality unless α, γ, and σ are known. We identify these parameters by substituting the right-hand side of expression (11.6) into a linearized version of equation (11.1) wherever a_{jt} appears:

$$a_{jt} = a_0 + \sum_{q=1}^{Q} \lambda_q a_{j,t-q} + \sum_{q=Q+1}^{2Q} \lambda_q c_{jt-q} + \lambda^x \mathbf{x}_{jt} + \lambda^b b_{jt-1} + \varepsilon_{jt}^a, \quad (11.1a)$$

$$c_{jt} = a_0 + \sum_{q=1}^{Q} \phi_q c_{j,t-q} + \sum_{q=Q+1}^{2Q} \phi_q a_{j,t-q} + \phi^x \mathbf{x}_{jt} + \phi^b b_{jt-1} + \varepsilon_{jt}^c, \quad (11.1b)$$

$$j=1,\ldots,N, \quad t=Q+1,\ldots,T.$$

We then estimate these equations jointly with the equilibrium price relationship that obtains when firms Bertrand-Nash compete in the product market (Berry 1994):[11]

$$\ln(p_{j,t}) = \ln\left\{c_{j,t} + \frac{(1-\sigma)/\alpha}{1-\sigma\cdot s_{j|g,t} - (1-\sigma)\cdot s_{j|g,t}\cdot s_{g,t}}\right\} + \varepsilon_{jt}^P. \quad (11.7)$$

That is, we estimate the demand parameters at the same time that we estimate the parameters describing the effects of international activities on each dimension of performance (λ_b and ϕ_b).

It remains to discuss the properties of the error terms in the system (11.1a), (11.1b), and (11.7). First, if ε_{it}^P reflects measurement error in costs or intermediate inputs, it will be correlated with the right-hand side variables in equation (11.7). Unlike Berry (1994), we assume this problem away by positing that the noise in equation (11.7) comes exclusively from measurement error in revenues (and thus in prices).[12]

Second, if the disturbance terms ε_{jt}^a and ε_{jt}^c are serially correlated, they will not be orthogonal to lagged endogenous variables ($b_{jt-1}, a_{jt-q}, c_{jt-q}$), and spurious correlation patterns may result. There are two standard ways to deal with this problem. One is to choose a sufficiently long lag length (Q), so that all persistence in the endogenous variables is absorbed by the explanatory variables, leaving ε_{jt}^{a} and ε_{jt}^{c} serially uncorrelated. This solution is simple and appealing, but in short panels like ours it means sacrificing most of the time series information in the data. Since time series variation is key for Granger causality tests, we choose the other standard approach. That is, we assume that the error terms are characterized by a standard error component specification and correct for the associated initial conditions problem.

Specifically, we write the disturbances as $\varepsilon_{jt}^{a} = \mu_{j}^{a} + \upsilon_{jt}^{a}$ and $\varepsilon_{jt}^{c} = \mu_{j}^{c} + \upsilon_{jt}^{c}$, where, $\mathrm{var}(\upsilon_{jt}^{k}) = \sigma_{\upsilon,k}^{2}$, $\mathrm{var}(\mu_{j}^{k}) = \sigma_{\mu,k}^{2}$, $\mathrm{cov}(\upsilon_{jt}^{k}, \upsilon_{jt-s}^{k}) = 0$, $\forall s \neq 0$, $\mathrm{cov}(\upsilon_{jt}^{a}, \upsilon_{jt}^{c}) = \sigma_{\upsilon,ac}$, $\mathrm{cov}(\mu_{j}^{a}, \mu_{j}^{c}) = \sigma_{\mu ac}$ and $\mathrm{cov}(\mu_{j}^{a}, \mu_{m}^{c}) = \mathrm{cov}(\mu_{j}^{k}, \mu_{m}^{k}) = 0$ $\forall j \neq m$ and $\forall k = a, c$. We then use Wooldridge's (2000) conditional likelihood function to deal with the initial conditions problem. That is, we write the joint density for T realizations on the vector ω_{jt} and the unobserved effects, $\mu_{j} = (\mu_{j}^{a}, \mu_{j}^{c})$, conditioned on ω_{j1} and x_j, as

$$
\begin{aligned}
& f(\omega_{j2},\ldots,\omega_{jT},\mu_{j}|\omega_{j1},x_{j},b_{j}) \\
& = f(\omega_{jT}|\omega_{jT-1},\omega_{j1},x_{jt},\mu_{j},b_{jT-1}) \cdot f(\omega_{jT-1}|\omega_{jT-2},\omega_{j1},x_{jt},b_{jT-2},\mu_{j}) \cdots \\
& \quad f(\omega_{j2}|\omega_{j1},x_{j1},b_{j1},\mu_{j}) g(\mu_{j}|\omega_{1j},x_{j}).
\end{aligned}
$$

We then express μ_j as a linear projection on ω_{j1} and the temporal mean of x_j, plus a residual plant effect

$$
\mu_{j} = \rho_{0} + \rho_{1}\omega_{1j} + \rho_{2}\bar{x}_{j} + \mu_{j}^{*}. \tag{11.8}
$$

We substitute this expression for μ_j into the density function above, conditioned on ω_{j1} and x_j, and integrate out the unobserved plant effects, $\mu_{j}^{*} = (\mu_{j}^{a*}, \mu_{j}^{c*})$. Since by construction μ_{j}^{*} is orthogonal to ω_{j1}, this approach eliminates the initial conditions problem, and the compound disturbance vector, $(\upsilon_{jt}^{a}, \upsilon_{jt}^{c}) + \mu_{j}^{*}$, has standard error components properties.[13] We use a full information maximum likelihood estimator for the system (11.1a), (11.1b), (11.6), (11.7), and (11.8), presuming that all disturbances are normally distributed.

Determinants of Activity

Our last methodological task is to develop a version of equation (11.2) that can be estimated. Structural estimation of the deep parameters is out of the question, given the complexity of the optimization problem and the number of parameters involved. Instead, we assume that the probability of choosing the kth activity bundle can be written as a reduced-form linear expression in the observable arguments of $g_b(\cdot)$. Also we drop all lags of more than one year and summarize the performance of competing firms with a cross-firm average of ω, excluding the jth firm, hereafter $\bar{\omega}^{-j}$:[14]

$$
b_{jkt} = \delta_{k}' b_{ji,t-1} + \alpha_{k}' \omega_{jt-1} + \eta_{k}' \bar{\omega}_{t-1}^{-j} + \varepsilon_{jkt}^{b} \qquad k = 1, 2^{K} \tag{11.2$'$}
$$

(Unlike in equation 11.2, firms' choices of activity bundles are not deterministic here, because we do not have access to the entire information on which they base their decision.)

As with equations (1a) and (1b), we adopt an error components specification for the disturbances, $\varepsilon_{jt}^{b} = \mu_{j}^{b} + \upsilon_{j}^{b}$, and we address the associated initial conditions problem using Wooldridge's (2000) technique. That is, we express each element of the vector μ_{j}^{b} as a linear function of the vector of initial states, b_{j0}, plus noise.

In principle, there are some efficiency gains to be reaped by estimating the system of equations (11.2′) jointly with those in the system (11.1a), (11.1b), and (11.7). However, as long as the residual plant effects in equation (11.2′) are not correlated with μ^{*} in equations (11.1a) and (11.1b), this is not necessary for consistency. We therefore opt to keep the model manageable and estimate the activity determinant equations separately.

Empirical Results

To implement our estimator, we use annual industrial survey data from three semi-industrial countries: Colombia, Mexico, and Morocco (see Roberts and Tybout 1997, for detailed descriptions of the data.) Ideally, we would like to study all of the activities described earlier, but each data set provides information on only a subset. The Colombian data reveal whether firms are importing intermediate goods and exporting; the Moroccan data identify exporters and firms with foreign owners (FDI); and the Mexican data identify exporters, importers of intermediate goods, and importers of capital goods. Thus although no single country spans the entire set of activities, across the three of them we observe a fairly large collection of conduits for international technology transfer.

We focus our empirical analysis on manufactured chemicals, for several reasons. First, treating all of the establishments in our data sets would be an overwhelming task. Second, among the sectors with sufficient observations to support inference, the chemical industries are most prone to engage in international activities. Third, these industries rely relatively heavily on scientists, technicians, and engineers, so when technology diffusion takes place, we are likely to find it at these firms. Fourth, in most chemical industries, imported final goods are sufficiently important to play the role of an outside good in our demand system. Finally, the chemical industries are well represented in all three countries. Cross-country comparisons thus allow us to examine whether particular types of production are prone to particular patterns of technology absorption.

We limit our analysis to plants that were present during all years of the analysis. (For Colombia, the sample period is 1981–91, for Mexico it is 1986–90, and for Morocco it is 1986–93.) Exclusion of entering and exiting plants obviously opens the door to selection bias in our findings, but it greatly simplifies the econometric modeling. We believe this is a price worth paying, because the omitted firms supply a very small fraction of the market. Furthermore, we are less concerned with precise parameter estimation than with determining whether significant patterns of association are present. It is highly unlikely that they would be manifest only among the new and dying plants that we leave out of our panels.

Determinants of Performance

Estimates of the demand system are reported in annex 1. The results for each four-digit chemical industry in Colombia, Morocco, and Mexico are presented in tables 11A.1–3. (Industries with fewer than 10 plants continually present are not treated.) To conserve space, we do not report parameters of the covariance matrices for the compound disturbances, but we do report the coefficients from equation (11.8), which relates the unobserved plant effects, μ^a and μ^b, to initial realizations of the performance variables, a_1 and c_1.

Under each industry heading, the left-hand column reports parameter estimates and the right-hand column reports the associated standard errors. Wald-statistics for the null hypotheses that product quality is unrelated to international activities and that marginal costs are unrelated to international activities are reported near the bottom of each table. All coefficient estimates that are at least twice their standard error are reported in bold, as are χ^2 statistics with p-values less than 0.05. The degrees of freedom for the Wald-statistics depend on the number of activity bundles considered, which varies across countries and industries. (The latter occurs within a country because some industries do not exhibit all possible activities.) Parameters, standard errors, and test statistics that describe the relationship between activities and performance are reported in shaded panels.

Demand Parameters Parameter estimates for the demand system appear in the top panel of each table. As hoped, α, the parameter that measures the sensitivity of indirect utility to price, is always quite significant and positive. This gives us some confidence that the price measure implied by our normalization rule contains information relevant to consumers. It is generally higher in Morocco than in the other countries, because the French accounting system there led to a unit price measure that was somewhat lower than the measure we calculated for Mexico and Colombia. As a result, the model assigned Morocco higher demand elasticities.

Estimates of σ are also quite accurate. They imply that for most industries, the standard error of $\varsigma + \nu$ is roughly twice the standard error of ν alone. For a given consumer, indirect utilities vary substantially across regions, as well as between foreign and domestic varieties. (It would be straightforward to use these figures to calculate measures of own-region or home market bias, but we did not do so.)

Finally, γ measures the effect of an increase in the price of imported goods on the relative utility attained from home goods. This parameter is not estimated as accurately as the other demand parameters, because it is identified solely by temporal variation, and the number of years we observe is limited. Nonetheless, it is positive in eight of the nine cases in which it is statistically significant at the 5 percent level (the exception is the detergent industry in Colombia). Overall, then, our estimates of the demand system conform well with our priors.

Determinants of Product Quality

Tables 11A.1–3 report estimates of the parameters that appear in equation (11.1a). Before considering the parameters of primary interest, let us recap our results on the control variables. First, conditioning on lagged performance and international activities, most industries analyzed in Colombia and Morocco showed no significant trend in relative quality. The exceptions were Colombian detergents and Moroccan detergents and paints, each of which exhibited a tendency to improve relative to imports over the sample period. Mexico, in contrast, hosted a number of industries that fell increasingly behind imported substitutes during the sample years. Rubber products and pharmaceuticals tended to improve, but five of the six remaining industries showed significant negative trends.

One might expect that plants beginning the sample period with large capital stocks would exhibit relatively high quality, since initial capital stocks should reflect pre-sample demand for their products. Indeed, in all of the industries in which initial capital stocks proved statistically significant, they were positively correlated with relative product quality. This relationship may seem spurious, since product quality is related to market share by identity (11.6), and firms with large market shares surely have large capital stocks. However, consumers care only about prices and product appeal, not productive capacity. So if our model is correctly specified and prices are properly measured, any relation between relative quality and initial capital stocks is indeed a consequence of interaction between quality and size.

The next control variables are lagged quality and lagged marginal cost. Lagged quality is significant in almost all cases, implying that our quality measure follows an autoregressive process, conditioned on other regressors. Given that product characteristics and reputation evolve slowly over time, this is what one would expect to find. In most cases, the AR(1) coefficient is significantly less than unity, but Colombian pharmaceuticals and Mexican fertilizers and pesticides yield roots close to one. (We did not conduct unit root tests for our system.) Coefficients on lagged marginal cost are usually unimportant, so for most industries marginal cost shocks have little effect on the subsequent evolution of product quality. In the four industry/country cases in which lagged marginal cost was significant, firms adjusted to cost shocks partly by reducing the future quality of their product.

The coefficients on initial quality realizations indicate that unobserved heterogeneity is often important and that Wooldridge's correction matters. In contrast, the coefficients on initial marginal cost realizations are usually insignificant, so our representation of unobserved heterogeneity in equation (11.8) is probably more general than it needs to be for most industries. Nonetheless, the covariance matrices for our compound disturbances imply that in many cases, persistent unobserved heterogeneity remains after conditioning on these variables.

Consider now the variables of primary interest—the dummies for the various activity bundles.[15] Wald-statistics reported at the bottom of each table test the joint null hypothesis that the previous period's activity bundles have nothing to do with current quality. Tellingly, all but three of these χ^2 statistics have a p-value greater than 0.05, so in 16 of the 19 country/industry cases, one cannot reject the null hypothesis that previous international activities have no effect on current product quality. Experiments with additional lags on activity bundles (not reported) left this basic message intact.

In the three instances in which activities are significantly related to quality, the dominant relationship is negative. In Moroccan plastics and Colombian fertilizers and pesticides, any combination of international activities lead to worse product quality than no international activities. In the other significant case, Moroccan paints and varnishes, firms that exported and firms with foreign ownership performed worse than firms without international activities, but firms that had both types of international relationship performed better. (In this case, none of the coefficients for these activity bundles was individually significant, so colinearity problems prevented us from drawing strong conclusions.) Overall, then, our methodology yields virtually no evidence that international activities Granger-cause improvements in product quality. (We return to the issue of whether we are missing significant linkages shortly.)

Determinants of Marginal Cost

The weak and occasionally negative association between product quality and international activities need not imply that firms pursuing these activities are misguided. It may be that international activities mainly help reduce costs rather than improve quality. Are such linkages picked up by our model?

Parameter estimates for our marginal cost equation (11.1b) are reported by industry and country in the lower panels of tables 11A.1–3. We begin our discussion of the results with the control variables. First, note that most industries show no significant trend in marginal costs once initial capital stocks, lagged costs, and activities are controlled for. The exceptions are plastics and soaps in Mexico and

Morocco, which trend significantly upward, and fertilizers in Mexico and pharmaceuticals in Colombia, which trend downward.

More interestingly, productive capacity (measured by initial capital stocks) seems to have little to do with our measure of marginal production costs. Except among Mexican rubber producers, where marginal costs rise with capacity, and Mexican pharmaceutical producers, where the opposite occurs, initial capacity is not a significant predictor of our cost measure. This suggests that scale economies are not dramatic and that production technologies are reasonably homothetic over the size range we observe in our panels.

As for the role of lagged marginal costs, all of the industries show clear evidence of serial dependence. Not surprisingly, all of the coefficients are greater than zero, and most are at least two standard deviations below unity. In half of the Mexican industries, lagged quality is also associated with current costs, usually positively, but no such relationship emerges in the other countries. Initial realizations on marginal costs are significant in most sectors, implying that unobserved heterogeneity is important. The (unreported) variance in residual plant effects is also often significant and correlated with the residual plant effect in the quality equation. However, the sign of this correlation varies from industry to industry.

Now consider the relation between international activities and marginal costs. Except in Morocco, our Wald-tests fail to reject the null hypothesis that the two are unrelated. In Morocco FDI alone is associated with significant cost reductions among paint and varnish producers, but exports and FDI together lead to cost increases. Interestingly, the plants that both export and have FDI are also the ones (weakly) predicted to have the highest quality product, so the combined effect of these activities may well be to enhance profits.[16] Exporting and FDI tend to increase costs among detergent and perfume producers, but there is no evidence of offsetting quality gains. Among Moroccan pharmaceutical producers, exporting and FDI significantly reduce future costs, alone or together. They also tend to reduce quality, however, so as with Moroccan paint and varnish producers, the net effect of these activities on profits is not immediately apparent.

Although none of the other country/industry panels shows a significant joint relationship between international activities and future marginal costs, in some instances particular activity bundles yield statistically significant effects. These occur exclusively among bundles that involve importing intermediate inputs. In Colombia fertilizer and pesticide producers reduce their future costs by importing intermediates, but they also reduce the quality of their product. In Mexico intermediate imports increase marginal costs, but they also increase quality among rubber producers. The same is true among Mexican fertilizer and pesticide producers, although the quality effect is not statistically significant. Among Mexican pharmaceutical producers, imported intermediates—in combination with exports or imported capital goods—reduce marginal costs and, if anything, tend to increase product quality. Only in the "other chemicals" sector do imported intermediates (in combination with imported capital and exports) significantly increase future costs without also tending to improve product quality.

Overall, then, most country/industry panels show no significant association between international activities and future marginal costs. However, a minority of industries exhibits this type of causal link. Furthermore, whether the activities increase or reduce costs, there is often an offsetting change in product quality, making the net effect of the international activities on profits ambiguous.

A Robustness Check Granger causality tests may miss important linkages if the transmission of knowledge does not coincide with the observed international activity. For example, when a foreign buyer places an order with a domestic firm, it may transmit blueprints and technical assistance years before the actual exports occur. To check whether this problem has undermined our inferences, we abandon our dynamic specification in favor of a model that simply tests for static correlation between international activities and performance. That is, we drop all a_{jt-q} and c_{jt-q} variables and replace b_{jt-1} with b_{jt} on the right-hand side of equations (1a) and (1b). If this model reveals no association, we have stronger evidence against the claim that international activities transmit knowledge. Statistical significance need not imply causal relationships from activities to performance, however. It may reflect causation from performance to activities (which we explore shortly) or transitory effects that do not involve knowledge transmission. For example, access to imported intermediates may simply enrich the menu of inputs for firms and allow them to produce better products or reduce their costs, as Ethier (1982) suggests.

Tables 11A.4–6 report the static version of our model for each industry/country panel. (We still use an error component specification, but we drop Wooldridge's correction for initial conditions, because lagged performance measures no longer appear on the right-hand side.) When lagged quality is dropped and international activities are no longer lagged, many sectors show a significant association between product quality and international activity. Furthermore, the association is now positive in Colombia for firms that import their intermediate inputs. By itself, this result has several interpretations. It may mean that importing intermediates allows a firm to learn something about technology and to permanently increase the quality of its product, as Grossman and Helpman (1991) suggest.[17] Alternatively, it may reflect the static benefits from importing that Ethier (1982) notes. This interpretation strikes us as more plausible, given that we picked up no evidence of Granger causation from imported intermediates to subsequent gains in product quality and that imported intermediates are unlikely to be preceded by ongoing contacts with foreign suppliers.

Mexico and Morocco also exhibit more significant associations in the static model. However, in contrast to Colombia, international activities still seem to hurt quality, if anything. Thus in determining the effects of international activities on performance, country conditions may be more important than technological features of industries that are common across borders.

Finally, in the marginal cost equation, coefficients differ from those in the dynamic model, but the evidence that international activities are associated with lower costs is, if anything, weaker. In sum, with the exception of the Colombian results on product quality, the timing issue does not appear to be the main reason why firms' product quality and marginal costs are unrelated to their international activities.

Determinants of Activity

Our last empirical exercise addresses the issues of how activities are chosen and whether performance causes international activities. We begin with a descriptive review of the activity patterns found in our data sets. Without controlling for anything, in tables 11B.1 and 11B.2 we report probabilities of different activity bundles and transitions between bundles based on simple cell counts. Clearly, regardless of which activity combinations we consider, the pursuit of one activity increases the likelihood that others will be pursued as well. Similar statements hold for transitions. If one of the international activities has been initiated, the probability that others will follow increases. Accordingly, empirical models that focus on a single activity as the key to international technology diffusion probably suffer from significant omitted variable bias.

Of course, patterns of association may not reflect complementarities among the activities. It is possible that certain firm characteristics—for example, location in a port city or ethnic ties to foreign countries—may make firms engage in multiple international activities, even when true complementarities are absent. To get at this possibility and to better understand firms' activity choices, we estimate the system of linear probability equations (11.2′), allowing for unobserved heterogeneity and using Wooldridge's (2000) correction for the initial conditions problem.

Table 11B.3 presents results for Colombia. We report estimates for three of the four possible activity bundles. (The bundle with no international activities is omitted. Parameter estimates for this option can be derived from the constraint that the sum of the probabilities of the different bundles must always be unity.) The first column reports results for all firms but does not include performance measures. Two points merit mention. First, activities are highly persistent, even after controlling for unobserved heterogeneity. This suggests that the start-up costs associated with the initiation of new international activities are nontrivial, and thus transitory policy or macro shocks may have lasting effects on activity patterns (Roberts and Tybout 1997). Second, we continue to find that the probability that a firm becomes an exporter is higher if it was already importing intermediates the previous year, and the probability of becoming an importer is higher for exporters than for non-exporters. However, tests reported at the bottom of table 11B.3 fail to reject the null hypotheses that these differences in conditional probabilities are zero. Thus although imported intermediates and exporting tend to go together, the dynamic interactions between them are not strong in this fuller specification.

Our second set of estimates (columns 3 and 4 in table 11B.3) adds performance measures to investigate whether firms with high quality, low costs, or both are more likely to engage in activities. Since our quality variable is normalized to have the same mean in each of the four-digit industries, we cannot simply pool all sectors. Instead, we restrict attention to pharmaceuticals, the four-digit industry with the largest number of firms.[18] Within this four-digit industry, the two patterns noted above continue to appear. Activities are highly persistent, and the presence of one activity increases the probability that another will be initiated. However, these dynamic interactions are not statistically significant.

Regarding the effects of quality and cost on the choice of activities, the signs of the coefficients are intuitive for the "both export and import" bundle. Firms with higher quality and lower costs are more likely to pursue this bundle of activities. However, neither coefficient nor the coefficients for the other bundles are significant, so the case for causation from previous performance to activity choices is weak. Table 11B.4 reports analogous results for Morocco, where we observe data on exporting and foreign equity but the degree of persistence, as measured by the magnitude of the coefficients on own lagged activities, is generally smaller. This is especially true for FDI, with past FDI activities increasing the probability of current FDI activity by only 10 percent in the first set of regressions, and not at all in the second set, which includes performance variables (reported only for plastics).

One puzzling exception to the pattern of insignificant effects of performance on the choice of activities is in the equation for the choice of both activities. Other things equal, firms with higher quality and lower costs are less likely to choose both activities. This mirrors the negative correlation between product quality and the joint pursuit of exports and FDI found in tables 11A.2–5. Taken at face value, it suggests that foreign investors are attracted to plants that have been performing relatively poorly.

Conclusion

This chapter has several basic messages. First, by imposing enough structure on the production function and the demand system, it is possible to measure product quality and marginal costs at the plant level and to relate the evolution of these variables to firms' activity histories. Doing so, we find strong firm-level persistence in both quality and marginal costs, as expected. However, in most of the industry/country panels we studied, past international activities do not help much to predict current performance, once past realizations on quality and marginal cost are controlled for. That is, activities do not typically Granger-cause performance. Interestingly, in the minority of cases in which significant associations emerge, international activities appear to move costs and product quality in the same direction. Thus the net effect on profits in these cases is not immediately apparent.

Concerning the determinants of international activities, several basic patterns emerge. Most fundamentally, activities are highly persistent, even after controlling for unobserved heterogeneity. This suggests that firms incur sunk threshold costs when they initiate or cease activities, so temporary policy or macro shocks may have long-run effects on the patterns of activities observed in a particular country or industry.

Second, activities tend to go together. Thus studies that relate firms' performance to one international activity and ignore the others may generate very misleading conclusions. The bundling of activities seems to mainly reflect unobserved plant characteristics, however, such as managerial philosophy, contacts, product niche, and location. Once these are controlled for, there is little evidence that engaging in one international activity increases the probability that others will occur in the future.

Finally, we find weak evidence (from Colombia) that firms that are already strong performers seek out foreign markets as a way to enhance their profits. These results do not generalize to Morocco, where plastics firms that have been performing poorly are more likely than other firms to become exporters and receive FDI.

Chapter 11 Annex A: Demand Parameters and Performance Equations

TABLE 11A.1 Colombian Chemical Industries: Dynamic Model

	Basic industrial chemicals		Fertilizers and pesticides		Plastics	
Demand parameter						
α	**2.151**	0.1	**5.363**	0.403	**16.16**	1.367
σ	**0.627**	0.017	**0.714**	0.018	**0.482**	0.004
γ	0.074	0.153	**1.69**	0.465	6.355	3.724
Product quality equation (11.1a)						
Intercept	**–0.449**	0.191	1.978	1.282	0.128	2.617
Trend	–0.018	0.012	–0.046	0.033	–0.818	0.325
ln (initial capital)	0.051	0.027	–0.197	0.158	0.690	0.600
X only	0.041	0.087				
MI only	0.112	0.071	**–0.904**	0.302	0.564	1.510
X and MI	0.064	0.09	**–0.811**	0.258	1.482	1.707
$a(t-1)$	**0.819**	0.067	**0.648**	0.077	0.183	0.17
$a(1)$	**0.124**	0.054	**0.329**	0.100	**0.358**	0.098
$c(t-1)$	–0.386	0.319	–0.056	0.462	–0.829	11.46
$c(1)$	0.624	0.336	**–3.481**	0.873	2.071	7.392
Marginal cost equation (11.1b)						
Intercept	–0.031	0.032	**0.175**	0.040	–0.002	0.013
Trend	0.001	0.002	0.001	0.003	0.000	0.002
ln (initial capital)	0.003	0.005	–0.017		0.005	0.004
X only	0.001	0.015				
MI only	0.009	0.012	**–0.069**	0.031	0.011	0.011
X and MI	0.004	0.016	**–0.063**	0.027	0.000	0.012
$c(t-1)$	**0.705**	0.067	**0.915**	0.063	**0.717**	0.074
$c(1)$	**0.33**	0.072	–0.134	0.093	0.058	0.05
$a(t-1)$	–0.004	0.01	–0.01	0.008	0.001	0.001
$a(1)$	0.002	0.008	0.012	0.007	–0.001	0.001
Quality effects		2.658		**10.634**		0.756
MC effects		0.676		5.936		1.293
–ln(L)		–572.82		–301.98		–72.99
Number of observations		324	117			108

TABLE 11A.1 (*Continued*)

Paints and varnishes		Pharmaceuticals		Detergents and perfumes	
3.388	0.161	**2.236**	0.074	**4.174**	0.219
0.458	0.018	**0.671**	0.009	**0.662**	0.008
–1.675	0.927	0.096	0.084	**–3.133**	0.348
0.538	0.619	0.135	0.154	**1.665**	0.659
–0.05	0.057	0.001	0.012	**0.189**	0.037
0.166	0.072	–0.012	0.02	–0.039	0.066
		–0.103	0.161		
–0.101	0.362	–0.04	0.071	0.05	0.152
–0.41	0.407	–0.04	0.103	0.096	0.214
0.378	0.05	**1.016**	0.038	**0.356**	0.069
0.752	0.068	–0.019	0.040	**0.694**	0.093
–2.199	2.05	**–1.572**	0.194	0.802	0.607
0.3	2.148	**0.799**	0.199	**–2.456**	0.996
0.031	0.019	**0.135**	0.039	–0.001	0.042
0.001	0.001	**–0.008**	0.003	–0.005	0.003
–0.003	0.002	–0.003	0.005	0.002	0.004
		–0.028	0.041		
–0.011	0.012	–0.009	0.018	0.007	0.014
0.008	0.014	–0.026	0.025	0.015	0.020
0.788	0.066	**0.526**	0.057	**0.632**	0.074
0.022	0.067	**0.235**	0.059	**0.156**	0.079
–0.002	0.002	–0.012	0.01	0.001	0.007
0.002	0.002	0.011	0.011	–0.008	0.008
	1.495		0.62		0.212
	4.323		1.306		0.612
	–47.97		–1,103.15		–543.54
	120		639		351

Source: Authors.

Note: Boldfaced figures indicate values that are significant at the 5 percent level.

TABLE 11A.2 Moroccan Chemical Industries: Dynamic Model

	Paints and varnishes		Pharmaceuticals	
Demand parameter				
α	**21.058**	2.614	**11.06**	0.996
σ	**0.867**	0.014	**0.761**	0.020
γ	**20.700**	8.110	7.844	4.069
Product quality equation (11.1a)				
Intercept	4.33	4.831	–1.738	2.298
Trend	**0.634**	0.215	–0.008	0.071
ln (initial capital)	0.273	0.311	**0.533**	0.185
X only	–1.217	1.164	**–1.219**	0.524
FDI only	–0.673	0.81	–0.809	0.463
X and FDI	2.088	1.447	–0.691	0.516
$a(t-1)$	**–0.558**	0.169	0.069	0.154
$a(1)$	**1.129**	0.141	**0.654**	0.127
$c(t-1)$	**19.86**	7.190	2.515	2.016
$c(1)$	–7.708	9.734	1.23	4.536
Marginal cost equation (11.1b)				
Intercept	0.099	0.086	–0.088	0.159
Trend	0.003	0.007	–0.002	0.006
ln (initial capital)	–0.005	0.009	0.02	0.014
X only	0.01	0.039	**–0.094**	0.038
FDI only	**–0.007**	0.024	**–0.067**	0.033
X and FDI	**0.117**	0.044	**–0.095**	0.035
$c(t-1)$	**0.633**	0.178	**0.534**	0.148
$c(1)$	**0.601**	0.266	0.15	0.342
$a(t-1)$	–0.009	0.006	–0.008	0.011
$a(1)$	0.005	0.005	0.01	0.01
Quality effects		**7.834**		5.753
MC effects		**9.573**		**8.554**
–ln(L)		–111.051		–216.877
Number of observations		84		76

TABLE 11A.2 (*Continued*)

Detergents and perfumes		Chemicals not elsewhere classified		Plastics	
1.479	0.132	**10.204**	0.789	**7.065**	0.393
0.256	0.030	**0.703**	0.025	**0.676**	0.028
7.977	3.031	4.482	3.236	**4.601**	1.329
–0.609	0.823	1.833	1.212	0.623	0.58
0.275	0.056	–0.025	0.082	0.041	0.031
–0.063	0.06	0.106	0.096	–0.007	0.076
–0.409	0.334	0.334	0.992	–0.328	0.224
0.098	0.308	0.357	0.477	–0.122	0.181
–0.103	0.277	1.534	0.862	**–3.105**	0.653
0.466	0.114	**0.532**	0.106	**0.475**	0.103
0.608	0.140	**0.234**	0.084	**0.478**	0.096
–1.069	0.987	–3.202	1.909	**–3.655**	1.022
1.768	1.061	0.741	1.848	**2.341**	0.981
0.032	0.073	0.036	0.055	0.095	0.058
0.015	0.006	–0.002	0.005	**0.007**	0.002
–0.01	0.007	0.002	0.007	–0.009	0.008
0.069	0.04	–0.008	0.061	–0.015	0.021
0.041	0.037	0.002	0.032	–0.023	0.017
0.099	0.033	0.047	0.056	–0.013	0.057
0.672	0.118	**0.314**	0.140	**0.208**	0.075
0.006	0.112	**0.327**	0.139	**0.491**	0.094
0.021	0.014	0.002	0.007	–0.01	0.007
–0.023	0.017	0.001	0.006	0.015	0.008
	2.115		3.354		**23.849**
	9.715		0.837		2.192
	35.459		–317.777		–918.04
	60		186		492

Source: Authors.

Note: Boldfaced figures indicate values that are significant at the 5 percent level.

TABLE 11A.3 Mexican Chemical Industries: Dynamic Model

	Basic industrial chemicals		Synthetic resins		Plastics	
Demand parameter						
α	**1.368**	0.110	**1.828**	0.166	**2.283**	0.101
σ	**0.586**	0.047	**0.532**	0.049	**0.763**	0.036
γ	–0.488	1.264	**1.927**	0.590	1.562	1.474
Product quality equation (11.1a)						
Intercept	–2.752	1.861	**1.730**	0.437	**3.450**	0.737
Trend	–0.010	0.255	**–0.356**	0.095	**–0.707**	0.212
ln (initial capital)	**0.234**	0.055	0.010	0.019	–0.049	0.030
X only	0.270	0.387	–0.052	0.137	0.120	0.199
MI only	**–0.401**	0.157	0.026	0.077	–0.065	0.152
MK only	–0.218	0.185	**–0.291**	0.141	0.032	0.268
MI and X	–0.288	0.151	0.135	0.078	–0.035	0.192
MK and X	–0.347	0.190	0.088	0.166	0.253	0.540
MI and MK	–0.221	0.238			–0.121	0.151
MI, MK, and X	–0.255	0.139	0.029	0.057	0.086	0.156
$a(t-1)$	–0.012	0.094	**0.854**	0.052	0.755	0.043
$a(1)$	**0.679**	0.090	**0.146**	0.048	0.353	0.047
$c(t-1)$	1.304	0.784	0.475	0.374	0.014	n.a.
$c(1)$	–0.739	0.752	–0.192	0.341	**–0.429**	0.154
Marginal cost equation (11.1b)						
Intercept	–0.028	0.115	**0.089**	0.035	**–0.105**	0.038
Trend	–0.002	0.018	–0.027	0.014	**0.045**	0.010
ln (initial capital)	0.002	0.012	–0.003	0.004	0.001	0.003
X only	0.024	0.035	–0.025	0.031	0.023	0.023
MI only	0.002	0.027	–0.005	0.018	0.000	0.018
MK only	–0.020	0.023	–0.003	0.032	0.022	0.034
MI and X	–0.011	0.022	0.019	0.018	–0.005	0.024
MK and X	–0.004	0.029	–0.037	0.038	0.103	0.062
MI and MK	0.063	0.040			0.000	0.018
MI, MK, and X	0.019	0.023	0.017	0.013	0.029	0.018
$c(t-1)$	**0.975**	0.071	**0.838**	0.085	**0.755**	0.043
$c(1)$	0.091	0.062	**0.184**	0.078	**0.173**	0.039
$a(t-1)$	–0.009	0.052	**0.027**	0.011	0.001	0.005
$a(1)$	0.003	0.036	**–0.026**	0.010	0.001	0.005
Quality effects	11.311		8.451		2.099	
MC effects	8.508		5.126		6.504	
–ln(L)		–212.78		–431.26		–345.26
Number of observations		236		156		580

TABLE 11A.3 (*Continued*)

Other chemicals		Rubber products		Fertilizers and pesticides		Pharmaceuticals		Soaps and perfumes	
2.261	0.131	**2.966**	0.224	**1.975**	0.238	**1.197**	0.070	**1.725**	0.121
0.729	0.074	**0.734**	0.090	**0.740**	0.084	**0.709**	0.083	**0.589**	0.034
1.691	0.580	–0.565	0.457	**3.713**	0.625	–0.363	0.447	**9.878**	1.782
1.344	0.488	0.019	0.305	**4.795**	0.732	–0.860	0.562	**7.670**	1.141
–0.322	0.103	**0.170**	0.082	**–1.345**	0.202	**0.206**	0.102	**–0.933**	0.269
0.026	0.015	**0.062**	0.028	–0.044	0.023	0.000	0.019	–0.048	0.032
–0.133	0.236	0.101	0.381			0.191	0.271		
0.077	0.054	**0.261**	0.122			0.032	0.091	–0.060	0.118
–0.074	0.129	0.247	0.284	**–0.586**	0.277	–0.107	0.193	0.137	0.517
0.059	0.066	0.141	0.110	0.178	0.159	0.058	0.062		
0.040	0.082	–0.118	0.194	0.026	0.156	0.056	0.093	–0.038	0.159
–0.135	0.086	–0.163	0.149			0.071	0.063	0.158	0.144
0.887	0.044	**0.935**	0.068	**1.096**	0.124	**0.809**	0.061	**0.652**	0.049
0.092	0.043	0.050	0.072	–0.046	0.128	**0.168**	0.058	**0.426**	0.051
–0.154	0.245	–0.555	0.481	–0.070	0.632	0.041	0.146	–0.121	0.324
0.317	0.222	0.542	0.483	0.617	0.378	–0.086	0.128	0.151	0.318
0.028	0.029	–0.062	0.039	**0.358**	0.096	**0.277**	0.092	**–0.345**	0.143
–0.003	0.010	0.025	0.010	**–0.069**	0.027	0.007	0.020	**0.125**	0.049
–0.003	0.003	**0.018**	0.005	0.006	0.004	**–0.024**	0.010	–0.008	0.006
–0.030	0.055	0.012	0.066			–0.027	0.148		
0.016	0.013	**0.054**	0.022			–0.092	0.049	–0.023	0.023
–0.011	0.030	0.028	0.049	–0.043	0.051	–0.035	0.106	–0.040	0.103
0.028	0.015	–0.008	0.020	**0.075**	0.028	**–0.073**	0.033		
–0.018	0.019	–0.006	0.035	–0.019	0.029	**–0.107**	0.050	–0.013	0.032
0.044	0.020	–0.049	0.027			–0.054	0.034	0.000	0.029
0.693	0.057	**0.610**	0.103	**0.589**	0.097	**0.606**	0.081	**0.695**	0.066
0.238	0.052	**0.374**	0.100	–0.074	0.068	0.096	0.071	**0.249**	0.064
0.013	0.010	0.017	0.013	**0.053**	0.019	**0.064**	0.031	**–0.016**	0.008
–0.009	0.010	–0.022	0.014	–0.066	0.019	–0.055	0.030	**0.020**	0.009
8.379		9.503		7.029		2.367		2.091	
11.606		12.389		9.844		7.186		1.218	
	–618.66		–329.45		–149.38		–372.34		–189.00
	356		168		76		244		188

Source: Authors.

Note: Boldfaced figures indicate values that are significant at the 5 percent level.

TABLE 11A.4 Colombian Chemical Industries: Static Model

	Basic industrial chemicals		Fertilizers and pesticides		Plastics	
Demand parameter						
α	**2.151**	0.100	**15.969**	1.760	**18.317**	1.777
σ	**0.627**	0.017	**0.500**	0.000	**0.491**	0.003
γ	0.074	0.153	2.083	26.857	3.540	5.221
Restricted product quality equation (11.1a)						
Intercept	**–0.449**	0.191	26.713	56.287	1.052	7.206
Trend	–0.018	0.012	–3.853	3.103	–0.723	0.475
ln (initial capital)	**0.051**	0.027	3.154	6.285	**3.062**	0.476
X only	0.041	0.087				
MI only	0.112	0.071	–21.496	38.73	**5.531**	1.610
X and MI	0.064	0.09	–11.446	35.52	–1.402	2.165
Restricted marginal cost equation (11.1b)						
Intercept	**0.819**	0.067	**0.433**	0.106	–0.004	0.019
Trend	**0.124**	0.054	–0.001	0.002	**–0.007**	0.002
ln (initial capital)	–0.386	0.319	**–0.046**	0.022	**0.038**	0.004
X only	0.624	0.336				
MI only	–0.031	0.032	–0.008	**0.069**	0.06	0.015
X and MI	0.001	0.002	0.003	0.064	0.008	0.02
Quality effects		2.658		0.405		**22.322**
MC effects		4.338		0.112		**22.562**
–log (L)		–458.158		613.069		22.083
Number of observations		396		143		132

TABLE 11A.4 (*Continued*)

Paints and varnishes		Pharmaceuticals		Detergents and perfumes	
2.557	0.215	**2.196**	0.075	**4.169**	0.216
0.294	0.020	**0.700**	0.009	**0.664**	0.008
–0.258	1.328	**–0.748**	0.116	**–3.120**	0.312
3.827	0.410	**–10.451**	0.789	**–12.029**	0.897
–6.152	0.446	**–0.087**	0.019	**0.361**	0.037
–0.111	0.371	**1.149**	0.200	**0.757**	0.186
		–0.931	0.230		
–2.212	4.174	**0.274**	0.096	**0.294**	0.097
–0.027	0.115	**0.652**	0.141	–0.001	
–0.684	1.35	**0.507**	0.032	**0.332**	0.029
0.325	0.575	–0.01	0.001	**–0.007**	0.001
–1.314	0.716	0	0.01	–0.01	0.009
		–0.069	0.045		
0.124	0.031	0.014	0.018	–0.007	0.015
–0.003	0.001	**–0.056**	0.027	–0.028	0.016
	0.284		50.673		5.086
	17.529		11.91		3.262
	41.823		–837.104		–512.619
	132		781		429

Source: Authors.

Note: Boldfaced figures indicate values that are significant at the 5 percent level.

TABLE 11A.5 Moroccan Chemical Industries: Static Model

	Paints and varnishes		Pharmaceuticals	
Demand parameter				
α	**22.29**	2.789	**11.712**	1.057
σ	**0.83**	0.016	**0.73**	0.017
γ	10.807	7.052	3.978	2.747
Restricted product quality equation (11.1a)				
Intercept	5.817	11.856	3.84	5.261
Trend	**0.572**	0.169	–0.076	0.065
Ln (initial capital)	1.026	0.902	**0.812**	0.354
X only	–0.347	1.491	–0.668	0.643
FDI only	1.023	1.439	**–0.934**	0.459
X and FDI	2.909	1.742	–0.156	0.494
Restricted marginal cost equation (11.1b)				
Intercept	0.013	0.147	0.312	0.121
Trend	**0.011**	0.004	–0.004	0.004
Ln (initial capital)	0.009	0.018	–0.007	0.013
X only	–0.025	0.042	–0.042	0.042
FDI only	0.023	0.039	–0.048	0.031
X and FDI	**0.100**	0.047	0.022	0.034
Quality effects		4.333		6.394
MC effects		**9.193**		**9.144**
–ln(L)		–93.598		–223.352
Number of observations		112		104

Table 11A.5 (*Continued*)

Detergents and perfumes		Chemicals not elsewhere classified		Plastics	
1.478	0.13	**10.869**	0.402	**8.718**	0.441
0.255	0.027	**0.262**	0.004	**0.547**	0.019
−0.048	2.062	5.285	6.8	**4.639**	1.04
−4.077	4.199	13.286	9.291	2.286	1.943
0.278	0.049	0.093	0.132	**0.097**	0.024
−0.31	0.356	**0.491**	0.178	**0.67**	0.14
−1.166	0.545	−0.07	1.782	−0.156	0.244
0.13	0.521	0.028	1.008	−0.101	0.243
−1.095	0.559	−1.747	1.496	**−1.333**	0.542
0.257	0.08	0.087	0.068	**0.242**	0.067
0.013	0.004	0.002	0.003	**0.006**	0.002
−0.024	0.011	0.013	0.009	−0.013	0.008
−0.012	0.056	**−0.106**	0.051	−0.014	0.018
0.081	0.05	−0.005	0.033	0.013	0.018
0.121	0.052	−0.047	0.049	0.047	0.04
	8.737		1.534		6.54
	11.281		4.977		2.221
	49.601		221.778		−1021.81
	80		248		656

Source: Authors.

Note: Boldfaced figures indicate values that are significant at the 5 percent level.

TABLE 11A.6 Mexican Chemical Industries: Static Model

	Basic industrial chemicals		Synthetic Resins		Plastics	
Demand parameter						
α	**1.475**	0.1	**1.987**	0.229	**7.707**	0.824
σ	**0.517**	0.02	**0.665**	0.028	**0.167**	0.008
γ	0.142	0.113	**–0.652**	0.19	–1.275	0.885
Restricted product quality equation (11.1a)						
Intercept	**–4.386**	0.498	**–8.92**	1.304	**6.68**	2.656
Trend	**0.106**	0.034	**–0.592**	0.074	0.143	0.184
ln(K)	**0.55**	0.054	**0.739**	0.125	0.682	0.671
X only	0.058	0.264			–0.41	7.907
MI only	0.195	0.157			**–1.995**	0.903
MK only	**–0.366**	0.148	–0.307	1.016	**–3.36**	1.327
MI and X	**–0.437**	0.125	–0.519	0.978	**–2.569**	0.918
MK and X	–0.212	0.157			–1.852	1.905
MI and MK	0.144	0.203	–0.169	0.359	–0.393	1.06
MI, MK, and X	–0.256	0.132			–0.507	0.814
Restricted marginal cost equation (11.1b)						
Intercept	0.274	0.04	0.356	0.027	0.167	0.022
Trend	0.029	0.004	–0.025	0.005	0.021	0.003
ln(K)	–0.013	0.019	**–0.02**	0.009	0.005	0.008
X only	–0.069	0.062			0.036	0.109
MI only	–0.037	0.035			–0.003	0.026
MK only	–0.051	0.033	**0.307**	0.088	–0.079	0.039
MI and X	–0.041	0.028	0.108	0.072	0	0.026
MK and X	0.006	0.035			–0.106	0.058
MI and MK	–0.036	0.045	–0.068	0.053	0.004	0.032
MI, MK, and X	–0.032	0.03			0	0.023
Quality effects		**24.524**	0.833			**15.586**
MC effects		6.268	**15.62**		7.351	
–Log(L)		–308.656	154.594		18.271	
Number of observations		295		95		195

TABLE 11A.6 (*Continued*)

Other chemicals		Rubber products		Fertilizers and pesticides		Pharmaceuticals		Soaps and perfumes	
10.77	1.021	**1.921**	0.12	**3.524**	0.23	**4.712**	0.359	**2.643**	0.093
0.111	0.006	**0.513**	0.016	**0.428**	0.022	**0.486**	0.022	**0.555**	0.008
1.244	2.583	**−1.254**	0.236	−0.067	0.093	**−1.742**	0.254	**−3.073**	0.134
23.366	8.731	**−14.229**	1.115	−0.116	0.591	**−2.419**	1.049	**−10.119**	0.456
1.186	0.804	**0.493**	0.078	**0.017**	0.02	**−0.156**	0.057	**−0.679**	0.032
−0.131	.	**0.965**	0.162	**0.698**	0.051	**0.752**	0.134	**0.444**	0.060
−3.432	9.913			0.013	0.143	0.322	0.523	0.12	0.142
−2.39	4.414	0.06	0.319	0.07	0.112	**−1.181**	0.372	0.142	0.144
−2.477	7.056	−0.043	0.674	0.01	0.155	0.557	0.448	−0.157	0.270
−3.76	3.527			**−0.29**	0.131	0.259	0.373	0.09	0.202
−2.788	6.024							−0.147	0.260
−1.956	4.316	−0.14	0.398	0.236	0.134	−0.221	0.53	**0.396**	0.145
−4.341	3.659	0.042	0.355	0.005	0.109	0.429	0.446	0.285	0.162
0.426	0.044	0.404	0.07	0.261	0.027	**0.375**	0.054	0.309	0.032
0.008	0.006	0.016	0.005	0.009	0.002	−0.006	0.004	−0.006	0.003
−0.037	0.014	0.002	0.025	0.01	0.008	0	0.015	−0.017	0.010
−0.046	0.149			−0.013	0.036	0.03	0.054	0.006	0.023
−0.085	0.061	0.044	0.04	**−0.061**	0.023	−0.047	0.038	0.008	0.023
−0.053	0.114	0.003	0.087	−0.009	0.037	0.049	0.046	0.043	0.044
−0.067	0.047			**−0.078**	0.028	0.073	0.038	0.053	0.033
−0.066	0.086							−0.022	0.042
−0.011	0.066	0.018	0.051	**−0.075**	0.028	−0.017	0.055	0.041	0.023
−0.04	0.049	−0.023	0.046	−0.026	0.028	0.057	0.046	**0.063**	0.026
	1.685		0.382		**14.883**		**17.184**		10.458
	4.301		2.455		**13.005**		9.796		9.071
	746.139		−93.021		−957.527		−312.373		−810.053
	305		235		445		210		725

Source: Authors.

Note: Boldfaced figures indicate values that are significant at the 5 percent level.

Chapter 11 Annex B: Activity Patterns and Determinants

TABLE 11B.1 Activity Dynamics in the Colombian Chemicals Industry: Descriptive Evidence

	Year $t+1$				
Year t	None	Import	Export	Both	Total
None	509	70	9	3	591
Import	58	522	1	31	612
Export	7	2	45	7	61
Both	4	47	7	508	566
Total	578	641	62	549	1,830
P[export]					0.343
P[import]					0.644
P[export|import]					0.480
P[import|export]					0.903
P[export at t + k | not export at t]					0.037
P[export at t + k | not export at t, not import at t]					0.020
P[export at t + k | not export at t, import at t]					0.052
P[Ho: export at t + 1 independent of import at t]					0.003
P[import at t + k | not import at t]					0.126
P[import at t + k | not import at t, not export at t]					0.124
P[import at t + k | not import at t, export at t]					0.148
P[Ho: import at t + 1 independent of export at t]					0.590

Source: Authors.

TABLE 11B.2 Activity Dynamics in the Moroccan Chemicals Industry: Descriptive Evidence

Year t	Year $t+1$				
	None	FDI	Export	Both	Total
None	660	38	15	2	715
FDI	33	122	2	8	165
Export	24	4	49	7	84
Both	4	16	11	55	86
Total	721	180	77	72	1,050
P[export]					0.162
P[FDI]					0.239
P[export|FDI]					0.343
P[FDI|export]					0.506
P[export at t + k | not export at t]					0.031
P[export at t + k | not export at t, no FDI at t]					0.024
P[export at t + k | not export at t, FDI at t]					0.061
P[Ho: export at t + 1 independent of FDI at t]					0.013
P[FDI at t + k | no FDI at t]					0.064
P[FDI at t + k | no FDI at t, not export at t]					0.056
P[FDI at t + k | no FDI at t, export at t]					0.131
P[Ho: FDI at t + 1 independent of export at t]					0.008

Source: Authors.

TABLE 11B.3 Activity Dynamics in the Colombian Chemicals Industry: Econometric Results

Dependent variable	Regressor	All firms		Pharmaceuticals only	
		Coefficient	Standard error	Coefficient	Standard error
Export and import	Lagged both	0.604	0.034	0.778	0.079
	Lagged export	0.113	0.045	0.234	0.092
	Lagged import	0.011	0.014	0.107	0.076
	Lagged neither	0.015	0.014	0.110	0.071
	Quality	—	—	0.007	0.009
	Marginal cost	—	—	–0.043	0.037
	Initial both	0.332	1.468	0.240	0.048
	Initial export	0.029	0.661	0.276	0.088
	Initial import	0.058	1.093	0.046	0.031
	Average quality	—	—	0.002	0.009
	Average marginal cost	—	—	0.045	0.054
Export only	Lagged both	0.063	0.017	0.113	0.063
	Lagged export	0.507	0.027	0.466	0.070
	Lagged import	0.033	0.013	0.077	0.061
	Lagged neither	0.017	0.009	0.061	0.060
	Quality	—	—	–0.006	0.005
	Marginal cost	—	—	0.002	0.021
	Initial both	–0.048	0.018	–0.069	0.029
	Initial export	0.160	0.032	–0.051	0.071
	Initial import	–0.031	0.014	–0.037	0.022
	Average quality	—	—	0.005	0.005
	Average marginal cost	—	—	–0.031	0.041
Import only	Lagged both	–0.032	0.043	–0.115	0.104
	Lagged export	0.019	0.069	–0.088	0.131
	Lagged import	0.546	0.028	0.574	0.104
	Lagged neither	0.093	0.019	0.102	0.095
	Quality	—	—	0.005	0.013
	Marginal cost	—	—	0.089	0.054
	Initial both	0.072	0.044	0.170	0.065
	Initial export	0.013	0.065	0.123	0.126
	Initial import	0.287	0.031	0.311	0.047
	Average quality	—	—	–0.002	0.013
	Average marginal cost	—	—	–0.090	0.072

TABLE 11B.3 Activity Dynamics in the Colombian Chemicals Industry: Econometric Results (*Continued*)

Dependent variable	Regressor	All firms		Pharmaceuticals only	
		Coefficient	Standard error	Coefficient	Standard error
P[export at t + k | not export at t, import at t] − P[export at t + k | not export at t, not import at t]		0.012	0.016	0.013	0.032
P[import at t + k | not import at t, export at t] − P[import at t + k | not import at t, not export at t]		0.025	0.055	–0.065	0.092

Source: Authors.

Note: — = not available.

TABLE 11B.4 Activity Dynamics in the Moroccan Chemicals Industry: Econometric Results

Dependent variable	Regressors	All firms		Pharmaceuticals only	
		Coefficient	Standard error	Coefficient	Standard error
Export and FDI	Lagged both	0.427	0.039	0.594	0.077
	Lagged export	0.081	0.028	0.018	0.047
	Lagged FDI	–0.067	0.027	0.026	0.052
	Lagged neither	–0.003	0.007	0.021	0.050
	Quality	—	—	–0.010	0.005
	Marginal cost	—	—	0.081	0.041
	Initial both	0.442	2.030	0.000	0.000
	Initial export	0.007	1.815	0.002	0.018
	Initial FDI	0.147	1.697	0.058	0.020
	Average quality	—	—	0.012	0.006
	Average marginal cost	—	—	–0.107	0.066
Export only	Lagged both	0.097	0.038	0.136	0.155
	Lagged export	0.404	0.038	0.579	0.100
	Lagged FDI	0.085	0.031	0.168	0.100
	Lagged neither	0.037	0.010	0.135	0.090
	Quality	—	—	–0.005	0.013
	Marginal cost	—	—	0.060	0.092
	Initial both	–0.016	0.042	0.000	0.000
	Initial export	0.366	0.044	0.323	0.048
	Initial FDI	–0.064	0.030	–0.036	0.044
	Average quality	—	—	0.020	0.015
	Average marginal cost	—	—	–0.192	0.122
FDI only	Lagged both	–0.262	0.050	–0.395	0.139
	Lagged export	–0.073	0.040	–0.004	0.047
	Lagged FDI	0.101	0.040	–0.004	0.056
	Lagged neither	0.010	0.010	–0.030	0.044
	Quality	—	—	0.003	0.011
	Marginal cost	—	—	–0.048	0.074
	Initial both	0.296	0.053	0.000	0.000
	Initial export	0.063	0.046	–0.025	0.046
	Initial FDI	0.597	0.036	0.649	0.046
	Average quality	—	—	–0.005	0.013
	Average marginal cost	—	—	–0.090	0.071

TABLE 11B.4 Activity Dynamics in the Moroccan Chemicals Industry: Econometric Results (*Continued*)

Dependent variable	Regressors	All firms		Pharmaceuticals only	
		Coefficient	Standard error	Coefficient	Standard error
P[export at t + k|not export at t, FDI at t] – P[export at T + k | not export at t, no FDI at t]		–0.016	0.036	0.037	0.050
P[FDI at t + k|no FDI at t, export at t] – P[FDI at t + k | no FDI at t, not export at t]		0.002	0.035	0.022	0.042

Source: Authors.

Note: — = not available.

Notes

1. Unlike the other authors mentioned here, Eaton and Kortum (1996) do not attempt to empirically isolate a conduit for knowledge transfer.

2. Tybout (2000) provides a more extensive literature review.

3. Few studies in the other literatures do better. Pack (1987) is an exception. He uses detailed information on machines and workers at a sample of plants to calculate high-quality productivity indexes.

4. Milgrom and Roberts (1990) make a similar point in their study of the adoption of new technologies and organizational strategies. They do not consider international trade in goods, ownership, or information, so their list of activities differs somewhat from those that are the focus here. Nonetheless, their basic analytical point translates to our setting, mutatis mutandis. When nonconvexities and complementarities characterize the profit function, it may well be optimal to adopt bundles of new activities at once.

5. That is, the kth element of b_{jt} takes a value of one in period t if the firm is engaged in the kth possible bundle of activities during that period and otherwise takes a value of zero. In a country in which there are K possible activities, b_{jt} has 2^K elements.

6. Variation in factor prices across plants will also undermine our basis for inference. We controlled for regional variation in factor prices using regional dummies and never found significant effects.

7. It is possible to avoid this assumption, but doing so leaves the demand parameters underidentified and necessitates Bayesian estimation. Katayama, Lu, and Tybout (2003) provide details.

8. Anderson, de Palma, and Thisse (1992) show that one can think of these error components as reflecting heterogeneous tastes over unobserved product characteristics. From that perspective, Ackersburg and Rysman (2001) suggest that μ_1 and μ_2 be made functions of the number of products, since the addition of more products to a market is likely to crowd product space and effectively reduce the dispersion in tastes across products. We experimented with this generalization and found no cross-nest dependence of the error component variances on the number of products available locally.

9. As σ goes to zero, the within-group correlation of utilities goes to zero; as σ goes to unity, within-group correlation goes to unity.

10. More precisely, we calculate M_t using a constant real exchange rate to convert real dollar imports to pesos in all years. Thus we assume that the real dollar cost of imports corresponds to the volume of intermediate goods used to produce them. Our approach to measuring imports also implies that foreign producers do not adjust their dollar price in response to exchange rate fluctuations—that is, we assume complete pass-through. This implication is consistent with our expressions for equilibrium market shares and prices, which are based on the premise that the price of the outside good does not respond to adjustments in the prices of the domestically produced varieties. While the empirical literature suggests that the dollar prices of imported goods are likely to respond some to exchange rate fluctuations, our assumption does not seem too far from reality.

11. This condition presumes that the (common knowledge) performance vector $(\omega_1, \omega_{2t}, \omega_{jt}, \ldots \omega_{Nt})$ is realized at the beginning of each period and that future states of the industry do not depend on current price and quantity choices, given Ω_{t-1} and B_{t-1}. Also, mixed strategies in the product market competition are disallowed. See Ericson and Pakes (1995) for a formal discussion of the relation between product market competition and industry dynamics.

12. Lu and Tybout (2000) discuss some alternative ways to allow for noise in equation (11.2′). These alternatives complicate the estimation procedure but can be feasibly implemented. We plan to explore their properties in future work.

13. Note that we assume that μ^* is independent of b_j, although μ may not be.

14. In principle, a multinomial probit version of the decision rule could be estimated using the simulated method of moments (Geweke, Keane, and Runkle 1997). However, this approach is difficult to apply in our data sets, because we do not have any firm-specific information on variables that affect the utility of different choices (Keane 1992). Extensive experimentation with this estimator thus proved fruitless.

15. We use the following notation: X (exporter), MI (importer of intermediate goods), MK (importer of capital goods), and FDI (firm with at least 5 percent foreign ownership).

16. It is straightforward to calculate the predicted effect of these marginal cost and quality changes on profits firm by firm, but we did not do so.

17. As Grossman and Helpman (1991, p. 166) note, "Imports may embody differentiated intermediates that are not available in the local economy. The greater the quantity of such imports, the greater perhaps will be the number of insights that local researchers gain from inspecting and using these goods."

18. Alternatively, we might have pooled all firms and added industry dummies. The problem with this approach is that it adds a large number of additional parameters to be estimated. Since several of the other industries have quite small populations of firms, this rapidly exhausts degrees of freedom. A drawback of focusing on the largest industry is that the term $\bar{\omega}_{t-1}^{-j}$, summarizing competitors' performance, exhibits very little cross-firm variation. We therefore drop it from the estimation. Even for this limited specification, the number of parameters to be estimated is quite large: $Kx(2^K - 1)$ elements in the vector of coefficients on lagged activities (δ, $2 * K * (2^K - 1)$ elements in the vector of coefficients on lagged performance, α, and the same number again for the initial conditions correction. For this reason we are unable to estimate the model for Mexico, where the increase from $K = 2$ to $K = 3$ activities would result in 144 parameters to be estimated, with at most 580 observations in the plastics industry.

References

Ackersburg, Daniel, and Mark Rysman. 2001. "Unobserved Product Differentiation in Discrete Choice Models: Estimating Price Elasticities and Welfare Effects." Department of Economics, Boston University.

Aitken, Brian, and Ann Harrison. 1999. "Do Domestic Firms Benefit from Foreign Direct Investment? Evidence from Panel Data." *American Economic Review* 89 (3): 605–18.

Anderson, Simon, Andre de Palma, and Jacques-Francois Thisse. 1992. *Discrete Choice Theory of Product Differentiation*. Cambridge, MA: MIT Press.

Aw, Bee-Yan, and Amy Hwang. 1995. "Productivity and the Export Market: A Firm-Level Analysis." *Journal of Development Economics* 47 (3): 313–32.

Berry, Steven. 1994. "Estimating Discrete-Choice Models of Product Differentiation." *Rand Journal* 25 (2): 242–62.

Berry, Steven, James Levinsohn, and Ariel Pakes. 1995. "Automobile Prices in Market Equilibrium." *Econometrica* 63 (4): 841–90.

Bigsten, Arne, Paul Collier, Stefan Dercon, Bernard Gauthier, Jan Gunning, Anders Isaksson, Abena Oduro, Remco Ooostendorp, Cathy Patillo, Mans Soderbom, Michel Sylvain, Francis Teal, and Albert Zeufack. 1999. "Exports and Firm-Level Efficiency in the African Manufacturing Sector." School of Business, University of Montreal.

Blomström, Magnus, Robert Lipsey, and Mario Zejan. 1994. "What Explains the Growth of Developing Countries?" In *Convergence of Productivity: Cross-National Studies and Historical Evidence*, ed. William J. Baumol, Richard Nelson, and Edward Wolff, 243–59. New York: Oxford University Press.

Chen, Tain-jy, and De-piao Tang. 1987. "Comparing Technical Efficiency between Import-Substituting and Export-Oriented Foreign Firms in a Developing Country." *Journal of Development Economics* 36 (3): 277–89.

Clerides, Sofronis, Saul Lach, and James Tybout. 1998. "Is Learning by Exporting Important? Micro-Dynamic Evidence from Colombia, Mexico and Morocco." *Quarterly Journal of Economics* 115 (3): 903–47.

Coe, David, and Elhanan Helpman. 1995. "International R&D Spillovers." *European Economic Review* 39 (4): 859–87.

de Long, J. Bradford, and Lawrence Summers. 1991. "Equipment Investment and Long Run Growth." *Quarterly Journal of Economics* 106 (2): 445–502.

Eaton, Jonathan, and Samuel Kortum. 1996. "Trade in Ideas: Patenting and Productivity in the OECD." *Journal of International Economics* 40 (2): 251–78.

Ericson, Richard, and Ariel Pakes. 1995. "Markov-Perfect Industry Dynamics: A Framework for Empirical Work." *Review of Economic Studies* 62 (1): 53–82.

Ethier, Wilfred. 1982. "National and International Returns to Scale in the Modern Theory of International Trade." *American Economic Review* 72 (3): 389–405.

Geweke, John, Michael Keane, and David Runkle. 1997. "Statistical Inference in the Multinomial Multi-Period Probit Model." *Journal of Econometrics* 80 (1): 125–65.

Grossman, Gene, and Elhanan Helpman. 1991. *Innovation and Growth in the Global Economy.* Cambridge, MA: MIT Press.

Haddad, Mona, and Ann Harrison. 1993. "Are There Positive Spillovers from Direct Foreign Investment? Evidence from Panel Data for Morocco." *Journal of Development Economics* 42 (1): 51–74.

Hobday, Michael. 1995. *Innovation in East Asia: The Challenge to Japan.* Cheltenham, UK: Edward Elgar.

Katayama, Hajime, Shihua Lu, and James Tybout. 2003. "Why Plant-Level Productivity Studies Are Often Misleading and an Alternative Approach to Inference." Working Paper 9617, National Bureau of Economic Research, Cambridge, MA.

Katz, Jorge. 1987. *Technology Generation in Latin American Manufacturing Industries.* London: MacMillan.

Keane, Michael. 1992. "A Note on Identification of the Multinomial Probit Model." *Journal of Business and Economic Statistics* 10 (2): 193–200.

Keller, Wolfgang. 1998. "Are International R&D Spillovers Trade-Related? Analyzing Spillovers among Randomly Matched Trade Partners." *European Economic Review* 42 (8): 1469–81.

———. 2000. "Do Trade Patterns and Technology Flows Affect Productivity Growth?" *World Bank Economic Review* 214 (1): 17–47.

Kim, L. 1977. *Imitation to Innovation: The Dynamics of Korea's Technological Learning.* Boston: Harvard Business School Press.

Lall, Sanjaya. 1987. *Learning to Industrialize.* London: MacMillan.

Lu, Shihua, and James Tybout. 2000. "Inferring Plant-Level Performance Measures from Revenues and Costs." Department of Economics, Pennsylvania State University. University Park, Pennsylvania.

McFadden, Daniel. 1974. "Conditional Logit Analysis of Qualitative Choice Behavior." In *Frontiers in Econometrics*, ed. Paul Zarembka. New York: Academic Press.

Milgrom, Paul, and John Roberts. 1990. "The Economics of Modern Manufacturing: Technology, Strategy and Organization." *American Economic Review* 80 (3): 511–28.

Pack, Howard. 1987. *Productivity, Technology and Industrial Development.* New York: Oxford University Press.

Rhee, Yung Whee, Bruce Ross-Larson, and Garry Pursell. 1984. *Korea's Competitive Edge: Managing Entry into World Markets.* Baltimore: Johns Hopkins University Press.

Roberts, Mark, and James Tybout. 1997. "The Decision to Export in Colombia: An Empirical Model of Entry with Sunk Costs." *American Economic Review* 87 (3): 545–64.

Tybout, James. 2000. "Manufacturing Firms in Developing Countries: How Well Do They Do, and Why?" *Journal of Economic Literature* 38 (1): 11–44.

Tybout, James, and M. Daniel Westbrook. 1995. "Trade Liberalization and Dimensions of Efficiency Change in Mexican Manufacturing Industries." *Journal of International Economics* 39 (1): 53–78.

Westphal, Larry. 2001. "Technology Strategies for Economic Development in a Fast Changing Global Economy." *Economics of Innovation and New Technology* 11 (4–5): 275–320.

Wooldridge, Jeffrey. 2000. "The Initial Conditions Problem in Dynamic, Nonlinear Panel Data Models with Unobserved Heterogeneity." Working Paper, Department of Economics, Michigan State University, East Lansing.

12

MARKET DISCIPLINE AND CORPORATE EFFICIENCY: EVIDENCE FROM BULGARIA

Simeon Djankov and Bernard Hoekman

The conventional wisdom in the international trade literature is that greater import competition forces domestic firms to improve their efficiency. In contrast, many empirical studies suggest that import competition has no or only a weak effect on corporate performance. These findings may come about because many tests of the impact of import competition on firm performance have tended to focus on all firms producing tradable products. No distinction is generally made between import-competing firms and enterprises that export a substantial proportion of their output. The latter should be much less sensitive to a reduction in import barriers, insofar as the revealed ability to compete on export markets indicates that these firms should already be efficient. While imports are often a key input for such firms, governments in most countries have implemented schemes aimed at providing exporters with inputs at world prices. A reduction in overall levels of import barriers will therefore also imply less of a change in incentives than is the case for import-competing firms.

While not distinguishing between exporters and nonexporters may be a factor underlying the weak relationship that has been found between trade liberalization and subsequent improvements in corporate performance, other factors may lead to an overstatement of the impact of trade liberalization. Frequently, governments pursuing structural economic reforms may take actions that go beyond trade liberalization. Examples are tighter fiscal discipline, privatization, and the introduction of antitrust regimes, all of which should result in greater market discipline

This article originally appeared in the *Canadian Journal of Economics* 2000, 33(1): 190–212.

for enterprises. Conversely, if a government does not pursue such complementary actions when undertaking trade liberalization, the effects of greater competition from imports may be attenuated. For example, if soft financing at the enterprise level is available or competition law enforcement is lax, import competition may do little to influence firms' performance. Indeed, a perverse relation may exist if firms obtain increased subsidies or soft loans *because* import competition has caused a decline in their productivity.

In this chapter we investigate the relationship between firm productivity and increased competition in Bulgaria during the 1991–95 period, focusing on a number of major changes in policies—in particular, opening to international trade, corporatization and demonopolization of state-owned industry, and changes in the hardness of budget constraints confronting firms. Enterprise-level data are used to directly estimate changes in total factor productivity (TFP) growth, which we use as the measure of enterprise adjustment to the more competitive environment. We find that changes in import competition, domestic market structure (industry concentration), and firm-specific budget constraints (as measured by changes in arrears) all have the predicted impact on corporate performance. Increased import pressure is positively associated with subsequent productivity growth of Bulgarian firms. However, this finding is robust only once the sample is split into firms that are primarily exporters and those that are not. What matters for the exporters is competition on the global market. The results suggest that empirical work analyzing the impact of trade liberalization should be based on firm-level data that allow exporters and nonexporters to be distinguished. The analysis also illustrates the importance of accounting for changes in other policies that may enhance or attenuate the effect of greater competition from imports. In our sample, increases in domestic concentration and arrears are associated negatively with TFP growth.

The Literature on Competition and Corporate Efficiency

There is a substantial theoretical literature that studies the relationship between competition and corporate efficiency. The general hypothesis is that increased competition stimulates improvements in productivity. Two lines of argument have been developed in support of this hypothesis. The first is derived from the literature on X-inefficiency (internal to the firm); the second centers on industry rationalization. The X-inefficiency literature assumes that managerial effort is undersupplied in the absence of vigorous competition. In a theoretical investigation of the implications of the X-inefficiency hypothesis, Horn, Lang, and Lundgren (1995) show how greater competition may induce an expansion of output by incumbent firms through improved internal technical efficiency without any reallocation of resources across firms. Earlier studies (Holstrom 1982; Nalebuff and Stiglitz 1983) argue that incentive schemes for managers will generate better results the greater the number of players (firms) involved. This arises because of greater opportunities for comparison of performance. Hart (1983) builds an explicit model to show the link between increased competition and improved manager performance.

A second line of argument is that increased competition may lead to a rationalization of oligopolistic industries as firms are forced to compete for market share (Schmidt 1994). Resource reallocation occurs across firms within and between sectors. Although the shake-out process may result in a transitional decline in measured efficiency insofar as firms with increasing returns to scale lose domestic market share, over time this may be offset by greater output as the size of the market expands due to exit and access to export markets. Much depends on the existence of scale economies and the ease of entry and exit. Since competition raises the probability of bankruptcy and hence job losses, it also generates stronger incentives for workers to improve productivity and higher labor turnover across firms within sectors (Dickens and Katz 1987).

Both strands of the theoretical literature lead to the same prediction: greater competition leads, with some lag, to productivity improvements in imperfectly competitive industries. Few empirical investigations using firm-level data have established a strong link between greater competition and subsequent improvements in corporate productivity. Most studies look only at the partial correlation of a specific measure of increased competition; to the best of our knowledge there are no studies on developing countries or transition economies that attempt to account for different types of government policies that may affect competitive pressures at a given point in time. The four policy changes that are most frequently analyzed in the literature are trade liberalization (often proxied by changes in import penetration), hard budget disciplines, changes in market structure through entry and exit, and privatization.

Import Competition

Trade liberalization programs in developing countries in the 1980s and in transition economies in the early 1990s have generated firm- or industry-level data that allow tests of the imports-as-market-discipline hypothesis. Most studies exploring the effect of increased import competition on enterprise behavior estimate the impact on relative changes in TFP across sectors and have come to ambiguous conclusions. Harrison (1994) finds that the reduction in tariffs and the subsequent increase in import penetration in Côte d'Ivoire following the 1982 trade liberalization had a positive, although not statistically significant, effect on TFP growth. Haddad, De Melo, and Horton (1996) find no effect of increased import penetration on TFP growth in Moroccan industries following the 1986 liberalization. In contrast to these studies, van Wijnbergen and Venables (1993) find a strong positive effect of increased import penetration on labor productivity in a large sample of Mexican manufacturing firms. Several studies on transition economies also find a positive association between import competition and changes in productivity. Earle and Estrin (2001) conclude that there is a strong positive effect of increased import competition on labor productivity of Russian firms, and Falk, Raiser, and Brauer (1995) find a positive association between import penetration and industry performance in a large sample of Polish firms.

A second group of studies uses the dynamics of entry and exit across sectors as a measure of the effect of imports as market discipline. Roberts (1996) finds some evidence for a positive effect of import competition on increased turnover in Colombia, but only in highly concentrated industries. Using a sample of Chilean firms, Tybout (1996) finds only a weak association between import penetration and entry-exit patterns.

A third group of studies focuses on changes in industry mark-ups (price-cost margins) in countries that have undergone recent trade liberalization. A reduction in mark-ups is interpreted as a sign that increased competitive pressure has induced managers to improve efficiency. Levinsohn (1993) and Grether (1996) find some support for the imports-as-market-discipline hypothesis in studies of Turkey and Mexico, respectively.

Financial Discipline

A second strand in the empirical literature suggests that the imposition of hard budget constraints (elimination of direct subsidies, prudent bank lending, refusal by the government to write off bad debt) is necessary to ensure productivity growth. Prudent bank lending has been shown to lead to improved allocation of resources in the real sector and hence to affect economic growth positively (King and Levine 1993). Without hard budget constraints, it may be rational for managers to spend more time lobbying the government for subsidies and related types of support than undertaking painful restructuring measures. The empirical evidence suggests that "enterprises subjected to financial discipline show more aggressive collection of receivables, a closer link between profitability and investment, and a reorientation of goals from output targets to profits" (World Bank 1996, p. 45). Soft budget constraints are not necessarily introduced through direct government subsidies. Schaffer (1998) finds, for a large sample of Hungarian, Polish, and Russian firms, that governments use tax arrears to keep loss-making enterprises afloat.

Changes in Competition Policies and Market Structure

A third group of studies focus on competition in the domestic product market as a determinant of corporate performance. Two studies of British manufacturing firms (Nickell, Wadhwani, and Wall 1992 and Nickell 1996) use a panel framework to show that market concentration has an adverse effect on the level of TFP. By the nature of their estimation, this also implies that an increase in market concentration is followed by a fall in productivity. In the transition context, Lizal, Singer, and Svejnar (1995) find that the deconcentration of industry that followed the collapse of central planning in the Czech Republic did not result in improved labor productivity. Controlling for changes in imports, Earle and Estrin (2001) find no effect of changes in market concentration on the performance of Russian firms, as measured in labor productivity growth, total layoffs, or the introduction of new product lines. These results are somewhat surprising and may be due to the absence of panel data sets and the lack of attention given to controlling for the possibility that firms differ in terms of the hardness of their budget constraints.

Privatization

A great deal of attention has been given to determining the effects of privatization of firm performance (see Shleifer and Vishny 1994 for an overview). Clearly, this may have significant implications for managerial incentives to improve efficiency, and any empirical evaluation of trade liberalization should take into account the effect of privatization programs if they are pursued by a government. However, in the case of Bulgaria, privatization was not pursued with any vigor. Consequently, we make no attempt to survey the large and rapidly expanding literature.

Economic Reforms in Bulgaria

Until the late 1980s, Bulgaria was closely connected to the Soviet Union through the Council of Mutual Economic Assistance (CMEA) system. There was only a very small private sector, unlike Poland or Hungary, where private firms were allowed to operate in agriculture and services. State enterprises, organized in large vertically integrated conglomerates, produced according to quotas set by the Central Planning Committee. Their activities were financed by the State Bank, the only financial intermediary in the country. About one-third of industrial output was exported in 1988, of which some 80 percent went to CMEA markets (World Bank 1991). Similarly, Bulgaria relied to a great extent on the CMEA for its imports. Following the collapse of the CMEA in 1990, Bulgaria lost its major export markets, was forced to adjust to a much higher oil import bill as Soviet oil subsidies were eliminated and Iraqi supplies dried up following the Persian Gulf War, and was confronted with the repercussions of the civil war in Yugoslavia. The latter implied not just the loss of another export market but also higher transport costs to Europe. The magnitude of the economic shocks in the early 1990s therefore went beyond the demise of central planning.

As a result of these various developments, the economy went into a sharp recession. In 1991 industrial production fell by 22 percent, GDP declined by over 10 percent, and

TABLE 12.1 Performance Statistics, 1991–96

Panel A: Macroeconomic Indicators						
Indicator	1991	1992	1993	1994	1995	1996
Real economic growth (percent)	–11.7	–5.6	–4.2	1.4	2.6	–4.2
Industrial production (percent change)	–22.4	–12.2	–6.4	2.2	8.6	–7.8
Exports (percent change from previous year)	–5.6	11.8	23.4	37.1	29.3	15.3
Imports (percent change from previous year)	–18.6	53.2	8.2	9.1	18.3	8.3
CPI change (percent change from previous year)	338.5	91.3	75.0	52.1	32.9	298.1
Unemployment (percent of labor force)	11.1	16.4	17.0	13.3	10.6	11.7
Bulgarian lev/US$ exchange rate	21.8	24.5	32.7	66.1	70.7	782.3
Panel B: Stock of Tax Arrears (percent of GDP, end-period)						
Country	1991	1992	1993	1994	1995	1996
Bulgaria	6.7	10.3	10.9	12.2	11.1	13.4
Moldova	—	—	—	9.6	8.2	10.4
Poland	4.1	4.4	5.3	4.5	3.6	3.9
Romania	1.8	2.3	1.5	4.2	4.6	5.1
Russian Federation	—	—	1.7	2.5	3.5	10.0

Source: EBRD (1997).

Note: — = not available.

unemployment rose rapidly (table 12.1, panel A). A "big bang" stabilization and structural adjustment program was implemented during the course of 1991. Under it most subsidies were slashed. By mid-1992 most retail prices were liberalized, leading to a large jump in inflation, reflecting the existing monetary overhang. By 1994 output began to expand again, led by strong export growth, and unemployment rates started to decline from their peak of 17 percent in 1993. The majority of manufacturing industries became confronted with stronger market discipline. Aggregate subsidies (budget transfers) to the industrial sector declined from 14.9 percent of GDP in 1990 to 1.1 percent of GDP in 1995 (EBRD 1997). Although most remaining direct subsidies from the budget were concentrated in nontradable goods or services industries (utilities, mining, construction), many firms continued to be financed through soft loans from state-owned banks and build-up of arrears (on taxes, wages, and creditors). Bulgaria was among the worst performing European transition economies in this regard (table 12.1, panel B).

Competition from imports increased significantly. In 1992 imports jumped 50 percent, following the abolition of trade monopolies, licensing, exchange control, and quantitative restrictions. Imports continued to expand steadily after 1992. A unified floating exchange rate regime and a relatively flat five-band tariff structure were adopted, with a maximum rate of 40 percent and an average collected rate (tariff revenues as a share of imports) of some 7 percent in 1994–95 (IMF 1997).[1]

Very little privatization occurred up to 1996; the manufacturing sector continued to consist almost exclusively of state-owned enterprises: as of end-1995 only 7 percent of manufacturing output was in private hands (Bulgarian Privatization Agency 1997). However, a 1992 Law on Demonopolization led to a significant reduction in concentration and vertical integration in many sectors, as enterprises were transformed into joint stock companies and often broken up into constituent parts. Some new domestic entry also occurred in manufacturing sectors—in the 1991–95 period about 12 percent of manufacturing (mostly food and clothing) was produced in new private firms (National Statistical Office of Bulgaria 1997). As a result, concentration in most industries declined (table 12.2).

The transformation of state-owned enterprises into joint stock companies was accompanied by the creation of boards of directors. Membership of the boards was drawn from line ministries and other government bodies. Although in principle the boards could lead to a strengthening of incentives for management to improve enterprise performance, it could also make it easier to engage in profit shifting and asset stripping by weakening the oversight role of the state (Schleifer and Vishny 1994). In the short run, the effect of breaking up enterprises and the resulting elimination of in-house upstream and downstream linkages could also lead to a reduction in efficiency, as these are replaced with arm's length relationships only with a lag and involve incurring a variety of transactions and search costs (Svejnar 1995). In the event, the absence of privatization in the manufacturing and banking sectors created an environment of lax financial discipline for many enterprises. Although the government reduced explicit budgetary subsidies significantly, it often brokered deals between banks and enterprises (both state owned) to ensure that production continued even if the enterprise generated negative value-added (Claessens and Peters 1997). There was also a partial write-off of bank debts in 1993 for a number of highly indebted enterprises (an earlier write-off in 1991 is not captured in the data). These write-offs provided a perverse signal to all firms. Matters were compounded by a reduction in the quality of the "regulatory oversight" to which state-owned

TABLE 12.2 Descriptive Statistics (all variables except the 1990 levels in log differences)

Sector	Tariff (percent) 1990	TFP growth				Import penetration				
		1991–92	1992–93	1993–94	1994–95	1990	1990–91	1991–92	1992–93	1993–94
Food products	26.3	–0.011	0.016	0.074	0.017	0.45	0.007	0.140	–0.114	–0.168
Tobacco products	37.2	0.064	–0.028	–0.136	–0.075	0.53	0.041	0.007	–0.065	–0.114
Textile mill products	9.6	–0.019	0.072	0.151	0.175	0.42	0.058	0.198	0.078	0.078
Apparel	13.2	0.131	0.015	0.146	0.187	0.68	–0.227	0.078	0.096	0.063
Wood products	5.6	0.142	0.011	–0.094	–0.023	0.24	0.013	0.163	0.054	–0.206
Furniture and fixtures	9.6	–0.024	0.119	0.058	0.018	0.35	0.125	0.012	–0.153	–0.307
Paper	16.9	0.214	0.185	0.209	0.242	0.51	–0.026	0.120	0.102	0.107
Printing and publishing	7.3	–0.032	0.116	0.006	0.076	0.58	–0.144	0.317	–0.143	–0.192
Chemicals	6.3	–0.016	0.043	0.134	0.123	0.42	–0.061	0.080	0.131	0.151
Petroleum refining	7.3	–0.031	0.008	0.170	0.277	0.27	0.185	0.086	0.081	0.083
Rubber	8.4	–0.108	0.137	0.082	0.164	0.24	0.166	0.302	0.052	0.094
Leather	7.1	0.054	0.098	0.092	0.095	0.51	0.139	–0.330	0.035	0.064
Stone, clay, and glass	12.6	0.137	–0.078	0.150	0.125	0.22	0.108	0.191	0.029	0.063
Primary metals	6.2	–0.120	0.040	0.005	0.092	0.33	0.327	0.028	0.048	–0.004
Fabricated metals	10.9	–0.034	0.215	–0.041	0.134	0.26	0.057	0.026	–0.063	0.012
Nonelectrical machinery	8.9	0.162	0.186	0.044	0.083	0.42	0.255	0.143	0.091	–0.032
Electrical machinery	12.8	0.044	0.082	0.016	0.063	0.46	0.056	0.109	0.040	–0.014
Transportation equipment	14.4	–0.051	0.040	0.088	0.115	0.67	–0.010	0.000	0.022	–0.049
Miscellaneous manufacturing	7.8	0.025	–0.018	–0.115	0.089	0.72	0.087	0.022	–0.184	–0.050

(*Continued*)

Sector	Arrears					Domestic concentration				
	1990	1990–91	1991–92	1992–93	1993–94	1990	1990–91	1991–92	1992–93	1993–94
Food products	0.22	–0.005	0.037	0.015	0.075	0.009	–0.118	0.000	–0.134	0.251
Tobacco products	0.06	–0.016	0.006	0.039	0.017	0.048	–0.021	0.000	0.236	0.045
Textile mill products	0.36	0.010	0.064	–0.164	0.079	0.042	–0.074	–0.053	–0.114	–0.318
Apparel	0.38	–0.002	–0.002	0.046	0.118	0.029	–0.071	–0.038	0.000	–0.167
Wood products	0.19	–0.009	0.005	0.032	0.022	0.066	–0.047	0.016	–0.065	0.394
Furniture and fixtures	0.18	0.003	0.066	0.083	0.038	0.048	–0.087	0.022	–0.169	–0.419
Paper	0.32	0.003	–0.048	0.042	0.021	0.129	–0.161	–0.088	–0.063	–0.187
Printing and publishing	0.76	0.002	0.133	0.056	–0.120	0.216	–0.028	–0.034	–0.025	–0.164
Chemicals	0.48	0.004	0.059	–0.244	0.022	0.199	0.114	0.040	0.087	0.080
Petroleum refining	0.55	0.005	0.001	–0.143	0.064	0.134	0.102	–0.118	–0.204	–0.284
Rubber	0.12	0.002	–0.005	0.002	–0.003	0.426	0.051	–0.096	–0.031	–0.030
Leather	0.09	–0.004	0.009	0.007	0.016	0.115	–0.120	–0.082	–0.066	–0.160
Stone, clay, and glass	0.85	0.002	0.216	0.011	0.058	0.157	–0.107	–0.074	–0.122	–0.221
Primary metals	0.79	0.006	0.133	–0.241	0.133	0.209	0.014	0.051	–0.023	0.045
Fabricated metals	0.68	0.002	–0.093	0.107	0.052	0.268	0.029	0.018	0.028	0.021
Nonelectrical machinery	0.39	0.004	0.052	–0.113	–0.114	0.087	–0.071	–0.104	–0.028	–0.255
Electrical machinery	0.58	–0.008	0.099	–0.182	0.057	0.035	–0.084	–0.111	–0.082	–0.323
Transportation equipment	0.41	0.005	0.095	–0.107	–0.090	0.127	0.054	0.058	0.094	0.068
Miscellaneous manufacturing	0.36	0.002	0.040	–0.130	0.107	0.123	–0.197	–0.184	–0.154	–0.288

Source: Authors.

Note: See text for definition of variables.

enterprises were subject, as efforts were made by "agents" (managers) and "principals" (members of the various ministries and government bodies that were chosen for the boards of directors) to retain or obtain control over good assets.

In 1996 a crisis emerged: real GDP fell by more than 4 percent, industrial output dropped by almost 8 percent, and inflation jumped to nearly 300 percent (table 12.1). This resulted in the fall of the Socialist government and led to subsequent attempts to sever the link between banks and enterprises. The effects of the 1996 crisis are not reflected in this study, as the data series ends in the fourth quarter of 1995. One can, however, observe a substantial build-up of arrears starting in 1992, which undoubtedly contributed to the events in 1996.

The Empirical Framework

The literature on TFP measurement relies either on Tornquist approximations of Divisia indexes or on production (or cost) function estimates. The latter approach is particularly useful when the underlying assumptions of the former (constant returns to scale, perfect competition, and profit maximization) are too demanding. Since we use data for the initial postreform period in Bulgaria, there are many reasons to suspect that Divisia indexes may be inappropriate for measuring productivity changes. Production function estimations are therefore used instead.

Following the derivation in Basu and Fernald (2002), assume each firm i in sector s has a production function for gross output:

$$Y_i = F^i \, (K_i, L_i, M_i, I_s, C_s, A_i, T_i), \tag{12.1}$$

where Y is gross output; K, L, and M are firm-specific inputs of capital, labor, and materials; and T indexes technology. As the primary focus of this chapter is to test the association between productivity growth at the enterprise level and changes in import competition (I_s), while controlling for changes in domestic market competition (C_s) and financial discipline (A_i), we augment the production function by including these variables as additional factors of production, with I and C being sector specific and A firm specific. We assume these variables have an impact on the productivity performance of firms by affecting the incentives for managers to address X-inefficiency.

The construction of each variable is described in the next section. The firm's production function F is homogeneous of degree g ($g \neq 1$) in K, L, M, I, C, and A. Firms are assumed to be price takers on factor markets but may have market power in output markets. The former assumption is reasonable, since wages were set centrally in the state sector during the sample period and most materials were bought abroad at world market prices. The value of the marginal product is then proportional to the price of the input, P_{Ji}:

$$P_i F_J^i = \mu_i P_{Ji}, \tag{12.2}$$

where F_J^i is the marginal product of input J for firm i. Firms may charge a markup μ_i over marginal cost, $\mu_i = P_i/MC_i$, where MC_i is marginal cost. Equations (12.1) and (12.2) imply that

$$dy_i = \mu_i[s_{Li}dl_i + s_{Ki}dk_i + s_{Mi}dm_i] + \gamma_1 di_s + \gamma_2 dc_s + \gamma_3 da_i + \frac{F_T^i T_i}{F^i} dt_i, \qquad (12.3)$$

where dy_i is output growth, changes in inputs in the first term are weighted by revenue shares (s_{Ji} is the input j's share in nominal gross expenditure), and $(F_T^i T_i / F^i)dt_i$ equals the technology change or TFP growth. Equation (12.3) implies that mark-ups for each enterprise may differ. Given the short time-series of the data available to us, we assume that mark-ups across all firms within a sector are the same. We do allow mark-ups to vary over the sample years, however, consistent with the methodology in Levinsohn (1993) and Harrison (1994).

We rewrite equation (12.3) as:

$$\Delta y_{i,t} = \alpha_i + \beta_{Lst}\Delta l_{i,t} + \beta_{Mst}\Delta m_{i,t} + \beta_{Kst}\Delta k_{i,t} + \gamma_1 \Delta I_{s,t-1} + \gamma_2 \Delta C_{s,t-1} + \gamma_3 \Delta A_{i,t-1} + \gamma_{4j} \sum Year_j + \varepsilon_{i,t}\ , \qquad (12.4)$$

where $\Delta y_{i,t} = lnY_{i,t} - lnY_{i,t-1}$ and similarly for $l_{i,t}$, $m_{i,t}$, $k_{i,t}$, $i_{s,\,t-1}$, $c_{s,\,t-1}$, and $a_{i,t-1}$ (all lower-case symbols represent logs). All β_t's are estimated over each industry (sector) s for each period, that is, a separate production function is fitted to each sector, using data for all enterprises in the sector. The heterogeneity of enterprise performance is captured in a firm-specific term (α_i).

Data for output, inputs, and arrears are at the level of the firm. The two sector-specific explanatory variables (import competition and concentration) are affected by government policies and are exogenous for individual firms.[2] I_s depends on demand factors and trade policies; C_s is driven by entry and exit decisions of other firms as well as by actions taken by the authorities to break up monopolies and corporatize state-owned firms.[3] A_i is largely a function of the willingness of creditors to tolerate a buildup of debt, but it can be affected by the political influence of the firm. Both the sector-specific explanatory variables and the firm-specific variable (arrears) are lagged one year to reduce possible endogeneity.

In addition, the effects of other changes in the economic environment have to be controlled for. We do not have good proxies for these changes, nor can we account individually for each of them. Instead, we include annual dummies in the estimating equation that pick up the net effects of changes in the economic environment at the aggregate level.

Because of the probable correlation between productivity effects and the independent variables, ordinary least squares (OLS) may give biased and inconsistent estimates. This simultaneity problem is endemic to the empirical literature on productivity measurement. One solution would be to use firm-specific instruments that cause changes in input demands but are uncorrelated with the productivity shock.

As Levinsohn (1993) and Feenstra (1996) note, however, suitable instruments are not readily available. A second approach, developed in Olley and Pakes (1996), uses semi-parametric estimation techniques to address the same issue. Their technique requires profit maximization and is therefore unlikely to be appropriate for Bulgarian data in the initial transition period. We address the issue by using *F*-tests to reveal whether OLS is appropriate and relying on the Hausman specification test to choose between random or fixed effects frameworks in cases where OLS should not be used. As discussed below, these tests suggest that a random effects model is most appropriate.[4]

As mentioned previously, our premise is that import competition should primarily affect firms that produce for the local market. In the empirical analysis, the sample of firms therefore is split into exporters and nonexporters. The basic idea underlying this is simple. If a firm sells its production abroad, its productivity growth should not be affected by changes in domestic market concentration and import penetration. In contrast, the association between import competition and subsequent productivity growth should be strong for firms that sell mostly domestically. Thus we expect to find a positive association between increases in import competition and subsequent improvements in productivity ($\gamma_1 > 0$). Conversely, we expect an inverse relationship between domestic concentration (defined to exclude import competition) and increases in enterprise arrears (our proxy for lax financial discipline) and productivity growth ($\gamma_2, \gamma_3 < 0$). We do not have theoretical predictions on the signs of the estimated coefficients on the annual dummy variables, as those pick up the net effect of a multitude of other policies that affect the incentives created for enterprise performance.

The Data

The data used are from the Bulgarian industry census and are reported on a quarterly basis to the National Statistical Office (NSO). The data set covers 1,337 medium-size and large manufacturing firms from the fourth quarter of 1991 through the fourth quarter of 1995 and includes detailed information on firm revenues, expenditures, profits, employment, export share in revenues, capital stock, and inventories. All firms are identified by their name and a six-digit sector code, following the 1987 Bulgarian Industrial Classification. We use sector-specific price information (as reported on a quarterly basis by the NSO) on inputs and output to convert the material input expenditures, sales revenues, and inventories of each firm into quantities. Fourth quarter 1995 relative prices are used as a base to avoid possible distortions arising from the adjustment in relative prices that occurred in the sample period. We use quarterly price deflators to minimize the distortionary effects of inflation on our derived measure of output. Once quarterly values of production and costs are deflated, the four quarterly observations are added together to construct annual measures of inputs and output. This is important, as each sector may have a different product cycle during the year.[5]

The firm-level data include information on the share of exports in total revenues. We define as exporters all firms that sold more than three-quarters of their production abroad during the first year of the sample 1991. (An alternative definition, whereby exporters are defined as firms that sell more than half of their production abroad, is also used below to assess the sensitivity of results to the definition of exporter status). We choose this definition instead of using the share of output exported during the sample period to deal with the problem that export status in a given year is endogenous to policy. Previous studies have shown that of the firms that are able to make the transition to becoming an exporter after a trade reform, many experienced an above-average contraction of their domestic market (Roberts and Tybout 1997). By defining exporters as enterprises that were exporting at the beginning of the sample period, we remove this endogeneity problem.

In the sample period, Bulgaria did not use generally accepted accounting practices. Instead, the old Soviet-style accounting system was used. The main difference was the focus on production rather than sales as the measure of performance, which implied that production for inventory (unsold production) was counted as sales. In many cases this was a considerable portion of overall firm output, as managers stored finished goods to use them as a bargaining chip with suppliers or customers at a later stage. Another item added to sales revenue was the direct subsidies received from the government. Shaffer (1998) argues that this accounting practice is defensible in the context of transition economies, where output prices may be held at artificially low levels. In such cases, the subsidy acts as a compensatory transfer from the budget. Since price fixing was abolished in Bulgarian manufacturing sectors in 1991, this reasoning is not applicable to our data. To adjust for these accounting peculiarities, firms' revenues were recalculated by subtracting inventory increases and subsidies from the sales.[6]

Another problem with the accounting data is the treatment of depreciation costs. The 1988 Law on Amortization stipulated depreciation schedules for machinery, buildings, and transport vehicles comparable to those of Western European countries. In reality, a large portion of firms' capital stock was old and should have been fully depreciated. Managers, however, were allowed to charge any depreciation rate in such cases. Most often the practice was to charge the rate of the last year from the schedule, but this was by no means universally applied. The prospect of privatization of some firms further obscured true depreciation costs. Managers of such enterprises, if and when they entertained the possibility of buying into the company or assisting other buyers in return for compensation or job security, had an incentive to underreport capital or to depreciate it faster, so as to influence downward the price at which they might subsequently be able to purchase the firm. Given that the data on capital stocks and depreciation are unlikely to provide a good measure of the flow of capital services, energy consumption is used as a proxy for capital utilization. This correction is attractive in the transition context, because it is a flow measure that does not depend on accounting measures of fixed assets (Burnside, Eichenbaum, and Rebelo 1995), but it has the drawback that the substitution of additional capital for reduced energy consumption is obscured.

As discussed earlier, a large-scale informal economy emerged in Bulgaria during the sample period. One effect of the opportunities for shifting profits from public enterprises to the private sector is that reported profits of state-owned enterprises are likely to be biased downward (for example, because of acceptance of buildup of arrears), while costs are likely to be biased upward (due to, for example, transfer pricing). To the extent that reported costs exceed real costs, and derived profits fall short of real profits, the estimated coefficients would be biased downward if estimated in levels. However, since the incentive structure for managers did not change significantly during 1991–96, except as a result of common economywide developments (which are controlled for), this problem should not affect *changes* in variables. This is the main reason for our choice of empirical tests—all regressions are run in (log)-differences rather than levels of both the left- and right-hand side variables. This should alleviate the biases in the reported data.

The measure of (the changes in) domestic market structure (C_S) is derived from registry data on all manufacturing enterprises in Bulgaria during the 1991–95 period. These data were also obtained from the NSO and cover 13,568 entities. This is an unbalanced panel, that is, the data record entry and exit of firms and cover enterprises both in and outside our sample. They thus fully account for the changes in domestic market competition arising either from the decline in incumbent firms' production, the entry of new firms, or the rationalization (breakup) of conglomerates.[7] We use the standard Herfindahl index (the sum of squared market shares of all firms in a given industry) as a measure of concentration.

Arrears, A_i, our proxy for the laxness of financial discipline at the firm level, are defined as the ratio of overdue payments to total debt in a given year. This definition implicitly corrects for firm size, as both the numerator and denominator vary with size. Data on arrears are not consistent with international accounting standards: overdue payments for 30 days or more are counted as arrears. This is a relatively short period when compared with the practice in the United States (120 days) or the European Union (90 days). This creates further problems once penalties are incurred, since they are treated as increases in the amount of outstanding arrears. As mentioned previously, there was also a partial write-off of bank debts in 1993 (an earlier write-off in 1991 is not captured in the data). Although the write-off affected only highly indebted firms, the extent to which it sent a signal to all firms and affected subsequent debt increases is not easily measured. In sum, the arrears data may be a volatile proxy for the hardness of budget constraints across firms.

Import penetration, (I_s), is used as a proxy for increased competitive pressures from abroad. The I_s variable is constructed using Bulgarian customs data. These data have the advantage of being in a format that can be easily mapped into the two-digit industry codes that correspond to the enterprise data.[8] While import penetration is often used in the literature as a proxy for trade liberalization, direct measures of changes in trade barriers are preferable. This is a tall order to fill in the Bulgarian context: prior to 1990 traditional trade policies were not relevant, as all trade was handled through state-run foreign trade organizations. Following the political changes in 1990 and the abolition of central planning, direct trade became possible. The extent of the trade liberalization that occurred was very significant; almost analogous to a move from autarky to free trade. This implies that the standard approach used in the literature to explore the effect of marginal changes (for example, a move from an average tariff of 40 percent to 25 percent) would not be appropriate to Bulgaria in any case. The use of import penetration implicitly accounts for the effect of the collapse of the CMEA, the disbanding of foreign trade organizations (state trading monopolies), and so forth.

Subsequent to the opening of the economy in 1990, traditional trade policy instruments such as the tariff became more relevant in Bulgaria, as such instruments were the only avenue available for firms seeking to reduce competition from imports. Although the initial trade policy environment was quite liberal—average tariffs being low, never having been relevant in the past—the structure of initial levels of tariffs and other charges on imports may nonetheless embody information regarding the "sensitivity" or ability of industries to seek or obtain protection. Initial tariff levels (reported in the first column of table 12.2) are used to investigate to what extent they are correlated with firm performance.

All variables were first constructed in levels and then converted into log-differences. The resulting variables are reported in table 12.2. Simple correlation analysis shows that for both the full sample and for nonexporters, there is no significant collinearity among changes in C_s, A_i, and I_s (table 12.3). There is a negative correlation between change in arrears and the level of the initial tariff and TFP growth and a positive correlation between I_s and TFP growth.

Results

Standard test statistics suggest that either fixed or random effects approaches should be used to estimate equation (12.5). An F-test indicates that ordinary least squares is inappropriate: the null hypothesis of identical firm effects is rejected: $F(1336, 3950) = 0.83$ with a critical value of 1.001 (table 12.4).[9] We rely on the Hausman test to determine whether a fixed or random effects specification is most appropriate. In all but one of the regressions the Hausman test indicates that random effects should be used.

TABLE 12.3 Pearson Correlation Table

Variable	TFP growth	Δ Import penetration	Δ Concentration	Δ Arrears	Tariff
Full sample					
TFP growth	1.00				
Δ Import penetratin (I_S)	0.22	1.00			
Δ Concentration (C_S)	−0.07	−0.05	1.00		
Δ Arrears (A_i)	−0.21	−0.00	0.02	1.00	
Initial tariff	−0.14	0.02	0.03	0.01	1.00
Import-competing firms					
TFP-growth	1.00				
Δ Import penetration (I_S)	0.32	1.00			
Δ Concentration (C_S)	−0.11	−0.06	1.00		
Δ Arrears (A_i)	−0.17	−0.01	0.01	1.00	

Source: Authors.

We start with a regression using the full sample (that is, both exporters and import-competing firms), focusing on the change in import penetration (table 12.4, column 1) and controlling for firm size and common (cross-industry) effects. We include a proxy for firm size in all regressions, as this may affect enterprise performance in a variety of ways. Large firms can be expected to have greater scope to build up arrears or benefit from import protection (due to lobbying power or because of bargaining agreements between politicians and firms [Shleifer and Vishny 1994]). Also, smaller firms are more likely to be driven out of the market by import competition more rapidly, since they cannot sustain operations through implicit subsidization across product lines. To take into account such factors, the log-level of the initial (1991) employment of each enterprise is used as a proxy for size.[10] Furthermore, we control for common effects (macroeconomic factors) through inclusion of annual dummy variables.

As can be seen from table 12.4, column 1, we obtain a positive coefficient estimate on the change in imports variable, but this is not statistically significant at the 5 percent level. If the regression is restricted to nonexporters (defined as firms that export less than 75 percent of their output in 1991), a substantially stronger impact of import competition is found. The coefficient estimate increases, and more important, becomes statistically significant (table 12.4, column 4).

We next add changes in concentration and arrears as explanatory variables. As predicted, coefficient estimates are negative: increases in either concentration or arrears are associated with a reduction in TFP growth. By the nature of the estimation procedure, this implies that a reduction in market concentration (arrears) is followed by a

TABLE 12.4 Estimation Results (random effects model, except where noted)

Independent variable	Full sample			Import-competing firms only[1]		
	I Random effects	II Random effects	III Fixed effects	I Random effects	II Random effects	III Random effects
Δ Import penetration	0.12 (1.84)	0.09 (1.63)	0.08 (0.72)	0.15 (2.44)	0.11 (2.61)	0.10 (2.10)
Δ Concentration		–0.12 (–2.43)	–0.16 (–2.51)		–0.14 (–3.12)	–0.18 (–3.22)
Δ Arrears		–0.03 (–2.05)	–0.02 (–1.91)		–0.03 (–2.19)	–0.03 (–2.07)
Δ Imports × Δ concentration			–0.45 (–1.75)			–0.48 (–1.12)
Size (1991 employment)	0.01 (2.14)	0.00 (1.73)	0.00 (1.70)	0.01 (1.79)	0.00 (1.53)	0.00 (1.42)
Year 1992 dummy	0.03 (2.14)	0.03 (2.21)	0.03 (2.16)	0.03 (2.32)	0.03 (2.09)	0.03 (2.04)
Year 1994 dummy	–0.02 (–2.62)	–0.02 (–2.58)	–0.03 (–2.71)	–0.02 (–2.41)	–0.03 (–2.19)	–0.02 (–2.16)
Year 1995 dummy	–0.09 (–5.13)	–0.10 (–5.62)	–0.12 (–5.69)	–0.11 (–5.42)	–0.14 (–6.26)	–0.14 (–6.31)
Industry-specific factor costs ($\beta_{Ls,t}$, $\beta_{Ms,t}$, $\beta_{Ks,t}$)	Yes	Yes	Yes	Yes	Yes	Yes
Sample size	5,348	5,348	5,348	4,312	4,312	4,312
Adjusted R^2	0.28	0.29	0.28	0.30	0.30	0.30
F-test ($A, B = A_i, B$)	0.83	0.84	0.84	0.87	0.88	0.88
Hausman test (random versus fixed) effects; critical value in parentheses)	48.13 (60.19)	52.36 (62.54)	79.64 (63.72)	47.68 (60.19)	49.62 (62.54)	51.02 (63.72)

Source: Authors.

Note: *t*-statistics are in parentheses; estimates are heteroskedasticity consistent (using the White correction).

1. Firms that exported less than 75 percent of their production in 1991.

rise in TFP, even when one accounts for the effects on import competition and arrears (concentration). Inclusion of these two domestic competition variables reduces the effect of imports, which become even less statistically significant. When the sample is limited to import-competing firms, import competition remains statistically significant, suggesting there are significant differences across the two types of firms.

Finally, we include imports interacted with concentration as a regressor. In general, inclusion of such interaction terms is a helpful robustness test and may also help to remove the heteroskedasticity of the error term.[11] Interacting imports and concentration also has an economic rationale, as it allows us to account for the possibility that imports may have different effects in competitive as opposed to imperfectly competitive market structures. In the case of Bulgaria, initial levels of concentration were very high and import penetration very low. Imports can be expected to expand more rapidly in products where markets are more competitive. In industries where there are economies of scale and firms retain an ability to set prices (capture rents), imports may expand less rapidly. If the interaction term is included in the full sample, the Hausman test indicates that random effects are inappropriate. Fixed effects were therefore used. Inclusion of the interaction term does not affect the results: imports remain insignificant, and the coefficient estimates for concentration and arrears are not affected.[12] The interaction effect is not statistically significant.

If the sample is restricted to import-competing enterprises, the inclusion of the interaction term does not have the effect of making changes in import penetration insignificant. Instead, the coefficient estimate for imports remains positive and significant at the 95 percent level.[13] These results illustrate the importance of distinguishing between import-competing and exporting firms. Only in the case of nonexporters is there a robust relationship between changes in import penetration and firm performance (as measured by the change in TFP).[14] The results also illustrate the importance of including additional measures of competitive pressures that reflect other (nontrade) policies implemented by the government. The estimated effect of changes in concentration and arrears on TFP performance is negative and significant whether exporters are included in the sample or not.

Work by Roberts and Tybout (1997); Clerides, Lach, and Tybout (1998); and Tybout and Westbrook (1995) has centered on the issue of whether export activity results in higher productivity growth. If the lagged log level of the export share in total sales is included as a regressor, we find that being an exporter is associated with higher TFP growth. Coefficient estimates of this variable are positive and statistically significant for all three specifications, the t-statistic being consistently about 3. Unfortunately, the data do not allow us to undertake an analysis of the type done by Tybout and his coauthors, which centers on testing the learning-by-exporting hypothesis. This requires information on when a firm enters a new export market, which is not available in our sample.

In all the regressions, the year dummies become increasingly negative in magnitude and increase in statistical significance. While this pattern in the year dummies may be due to a variety of common macro or "environmental" factors, a likely explanation is that it reflects the changing incentive structure confronting managers of enterprises. At the start of the period, the government was pursuing a large number of reforms, including not only price liberalization and abolition of foreign trade monopolies but also corporatization and rationalization of state-owned firms as a precursor to divestiture (privatization). This created incentives for managers of firms to undertake reforms in the expectation that this would enhance their ability to benefit from (or survive) the privatization process. In the event, the government never proceeded beyond rationalization of industries; privatization never got off the ground. Consistent with the models of Schleifer and Vishny (1994), this provided incentives for managers to divert resources and engage in asset stripping. The increasing magnitude of the coefficient estimate over time may reflect the fact that it took time for managers to learn how to manipulate the system in order to shift resources out of the firm, complemented by the gradual deterioration of the quality of the monitoring of management performance by the various government ministries. The contributions in Bogetic and Hillman (1995) describe in some detail the extent to which enterprises engaged in asset stripping and pursued "de facto" privatization of resources in Bulgaria. The end result was the financial crisis that occurred in 1996.

A weakness of the empirical approach pursued in this chapter is that the major explanatory variables are all endogenous. Although the preferable approach of using direct measures of the underlying (exogenous) policy determinants of these changes is impossible to employ in the Bulgarian context (and for that matter, for any transition economy that pursues a "big bang" liberalization approach), we are relatively comfortable with the use of the concentration and arrears variables. The change in concentration was to a very large extent driven by the policy of corporatization and breaking up conglomerates created under the central planning model.[15] The level and change in arrears is a direct measure of the ability of firms to obtain soft financing from the banking sector. Import penetration is a direct measure of the effect of a variety of policies to open up the economy to external trade.

Information is available on the initial trade policy stance and changes in this stance over time. To assess the robustness of the import competition–based estimations discussed above, we therefore rerun the regressions using the initial (1991) level of tariffs and surcharges. The trade policy variable is insignificant in the full sample (table 12.5). If the regression is restricted to nonexporters, the trade policy variable is negative, as predicted, but only marginally significant. In contrast to the earlier results, concentration and arrears become statistically insignificant once we include the interaction term between tariffs and concentration. The results suggest that those firms that initially had above-average rates of protection performed worse

TABLE 12.5 Estimation Results: Random-Effects Model

	Full sample			Import-competing firms only[1]		
Independent variable	I	II	III	I	II	III
1990 level of tariffs/ surcharges	–0.02 (–1.63)	–0.01 (–1.37)	–0.01 (–1.03)	–0.03 (–2.07)	–0.02 (–1.76)	–0.02 (–1.81)
Δ Concentration		–0.11 (–2.68)	–0.10 (–1.32)		–0.12 (–2.94)	–0.08 (–1.28)
Δ Arrears		–0.02 (–1.98)	–0.02 (–1.62)		–0.02 (–2.12)	–0.03 (–1.57)
1990 tariff level × Δ concentration			0.07 (1.21)			0.06 (1.13)
Size (1991 employment)	0.01 (2.04)	0.00 (1.79)	0.00 (1.78)	0.01 (1.92)	0.00 (1.81)	0.00 (1.80)
Year 1992 dummy	0.03 (2.29)	0.03 (2.32)	0.03 (2.26)	0.03 (2.42)	0.03 (2.18)	0.03 (2.05)
Year 1994 dummy	–0.02 (–2.89)	–0.02 (–2.76)	–0.02 (–2.68)	–0.02 (–2.72)	–0.02 (–2.29)	–0.02 (–2.50)
Year 1995 dummy	–0.10 (–6.13)	–0.11 (–6.28)	–0.10 (–6.34)	–0.10 (–6.20)	–0.11 (–6.02)	–0.10 (–5.98)
Industry-specific factor costs ($\beta_{Ls,t}$, $\beta_{Ms,t}$, $\beta_{Ks,t}$)	Yes	Yes	Yes	Yes	Yes	Yes
Sample size	5348	5348	5348	4312	4312	4312
Adjusted R^2	0.26	0.27	0.27	0.31	0.31	0.32
F-test ($A, B = A_i, B$)	0.81	0.82	0.81	0.84	0.84	0.84
Hausman test (random versus fixed effects; critical value in parentheses)	50.61 (60.19)	57.42 (62.54)	47.95 (63.72)	58.32 (60.19)	55.16 (62.54)	59.61 (63.72)

Source: Authors.

Note: t-statistics are in parentheses; estimates are heteroskedasticity consistent (using the White correction).

1. Firms that exported less than 75 percent of their production in 1991.

after the opening of the economy in 1990–91. These findings are consistent with the earlier ones using changes in import penetration as a regressor.

During 1991–95 little change occurred in the level of tariffs. Import surcharges ranging from 3 to 40 percent ad valorem and averaging 15 percent were imposed in August 1994 on some final consumer goods, but these covered only 48 narrowly defined (at the five-digit level) product lines, mainly dairy products and chocolate foods. After 1995 significant surcharges were implemented for a wider range of imports, in part to deal with a balance of payments crisis, but this period is out of sample. It is therefore not very informative to use changes in tariffs or surcharges as an explanatory variable during the 1991–95 period. Regressing TFP growth on changes in tariffs and surcharges results in insignificant coefficient estimates (not reported).

The analysis in this chapter is limited to a balanced panel of firms. In principle, this can lead to selection bias. However, in the case of Bulgaria, and transition economies more generally, it is not clear in what direction the bias goes. In contrast to a market economy, where entry and exit involves the creation and destruction of distinct (and usually independent) entities, in the case of Bulgaria during the period under consideration, entry and exit of this type was quite rare. Instead, the government pursued a policy of corporatization and restructuring of state-owned enterprises that involved breaking up existing conglomerates. It is this process that primarily drives the changes in market structure we observe in the data. The problem is that politics plays a large role in determining whether in this process an enterprise that is subjected to rationalization ends up with a better or worse "endowment." We have no information on the extent to which firms in the sample were subject to downsizing, nor do we have specific information regarding the "quality" of the assets lost or obtained in this process. What we can say with greater confidence is that there is no difference between exporters and nonexporters in this regard. As one of our main motivations is to investigate the importance of making such a distinction, we would argue that sample selection bias is not a serious issue.[16]

Conclusions

Relatively few empirical studies have found robust support for the predicted positive relationship between liberalization of trade and the subsequent performance of domestic industry. In this chapter we argue that this may reflect the fact that no distinction is made between exporters and nonexporters and that account is not taken of the combined effect of other sources of—and changes in—market disciplines. In principle, one expects trade liberalization to have a greater impact on nonexporting, import-competing enterprises than on those that export a significant share of their production. One also expects a greater impact of trade liberalization in environments characterized by hard budget constraints, active enforcement of competition policy, and governance structures that create incentives for managers to maximize operating efficiency. In the context of developing countries in particular, governments may pursue only piecemeal reform, with the effect of attenuating the impact of trade liberalization. Trade liberalization alone may not have a significant effect on corporate performance in the absence of complementary policy measures to promote greater competition and induce the necessary reallocation of resources within and across firms. Alternatively, governments may implement policy actions in multiple areas that all have the effect of increasing market disciplines. Account of such changes, or the absence thereof, should be taken in empirical assessments of the effects of trade liberalization.

Using data for Bulgaria during 1991–95, we find empirical support for the premise that changes in market discipline that emerged during the early 1990s—through trade liberalization, demonopolization and restructuring efforts, and hard budget constraints are associated with subsequent productivity growth in the enterprise sector. For the full sample of firms, we find that changes in import penetration do not have a statistically significant effect on firm productivity. It is only when we distinguish between firms on the basis of initial export intensity status and remove those firms that exported 75 percent or more of their output in 1991 that import competition becomes a significant explanatory variable. Arrears and concentration are not sensitive to this distinction. These variables are equally important for both types of firms.

Notes

1. Export taxes of 20 percent were applied on a small number of foodstuffs and materials used by the construction industry in 1993–94. These were phased out by the end of 1995.

2. Imports, concentration, and arrears are all endogenous. Although using lagged values should help reduce problems of contemporaneous correlation, relying on such endogenous variables as regressors does not provide information on, or tests of, the true underlying determinants of changes of productivity. As discussed below, however, this is the best that can be done given data constraints.

3. See Djankov and Hoekman (2000) for an analysis of the activities of the Bulgarian antitrust office.

4. This is consistent with our intuition, as fixed-effects estimation assumes firm productivity growth to be constant over time. This assumption is objectionable, since changes in productivity due to increased competition is the phenomenon we seek to explore. The random-effects model avoids the imposition of constant productivity growth over time, but it has the drawback that productivity shocks at the firm level are assumed to be uncorrelated over time. This may not be a reasonable restriction if there is convergence or divergence in corporate performance.

5. It would, of course, be preferable to use firm-specific price indices as deflators, but these were not part of the standard questionnaire sent to firms by the NSO. The use of sectoral price deflators obscures the variation of firm performance within a sector. This would be a serious problem if we were studying the convergence or divergence in productivity within sectors. It is less objectionable here, as the primary focus is on the performance of firms across sectors. The results should nonetheless be treated with caution.

6. At the suggestion of a referee, we also ran all the regressions reported below without subtracting subsidies from enterprise revenues. The results are not sensitive to whether or not subsidies are included. This is a reflection of the limited use that was made of direct subsidies.

7. See Djankov and Hoekman (1998) for a discussion and descriptive statistics on breakups and closure of plants.

8. Partner country export data were used to check the accuracy of these data. The two data sets were very similar.

9. In any event, it is also problematic on a priori grounds to assume homogeneity of firm characteristics.

10. We cannot use revenues (since that would create simultaneity problems in the regression) or fixed assets (since the data, as discussed earlier, are suspect).

11. As we employ the White correction in the estimation, any improvement in this regard is likely to be marginal.

12. This result may in part reflect the change from random- to fixed-effects estimation. Imports are already insignificant in a fixed-effects-based estimation once concentration and arrears are included in the regression.

13. These results are not very sensitive to the definition of what is an exporter. If we run the regressions using a criterion of exporting at least 50 percent of production to define exporters, the results are very similar. In the case in which the sample is restricted to nonexporters, the coefficient estimate on change in import penetration (controlling for arrears, concentration, and the interaction effect) is 0.11 ($t=2.12$).

14. Very similar results are found if instead of breaking the sample into two groups, the regression is run on the full sample (including changes in import penetration, arrears, concentration, and the imports'concentration interaction term on the right-hand side) but we control for exporters by interacting export status (using a dummy variable equal to 1 if 75 percent of output is exported during 1991) with the change in imports.

15. Imports can, of course, also be expected to play a role, but this does not appear to be the case in Bulgaria. The correlation between changes in concentration and changes in imports is effectively zero (see table 12.3).

16. Djankov and Hoekman (1998) analyze the extent to which the results of the panel regressions change if the sample includes spinoffs and plant closures. If such an unbalanced panel of Bulgarian firms is used, the magnitude of productivity improvement (as measured by a decline in price-cost margins, not TFP growth) is somewhat greater than for the balanced panel. This also suggests that possible selection bias is not a major source for concern, insofar as the bias works against the firms in the balanced sample.

References

Basu, Susanto, and John Fernald. 2002. "Aggregate Productivity and the Productivity of Aggregates." *European Economic Review* 46: 963–91.

Bogetic, Zelko, and Arye Hillman, eds. 1995. *Financing the Government in Transition: Bulgaria.* Washington, DC: World Bank.

Bulgarian Privatization Agency. 1997. *Annual Report* 1996. Sofia: Bulgarian Privatization Agency.

Burnside, Craig, Michael Eichenbaum, and Sergio Rebelo. 1995. "Capital Utilization and Returns to Scale." In *NBER Macroeconomics Annual*, ed. Ben Bernanke and Julio Rothemberg. Cambridge, MA: National Bureau of Economic Research.

Claessens, Stijn, and R. Kyle Peters. 1997. "Enterprise Performance and Soft Budget Constraints: The Case of Bulgaria." *Economics of Transition* 5: 305–22.

Clerides, Sofronis, Saul Lach, and James Tybout. 1998. "Is Learning by Exporting Important? Micro-Dynamic Evidence from Colombia, Mexico, and Morocco." *Quarterly Journal of Economics* 113: 903–47.

Dickens, William, and Lawrence Katz. 1987. "Inter-Industry Wage Differences and Industry Characteristics." In *Unemployment and the Structure of Labor Markets*, ed. Kevin Lang and Jonathan Leonard. New York: Blackwell.

Djankov, Simeon, and Bernard Hoekman. 1998. "Trade Reorientation and Post-Reform Productivity Growth in Bulgarian Enterprises." *Journal of Policy Reform* 2: 151–68.

———. 2000. "Competition Policy in Post-Central Planning Bulgaria." *Antitrust Bulletin* 65 (Spring): 227–48.

Earle, John, and Saul Estrin. 2001. "Privatization and the Structure of Enterprise Ownership." In *Russia's Post-Communist Economy*, ed. B. Granville and P. Oppenheimer. Oxford: Oxford University Press.

EBRD (European Bank for Reconstruction and Development). 1997. *Transition Report.* London: EBRD.

Falk, M., M. Raiser, and H. Brauer. 1995. "Output Decline and Recovery in Poland." Kiel Institute for World Economics Working Paper 723, University of Kiel, Germany.

Feenstra, Robert. 1996. "Estimating the Effects of Trade Policy." In *Handbook of International Economics*, Vol. III, ed. Gene Grossman and Ken Rogoff. Amsterdam: North Holland.

Grether, Jean-Marie. 1996. "Mexico 1985–1990: Trade Liberalization, Market Structure, and Manufacturing Performance." In *Industrial Evolution in Developing Countries*, ed. M. Roberts and J. Tybout. Oxford: Oxford University Press.

Haddad, Mona, Jaime de Melo, and Brendan Horton. 1996. "Morocco 1984–1989: Trade Liberalization, Exports, and Industrial Performance." In *Industrial Evolution in Developing Countries*, ed. M. Roberts and J. Tybout. Oxford: Oxford University Press.

Harrison, Ann. 1994. "Productivity, Imperfect Competition and Trade Reform." *Journal of International Economics* 36: 53–73.

Hart, Oliver. 1983. "The Market Mechanism as an Incentive Scheme." *Bell Journal of Economics* 14: 366–82.

Holstrom, Bengt. 1982. "Moral Hazard in Teams." *Bell Journal of Economics* 13: 324–40.

Horn, Henrik, Harold Lang, and Stefan Lundgren. 1995. "Managerial Effort Incentives, X-Inefficiency and International Trade." *European Economic Review* 39: 117–38.

IMF (International Monetary Fund). 1997. *Yearbook of Government Finance Statistics*. Washington, DC: IMF.

King, M., and R. Levine. 1993. "Finance and Growth: Schumpeter Might Be Right." *Quarterly Journal of Economics* 108: 717–37.

Levinsohn, James. 1993. "Testing the Imports-as-Market-Discipline Hypothesis." *Journal of International Economics* 35: 1–22.

Lizal, L., P. Singer, and J. Svejnar. 1995. "Manager Interests, Breakups and Performance of Enterprises in Transition." In *The Czech Republic and Economic Transition in Eastern Europe*, ed. J. Svejnar. San Diego, London, and Toronto: Harcourt Brace, Academic Press.

National Statistical Office of Bulgaria. 1997. *Statistical Yearbook*. Sofia: Government Printing Office.

Nalebuff, Barry, and Joseph Stiglitz. 1983. "Prizes and Incentives: Towards a General Theory of Compensation." *Bell Journal of Economics* 14: 21–43.

Nickell, Stephen, Sushil Wadhwani, and Martin Wall. 1992. "Productivity Growth in UK Companies 1975–86." *European Economic Review* 36: 1066–85.

Nickell, Stephen. 1996. "Competition and Corporate Governance." *Journal of Political Economy* 104: 721–46.

Olley, Stephen, and Ariel Pakes. 1996. "The Dynamics of Productivity in the Telecommunications Equipment Industry." *Econometrica* 64: 1263–98.

Roberts, Mark. 1996. "Colombia 1977–1985: Producer Turnover, Margins, and Trade Exposure." In *Industrial Evolution in Developing Countries*, ed. M. Roberts and J. Tybout. Oxford: Oxford University Press.

Roberts, Mark, and James Tybout. 1997. "An Empirical Model of Sunk Costs and the Decision to Export." *American Economic Review* 87: 545–64.

Schaffer, Mark. 1998. "Do Firms in Transition Economies Have Soft Budget Constraints? A Reconsideration of Concepts and Evidence." *Journal of Comparative Economics* 26: 157–79.

Schmidt, Klaus. 1994. "Managerial Incentives and Product Market Competition." Discussion Paper 430, University of Bonn, Germany.

Shleifer, Andrei, and Robert Vishny. 1994. "Politicians and Firms." *Quarterly Journal of Economics* 46: 995–1025.

Svejnar, Jan. 1995. "Enterprises and Workers in Transition." *American Economic Review* 86: 123–27.

Tybout, James. 1996. "Chile 1979–1986: Trade Liberalization and Its Aftermath." In *Industrial Evolution in Developing Countries*, ed. M. Roberts and J. Tybout. Oxford: Oxford University Press.

Tybout, James, and M. Daniel Westbrook. 1995. "Trade Liberalization and Dimensions of Efficiency Change in Mexican Manufacturing Industries." *Journal of International Economics* 31: 53–78.

van Wijnbergen, Sweder, and Anthony Venables. 1993. "Trade Liberalization, Productivity, and Competition: The Mexican Experience." Discussion Paper 345, London School of Economics.

World Bank. 1991. *Bulgaria: Crisis and Transition to a Market Economy*. Washington, DC: World Bank.

———. 1996. *World Development Report 1996: From Plan to Market*. Washington, DC: World Bank.

13

INNOVATION IN MEXICO: NAFTA IS NOT ENOUGH

Daniel Lederman and William F. Maloney

Although Mexicans have traditionally lamented their distance from divine protection and proximity to the United States, dramatically increased commercial interactions with arguably the most dynamic generator of new technologies in history could potentially yield the country large growth dividends through technological transfer.[1] Yet the main message of this chapter is that trade liberalization and the North American Free Trade Agreement (NAFTA) are helpful but not enough to help Mexico catch up to the level of technological progress of its North American partners, especially the United States. In fact, the evidence reviewed here suggests that, given its level of development, Mexico suffers from low levels of innovation effort, and it severely underperforms relative to successful economies, such as the Republic of Korea and the United States. In addition, its national innovation system (that is, how the private sector, universities, and public policies interact to produce economically meaningful innovation) is inefficient. Without addressing these deficiencies, it is unlikely that NAFTA alone will be sufficient for Mexico to catch up with the pace of innovation in North America.

The literature suggests that roughly half of cross-country differences in per capita income and growth are driven by differences in total factor productivity (TFP), generally associated with technological progress (see Dollar and Wolf 1997; Hall and Jones 1999). Furthermore, much of the widening gap between rich and poor countries results not from differences in capital investment but from differences in technological progress. This fact moves to center stage an abiding question in economic development: given the great potential gains to be realized by adopting technologies from the industrial economies, why do developing countries fail to do so? Perversely, the countries generating new technologies at the frontier appear to have faster TFP growth in manufacturing and agriculture than do poor countries that could, in theory, adopt existing technologies invented in the advanced economies (Martin and Mitra 2001).

Recent work in innovation stresses that adopting existing technology has its cost. Firms and countries need to develop an absorptive or national learning capacity, which is hypothesized to be a function of spending on research and development (R&D).[2] Although learning is often considered relevant only for basic science dedicated to expanding the knowledge frontier, Cohen and Levinthal (1989), among others, stress that learning (that is, knowing where the frontier is and figuring out what adaptations are necessary to move toward it) is the "second face" of R&D.

Pavitt (2001) argues that investment in pure research is also important for developing countries. First, countries that are most familiar with the frontiers of basic science will best train the applied problem solvers in the private sector. Second, even basic research does not flow easily or without cost across borders, so developing countries cannot simply rely on what is being generated in the advanced countries. Lederman and Sáenz (2003) present econometric evidence suggesting that innovation outcomes (patents per capita) are an important explanation of the levels of development observed around the world.

In this chapter we review indicators of Mexico's technological progress and innovation efforts since the 1960s and ask what the global and Mexican experiences suggest about how greater openness to trade and FDI should affect specific economies. We argue that Mexico's weak national learning and innovative capacity represents a critical missing complement to its liberalization efforts. This is a result partly of weak efforts in education and innovation and partly in how these efforts are coordinated, through a largely dysfunctional national innovation system. The principal conclusion is that exploiting the opportunities offered by greater integration with Canada and the United States requires Mexico to make substantial policy improvements in both the level and efficiency of innovation efforts. NAFTA is not enough.

Mexican Innovation and Technological Progress Since the 1960s

One of the most heavily used indicators of progress in innovation is the level and growth rate of TFP. This is generally understood to be the part of the growth of gross domestic product (GDP) that is not explained by the accumulation of raw labor, physical capital, and perhaps human capital, ideally after controlling for capacity utilization. Since the pioneering work of Solow (1956, 1957), this indicator has been thought to be a good proxy for technological progress.

Figure 13.1 shows average annual TFP growth rates in Mexico and selected other countries and regions between 1960 and 1999.[3] Mexico's TFP performance

FIGURE 13.1 Growth Rates of Total Factor Productivity in Selected Countries and Regions, 1960–99

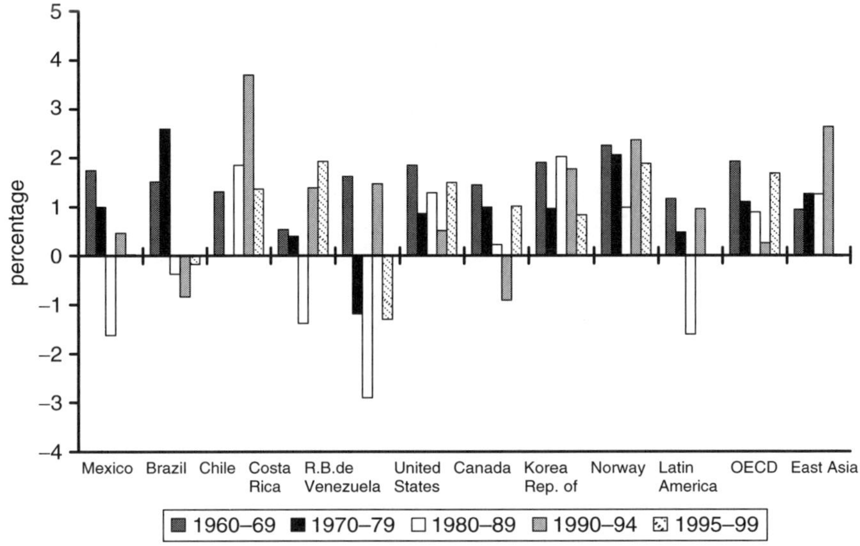

Source: Loayza, Fajnzylber, and Calderón (2002).

was highest in the 1960s, and, as in most of the other economies except those of Brazil and the East Asia and Pacific region, it declined in the 1970s and declined with the rest of Latin America in the 1980s (largely as a result of the debt crisis).[4] The slight recovery in the 1990s may have resulted from economic reforms implemented in the late 1980s and early 1990s in most Latin American and Caribbean countries, although recent econometric estimates suggest that, on average, these effects were modest in the region.[5] Figure 13.1 suggests that Chile, the most advanced reformer, performed far better than either the Latin American or the Asian regional averages for the past two decades. Given the overall similarity in policy packages, there would seem to be nothing in the economic model adopted that intrinsically dictated lower rates of TFP growth in Mexico.

An explanation for these patterns was put forward by Acemoglu, Aghion, and Zilibotti (2002), who argue that there are two stages of technology adoption. The first is based on fomenting the accumulation of technology embodied in capital formation, even if this requires some static efficiency losses through interventionist policies, arguably including the import-substituting industrialization adopted in Latin America. The second stage, centered on innovation, requires greater structural flexibility and fewer distortions. In their view, Brazil, the Republic of Korea, Mexico, Peru, and Taiwan (China) all successfully pursued the first stage, but the

Asian economies were able to make the transition to efficient innovative economies whereas Latin America was not. The case of Chile, which liberalized about 10 years before Mexico, supports this diagnosis and offers some reason to suppose that Mexico will experience a similar rebound in TFP in the coming years, all other things equal. In light of the successful growth experiences of the relatively open economies of Finland, Ireland, Israel, and Spain over a similar period, it is difficult to argue that the extreme "closedness" of the region was necessary or desirable, especially given the difficult political economy problems of moving to a more innovative structure later.[6]

Productivity growth in Norway, another petroleum exporter, was quite strong by international standards throughout this period, suggesting that net exporters of natural resources do not necessarily have lower potential for productivity growth. On the contrary, Lederman and Maloney (2002) find that countries rich in natural resources tend to experience faster economic growth, even after controlling for the contributions of human and physical capital accumulation.

Another commonly used innovation proxy is the number of patents granted (figure 13.2). It is generally believed that patent statistics broadly reflect the flow

FIGURE 13.2 Patents per Million Workers in Selected Regions, 1960–2000

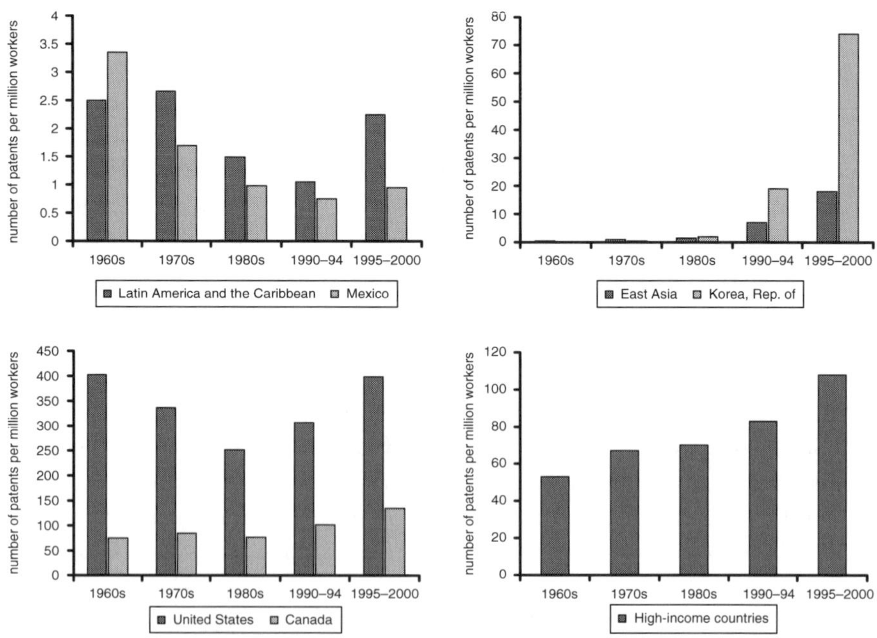

Source: U.S. Patent Office.

of innovations covering either adaptations of existing patents or brand-new inventions (Griliches 1990; Patel and Pavitt 1995). Measures of the number of patents granted to researchers from around the globe, however, are not without flaws. One important consideration is the fact that the costs of applying for patents, the level of intellectual property protection, the pecuniary benefits from patents, and other institutional features vary greatly across countries. Thus patents granted in one country are not strictly comparable to those granted by others. We use the number of patents granted to Mexican residents by the United States Patent and Trademark Office (USPTO) as a proxy for the flow of innovation.[7] The data from the USPTO are attractive because of their global and long time coverage and especially because it is commonly understood that the United States offers perhaps the most advanced levels of intellectual property protection in the world (Maskus 2000). Although the costs of the application process are likely to be higher in the United States than in most other countries, the benefits are also likely to be higher.

The evidence suggests that over time, Mexico's patenting activity follows a pattern similar to its TFP growth rate. The number of patents granted to Mexican innovators peaked in the 1960s, declined continuously until the first half of the 1990s, and picked up again after the implementation of NAFTA in the second half of the 1990s. This resurgence was quite modest by historical standards, however, and was insufficient to make a significant dent in the observed gap with respect to Canada and the United States. Mexico is far behind East Asia (especially the Republic of Korea) and lags Costa Rica and República Bolivariana de Venezuela.

This pattern remains even when differences in the level of development are accounted for. Figure 13.3 benchmarks performance by researchers in Mexico, comparing their performance with average performance in countries with the same levels of GDP, the same size labor force, and the same value of merchandise exports to the United States since the 1960s.[8] It shows how far Mexico is from the average of similar economies (the zero line); the embedded table reports Mexico's average absolute levels by decades. A negative number on the vertical axis is evidence of underperformance. Mexico's performance in patents has fallen from more than twice the average in the 1960s to substantially below the average in the 1980s, just prior to the debt crisis and the structural reforms.

Another indicator of innovation is the number of scientific publications, which can be interpreted as a measure of outcome of basic rather than applied research. Mexico shows some upward progress from the 1980s, to roughly half of its expected value.

In summary, in terms of TFP growth, patenting activity, and scientific publications, Mexico lags behind many countries at its level of development and sorely behind the newly industrial countries of Asia that were once its comparators.

FIGURE 13.3 Number of Scientific Publications and Patents in Mexico Relative to Comparative Countries, 1960–2000

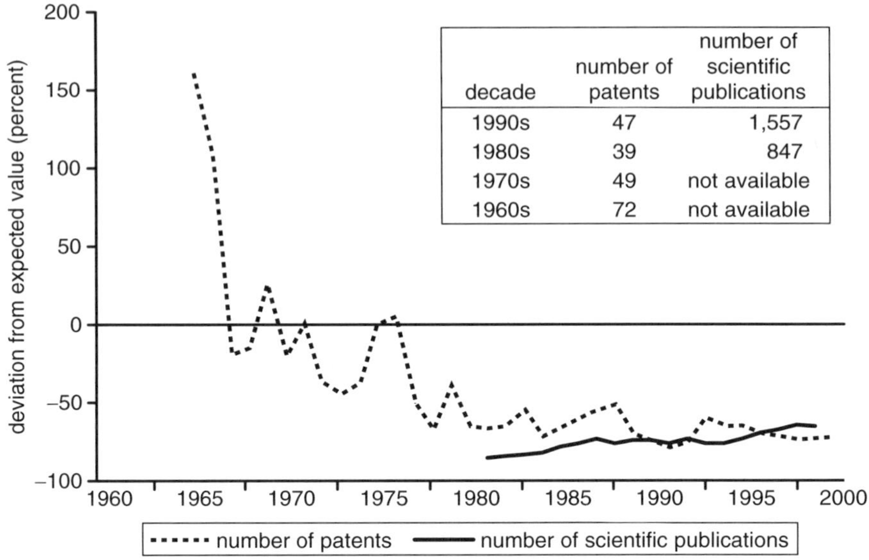

Source: Authors' calculations.

How Trade and Foreign Direct Investment Affect Innovation and Technological Progress

What does the international evidence say about how the increased trade and FDI encouraged by NAFTA might affect growth through the transmission or generation of technology? The bulk of the literature suggests modest effects, working primarily through physical capital accumulation.[9] Using a panel of countries since 1960, Loayza, Fajnzylber, and Calderón (2002) find that a 1 percent increase in the portion of the trade-to-GDP ratio that is related to trade policies leads to an increase in the growth rate of GDP per capita of 0.025–0.010 percentage points a year.[10] Consistent with the findings of Baldwin and Seghezza (1996) and Levine and Renelt (1992), Wacziarg (2001) finds that the most statistically robust channel through which trade positively affects economic growth is investment, with domestic investment growth accounting for more than 60 percent of the positive effect of trade on growth.[11] He speculates that these results are consistent with theories that focus on the procompetitive effects of trade, because the survival of existing firms and the entry of new ones after trade liberalization probably require large fixed capital costs.

An emerging body of literature also finds an impact of trade and investment variables on TFP growth, although the exact channel remains ambiguous. The overall impact of NAFTA on productivity in Mexico was positive: the agreement was associated with convergence in rates of TFP growth among the manufacturing sectors in Mexico and the United States. López-Córdova (2002) estimates that the whole package of NAFTA–related phenomena (lower Mexican tariffs, the preferential tariff margin in the United States, higher import-to-output ratio, and participation of foreign producers) increased TFP by 10 percent. Schiff and Wang's (2002) estimate of the same package is 5.6–7.5 percent. These estimates are broadly consistent with estimates of a very large impact of the free trade agreement in Canada. Trefler (1998) argues that overall manufacturing TFP rose by 0.2 percent a year (1.0 percent for the firms most affected by trade), primarily as a result of plant turnover and rising technical efficiency within plants. Hence it is likely that the rates of TFP growth in Canada and Mexico would have been even lower than those shown in figure 13.1 if NAFTA and its predecessor, the Canada–U.S. Free Trade Agreement, had not been implemented.

Imports of capital or intermediate goods, which are thought to embody or transmit knowledge of new production techniques or skills, appear associated with a positive effect on the levels of TFP.[12] Coe, Helpman, and Hoffmaister (1997) find that the overall level of imports is important for international technology diffusion for 77 developing countries. Looking at industry-level data from eight OECD countries, Keller (2002) finds that roughly 50 percent of TFP growth in manufacturing industries results from their own R&D spending, 30 percent results from other domestic industries, and 20 percent results from R&D expenditures in foreign industries, some of which is transmitted through trade. He speculates that the importance of R&D expenditures in foreign industries may be much higher in developing countries, where local R&D effort is much lower than in high-income countries. For Latin America, Schiff and Wang (2004) find modestly positive effects of the technology embodied in intermediate inputs on TFP for certain high–R&D industries in Latin America.[13] Looking specifically at NAFTA, Schiff and Wang (2002) find that the roughly 14–18 percent increase in total imports from NAFTA partners to Mexico led to an increase in TFP levels in manufacturing industries of 5.1–7.0 percent. The 3 percent diversion of imports from other OECD countries, whose imports have no impact on TFP, led to another 0.47 percent increase in TFP.[14]

The interpretation of these results is not straightforward. Seemingly in contradiction to the above studies, Eaton and Kortum (1996) find that bilateral imports do not help predict bilateral patenting activity, their indicator of international technology diffusion. Based on firm-level data from Mexico, López-Córdova (2002), like Muendler (2002) for Brazil, finds a negative impact of imported inputs on manufacturing TFP.[15] Schiff and Wang (2002) express doubt about the meaning of their own estimates in the Coe-Helpman-Hoffmaister tradition. The fact that input trade with the United States is a good vehicle for technology transfer to Mexico but that trade with other high-income OECD countries apparently has no effect on TFP is counterintuitive. The result is strikingly consistent with Keller (2002), who finds that the impact of trade in intermediate goods decreases with geographic distance between trade partners. In fact, employing Keller's elasticity, the U.S. impact on Mexico should be, and is, roughly 10 times as large as that with respect to the OECD. However, space-dependent depreciation of technology embodied in inputs seems unlikely and, as Schiff and Wang (2002) suggest, these results may be picking up greater collaborative and subcontracting relationships across the border rather than the effortless transfer of production knowledge embodied in the intermediate inputs themselves. This in no way undermines the benefits of an open trade stance with respect to the United States, but it does suggest that the extraordinarily large TFP-enhancing effects of trade with the United States reflect nontrade channels of influence, which may be related to personal and business interactions among businesspeople, firms, and researchers.

Firms may also learn by exporting in the sense that participation in foreign markets may help them identify the latest production, management, and marketing techniques. Thus exporters in Mexico could have enhanced their learning capacity during the post–NAFTA and trade liberalization period. Numerous cross-sectional studies have shown that Mexican exporters tend to be more technically efficient, presumably because of technological development related to the import of technologies from abroad (Alvarez and Robertson 2001; Meza González and Yaguë 2002). However, using the only micro-level panel data spanning the NAFTA period that allow for the determination of causality—whether exports make a firm more efficient or more efficient firms export—López-Córdova (2002) finds no impact of exporting on TFP growth and a negative correlation with productivity levels. A World Bank study (2000) finds that the number of years of experience in exporting does seem to be associated with rising TFP levels, although the estimates did not control for unobserved firm-specific characteristics and the act of beginning to export itself did not appear to stimulate productivity growth.

The absence of a positive finding is consistent with the panel regressions by Clerides, Lach, and Tybout (1998) for Mexico's early period of liberalization, as well as for Colombia and Morocco. The researchers found little evidence in any country that firms' cost structures change after breaking into the export market. They argue that higher productivity is likely to be the result of a selection of the better firms into exporting—that is, the Schumpeterian reallocation effect.[16] These results are also consistent with Kraay, Soloaga, and Tybout's (chapter 11 of this volume) analysis of firms in the chemical industry in Mexico and Colombia. The researchers were unable to establish Granger causality between engaging in international activities—be it imports or exports—and indicators of productivity gains.[17] The disappointing results regarding the lack of a robust positive effect of exporting on TFP growth are found elsewhere, in both emerging and advanced economies.[18]

There can be little doubt that FDI increases the capital stock in the host country and contributes the technology embodied in that capital. It also may have reallocation effects, to the degree that it displaces inefficient firms. However, the macroeconomic evidence does not suggest strong causality between FDI and TFP growth, and the microeconomic evidence for technological spillovers to other firms is thin and negative.[19] López-Córdova (2002) finds a negative direct effect of FDI on the same industry's TFP. This is consistent with numerous other panel studies of other developing and industrial countries.[20] Other literature on Mexico is sparse. Blomström and Wolff (1994) find that both the rate of local firms' labor productivity growth and their catch-up rate to the multinationals are positively related to the industry's degree of foreign ownership. They find that the rate of convergence of industry labor productivity with the U.S. rate of growth is higher in industries with a higher share of multinationals. They point out, however, that it is difficult to distinguish a rise in within-firm productivity from increased competition that forces out less efficient firms, raising the average rate of growth.

In summary, our reading of the international evidence is that NAFTA may have helped spur trade, FDI, and economic growth, but the trade channel's benefits were driven mainly by factor accumulation effects and displacement of inefficient firms by more efficient firms. FDI stimulated by NAFTA aided Mexico's economic recovery, but it did not necessarily augment the technological progress component of TFP. The evidence for direct knowledge spillovers through these mechanisms seems generally weak.

National Learning Capacity: The Missing Complement

These findings seem somewhat counterintuitive, given how important trade and, to a lesser extent, FDI appear to have been in the "Asian miracles," which did experience rapid rates of TFP growth. The lackluster productivity growth of Mexico offers important insights. Trefler (1999a, 1999b) argues that NAFTA was associated with closing the gap with the United States in some manufacturing activities but in exacerbating it in others, such as computers and industrial machinery. He argues that a critical explanation is low levels of Canadian R&D and deficient basic science and that the presumption that Mexico can rely on basic science from the United States is misguided. By the time a seminal innovation is transferred from the United States, its most valuable applications have already been exploited by U.S. companies. To support this point, Trefler cites evidence showing that a 5 percent increase in R&D in the United States is associated with 6.7 percent increase in U.S. productivity but only a 2.4 percent increase in Canadian productivity.

Both partners' experiences suggest that, although they are helpful, trade and FDI are not enough to put Mexico on a rapid growth path; important complementary efforts in innovation are necessary. An emerging strand of the growth literature stresses the importance of "national learning" or "innovative" capacity to adopt new technologies and implicitly take advantage of the new trade opportunities for convergence. Acemoglu and Zilibotti (2001) argue that most technologies developed in advanced countries are not as productive in developing countries because the host countries' low human capital is not appropriate for using innovative production processes. Lloyd-Ellis and Roberts (2002) argue that education and technological progress are not only complements but dynamic complements, with the return to each determined by the growth of the other. Technology transfer from the United States to Mexico, for example, will not lead to the equalization of productivity levels between these countries as long as Mexico's human capital is much lower than that in the United States.

Howitt and Mayer-Foulkes (2002) offer a "convergence club" theory that explains why human capital—and innovative effort more generally—are essential for convergence among countries. They trace three possible productivity-growth paths for countries exposed to identical technological progress. Countries with high human capital relative to the technological frontier will experience the fastest rates of TFP growth.[21] That is, the most dynamic economies will tend to be those that have the necessary human capital and the required learning capacity for pushing the technological frontier forward. Countries with lower learning capacity will tend to rely on the adoption of technologies previously invented in the most dynamic countries (as Trefler [1999] suggests is the case in Canada), and they will appear to show slower levels of TFP growth than the leading countries. Economies that have inadequate human capital or learning capacity to adopt technologies stagnate.[22] Microeconomic evidence from Mexico suggests that firms that spend more on R&D, use highly skilled workers, and train are more likely to adopt new technologies (López-Acevedo 2002). But the question remains whether, in the aggregate, Mexico is making the necessary complementary effort in innovation that theory suggests is necessary for it to converge with U.S. levels or whether an excessive reliance on passive transfers through trade and FDI will leave the country stuck in a low-growth trap.

Where Does Mexico Stand in Innovation Effort?

As a way of quantifying Mexico's innovation effort, we benchmarked two common indicators of innovation inputs: expenditure on R&D and payments for licensing of new foreign technologies, both with respect to GDP and the labor force.[23] Expenditure on R&D extends beyond investment in cutting-edge technologies to expenditure

FIGURE 13.4 Ratio of R&D to GDP in Selected Countries

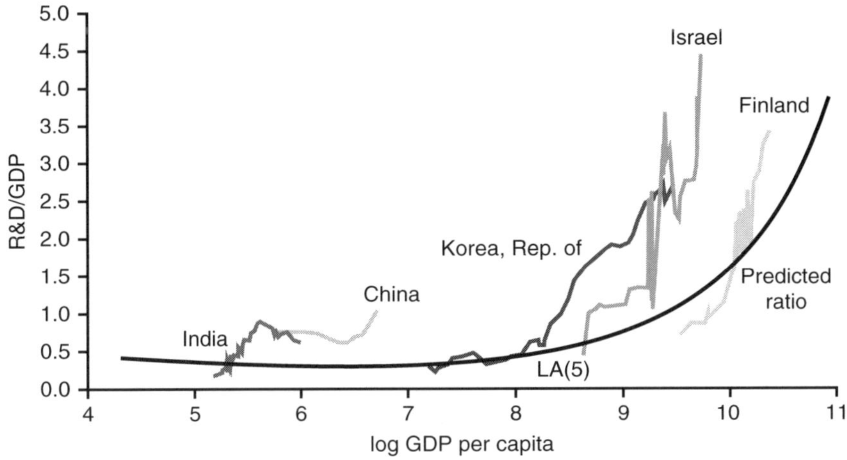

Source: Lederman and Maloney (2003).

on adoption and adaptation of technologies. It shows substantial variance across country growth trajectories. Not only does the share of GDP dedicated to R&D in the average country increase with income per capita, several high-growth comparator countries—Finland, the Republic of Korea, and Israel—had dramatic takeoffs relative to this benchmark, a path that China and India appear to be attempting to follow (figure 13.4). The average effort of five middle-income Latin American countries (Argentina, Brazil, Chile, Costa Rica, and Mexico) is substantially below trend.

Since the late 1960s, Mexico's R&D effort has been below the level of countries at the same level of development. In contrast, licensing expenditures since the late 1970s have met or exceeded the predicted level (figure 13.5).[24] The debt crisis of the 1980s was associated with a rapid decline in R&D effort. It is therefore possible that one of the channels through which the process of adjustment during the 1980s hampered productivity growth may have been through the reduction of R&D investments. During the NAFTA period, R&D spending rose again, although not to the average. From this viewpoint Mexico needs to do more to stimulate innovative activity and address impediments to it. This becomes even more obvious when we look at high-innovation countries. Among these countries, only Canada experienced neither a boom nor a continuously high level of R&D effort, a result consistent with Trefler's (1999a, 1999b) argument that Canada risks falling

FIGURE 13.5 Innovation Inputs in Mexico

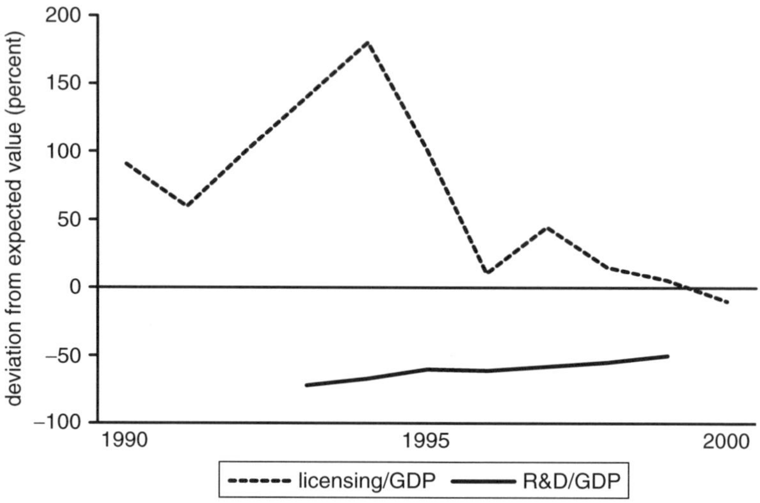

Source: Authors' calculations.

into a lower growth equilibrium if it does not increase its R&D effort. The cases of Finland and the Republic of Korean are remarkable in that they went through periods of very rapid improvements in their relative R&D effort; by the early 1980s both were well above the median.

How Much Should Mexico Spend on Innovation?

Caution is necessary in interpreting these benchmarking results and those like them, for two reasons. First, standard growth theory suggests that levels of investment—in physical and knowledge capital—determine the steady-state level of income. Hence in some sense, the above benchmarking is simply asking what combination of various kinds of investments a country used to attain its current level of income. At the most basic, very high use of R&D in China or India may suggest only that these countries are inefficiently using those factors. As Maloney and Rodriguez-Clare (2005) argue, the accumulation of knowledge capital and physical capital are closely related (TFP and the capital/output ratio are highly correlated), suggesting strong complementarities between the two. It may make little sense for a country to embark on a research agenda without a well-developed private sector to complement it. Furthermore, as Maloney (2005) points out, Latin America attained its levels of income with low levels of R&D and education and moderate levels of licensing relative to what was expected given the level of GDP and the size of the population in these countries. In contrast, in Finland and the newly industrialized countries of Asia, R&D, education, and licensing all exceeded expectations. But this does not suggest that one recipe is preferred to the other: Italy and Spain followed the Latin recipe and converged very quickly to EU levels of income. That said, their recent slowdowns suggest that while process and managerial innovation can generate substantial TFP gains, over the longer term a greater emphasis on R&D–type activities in the production mix is probably necessary.

Another way of posing the question is to ask whether the returns to R&D can justify the dramatic increases in R&D expenditures seen in Finland or the Republic of Korea. Most estimates of the impact of R&D spending on TFP in selected U.S. firms and industries are very high, ranging from 30 to 120 percent, which, when compared with a 7 percent return on capital, implies that the United States should invest several times more in R&D than it currently does.[25] Lederman and Maloney (2003) also find estimates of social rates of return to R&D for a global sample, including poorer countries in this range. However, they find that there appears to be a *U*-shaped pattern relating returns to level of development, a pattern in which countries such as Mexico show much lower returns than countries such as Finland. Why this should be the case is partly the subject of the next section.

Lederman and Maloney also find that not only do natural resource–abundant economies (captured by the ratio of net exports of natural resources) such as Mexico appear to grow more rapidly than others, the interaction with R&D spending is significantly positive.[26] That is, a much higher commitment to innovation may have allowed Australia and Scandinavia to exploit their natural resources more effectively than Mexico and other Latin American economies.

Why Is Innovation Effort So Low in Mexico?

High social returns may not translate into high private returns—and hence innovative effort—because market imperfections tend to reduce the equilibrium marginal private returns to R&D.[27] Knowledge is especially susceptible to market failures that lead to an underinvestment in R&D and other innovation-related activities. These market failures take several forms.

Nonappropriability Innovators may be unable to exclude others from using their ideas. This is implicit in the finding that rates of return to R&D tend to be roughly four times the private rate in the studies cited above. Recognition of this failure has led to an emphasis on public interventions necessary to ensure the socially optimal level of innovation. These interventions include subsidies to R&D,

temporary monopoly rights granted through patents, and other intellectual property regime instruments.

The same problem slows the transmission of existing ideas. A firm that incurs the costs of tapping into the global stock of knowledge will soon find its discoveries adopted by other firms that free ride on the investment. Historically, this has given rise to institutions ranging from agricultural extension services to technology parks to institutions designed to act as "antennae" for new ideas at the sectoral and national levels.

Lumpiness and Scale Economies That Dictate Specialization R&D and innovation are characterized by economies of scale and lumpiness. To be effective, resources often need to be concentrated in a manner that is beyond the capacity of the individual firm. Combined with the fact that even patents are not effective in resolving the nonappropriability problem in "pure science," this provides a rationale for institutions dedicated to R&D and innovation efforts, such as research centers and universities. Innovation and knowledge developed by these institutions tend to be nonexclusive and are made available to all interested parties.

Innovation, Diffusion, and Application That Require Collaboration among Many Institutions and Firms Although innovation is sometimes the product of a single firm, the more common pattern is one of joint efforts among various firms, among firms and R&D–related institutions, or among various R&D-related institutions. Progress does not proceed linearly from pure science to applied technologies, it moves in both directions, with feedback from frontline users of technology to researchers essential for refining products and production processes (Nelson and Rosenberg 1993). As a result of specialization, the full supply chain of knowledge is not totally integrated, either vertically or horizontally. Technological advance is not necessarily evenly diffused throughout the supply chain. Success may require cooperation among actors, cooperation that is subject to coordination failures and transactions costs. In many industrial countries these issues have given rise to national institutions devoted to formenting collaboration or eliminating impediments to technological collaboration between institutions.

The development and necessary interaction and coordination of these market and nonmarket institutions has led to the concept of national innovation systems and the emergence of an extensive body of relevant literature that we can only touch on here (Nelson and Rosenberg 1993; OECD 1999, 2001). It is through networks of public and private firms interacting in a concerted way to generate and adopt technologies that nations learn. This national learning capacity is what permits nations to adopt and innovate in their initial areas of comparative advantage and helps create new areas (see Furman, Porter, and Stern 2002; Nelson 1993; Romer

1990; Wright 1999). National learning capacity can be seen as the factor determining which convergence club countries find themselves in, according to Howitt and Meyer-Foulkes (2002). A broader concept of a national innovation system includes other supporting institutions, such as credit and labor markets and even the degree of market friendliness of the economy (that is, whether there is an incentive to develop an idea and how difficult it is to take an idea to market).

Mexico has numerous innovation-related public institutions, including institutions associated with the Ministry of Education, the National Council for Science and Technology (CONACYT), and the national petroleum company (PEMEX), that address particular market failures. As is generally the case, however, market forces will not guarantee that the various components of these nonmarket institutions function coherently in a fashion that remedies the market failures they were intended to address. Finland, consistently ranked among the top two most competitive economies in the world, is extraordinarily mindful of the need to ensure the smooth collaboration of elements of its national innovation system—in a way that Mexico is not. The challenge for public policy is thus not only to help establish research and educational agencies but to see that market and extra-market incentives ensure that these agencies are adequately integrated with one another and with the productive sector.

The Efficiency of Knowledge Generation

The weak functionality of the Mexican national innovation system may contribute to low private and social rates of return and the apparent low efficiency in the use of innovation investment, measured in terms of how well it converts R&D financing into patenting.

Table 13.1 presents econometric estimates of patenting elasticities with respect to total R&D investment.[28] Also included in the regression in order to standardize by trade and endowment patterns are the log of exports to the United States, net exports of natural resources, and the log of the pre-sample mean of patents (1963–85). Every regression includes a dummy for Latin America and a dummy for Mexico that are interacted with the R&D terms as a measure of whether the transformation rates are higher than the sample mean. The first three columns suggest that there is a strong and statistically significant relationship between R&D expenditures and patents. However, the Latin American dummy with R&D generates a strongly negative coefficient, suggesting both that the region performs substantially worse than the norm in converting R&D expenditures into innovation and that Mexico performs even worse than the region as a whole. More generally, these results suggest that some of the spillovers among firms in the OECD are not occurring in Latin America.

TABLE 13.1 Determinants of Patent Counts

Methodology: Negative binomial, presample mean estimator *Presample: 1963–84*				
Number of observations	512	512	512	512
Number of countries	53	53	53	53
ln(average patents 1963–84)	0.32*	0.40*	0.38*	0.26*
ln(R&D expenditure)	0.78*	0.48*	0.48*	0.24*
ln(Ustrade)		0.37*	0.35*	0.32*
Nrleamer		−0.03**	−0.03*	−0.07*
ln(quality)*ln(R&D)				0.17*
ln(colaboration)*ln(R&D)				
ln(years of education)*ln(R&D)				
Tertiary enrollment*ln(R&D)				
Secondary enrollment*ln(R&D)				
LAC*ln(R&D)	−0.03*	−0.04*	−0.04*	0.00
MEX*ln(R&D)	0.00	−0.04*	−0.04*	−0.02*
Openness*ln(R&D)			0.02**	0.02***
Trade residual*ln(R&D)				
ICT*ln(R&D)				
Time trend	0.05*	0.04*	0.03*	0.03*
Pseudo *R*-squared	0.17	0.19	0.18	0.21
Log likelihood	−2,889.4	−2,828.5	−2,946.6	−2,766.2

Figure 13.6 expands this exercise by plotting the interactive dummies for a number of other regions and countries, leaving the OECD as reference. Finland, the Republic of Korea, and Taiwan (China), to mention a few, appear to be more efficient than Latin America and Mexico.

The negative Latin American effect disappears altogether when the quality of scientific and research organizations (universities and public research institutes) and the quality of collaboration between the productive sector and universities are introduced. When tertiary enrollment rates are added, Mexico's dummy also disappears. It is noteworthy that secondary enrollment and the level of development of information and communications technology do not explain the region's innovation inefficiency, suggesting that a focus on computer and communications infrastructure to the detriment of more traditional institutions of learning seems inadvisable.[29] In summary, Mexico's inefficiency results from a combination of low enrollment rates in universities and poor quality research and linkages between the universities and the productive sector.

TABLE 13.1 (*Continued*)

512	512	328	512	328	295
53	53	52	53	52	49
0.33*	0.28*	0.37*	0.23*	0.30*	0.28*
0.35*	0.31*	0.42*	0.15*	0.29*	0.34*
0.30*	0.38*	0.27*	0.37*	0.22*	0.19*
−0.08*	−0.10*	−0.07*	−0.11*	−0.09*	−0.08*
			0.12*	0.09*	0.09*
0.11*			−0.01	0.03**	0.02
	0.10*		0.08*		
		0.08*		0.07*	0.07*
		0.02		−0.01	−0.01
−0.01	−0.01*	−0.02*	0.01	0.00	0.00
−0.03*	−0.03*	−0.03*	−0.02**	−0.01	0.00
0.01	0.00	0.02*	0.00	0.02*	0.03*
					0.00
0.03*	−0.01	−0.03**	0.00	−0.01	−0.02
0.20	0.21	0.23	0.22	0.24	0.25
−2,770.7	−2,731.7	−1,722.0	−2,700.3	−1,686.5	−1,529.6

Source: Bosch, Lederman, and Maloney (2003) and authors' estimates.

*Significant at the 1 percent level.

**Significant at the 5 percent level.

Structural Change in Trade but Not in Innovation

As an additional measure of how well the national innovation system is working, we can determine how well Mexico's innovation effort corresponds with the emerging areas of comparative advantage—that is, the extent to which the national innovation system is "supporting" sectors with potential dynamism. NAFTA stimulated a structural change in which Mexico became a net exporter of machinery, including telecommunications equipment, road vehicles and parts, and office and data processing equipment (De Ferranti and others 2002). In addition, the incidence of intraindustry trade rose remarkably fast after 1994. What is not clear is whether these are simply upscale *maquiladora* operations or whether they have the potential to become knowledge clusters that both provide employment opportunities and drive future growth.

FIGURE 13.6 Efficiency of Research and Development in Selected Countries, 1985–2000

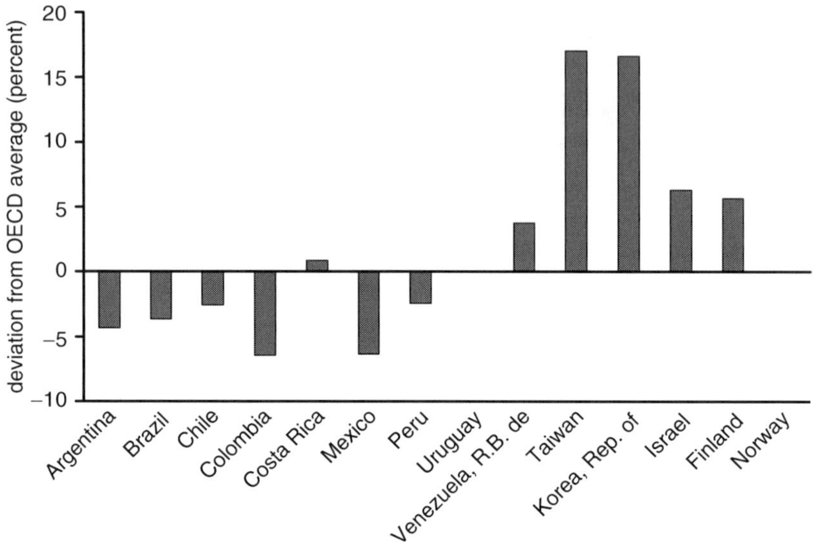

Source: Lederman and Maloney (2003).

The Innovative Revealed Comparative Advantage (IRCA) index reveals how patenting in various sectors in Brazil, the Republic of Korea, Mexico, and Taiwan (China) relative to each country's total patenting compares with the world's share of total patenting in that sector (figure 13.7) (Patel and Pavitt 1995). A value above unity suggests that a country has an IRCA in a sector. The principal sectors in which Mexico shows an IRCA are traditional processed foods, soaps and paints, and primary ferrous products. It has no IRCA in computers or automobiles. In contrast, Brazil's aircraft index has been rising since the privatization of EMBRAER (Brazil's small airplane and parts producer) in 1994. More dramatically, the Republic of Korea shows a notable increase in the index for electronic equipment, a sector in which it had a clear IRCA by 2000. Taiwan (China) also experienced substantial changes in its innovative structure and now has a clear IRCA in transportation industries (motorcycles and bicycles, miscellaneous transportation equipment) and various electronics products (radio and TV receiving equipment, household appliances).

Mexico's inability to generate an IRCA in the areas in which it appears to be exporting heavily suggests that the national innovation system is not supporting the productive sector. To the degree that Mexico is simply relying on temporarily

FIGURE 13.7 IRCA Index in Selected Industries and Countries, 1980–2000

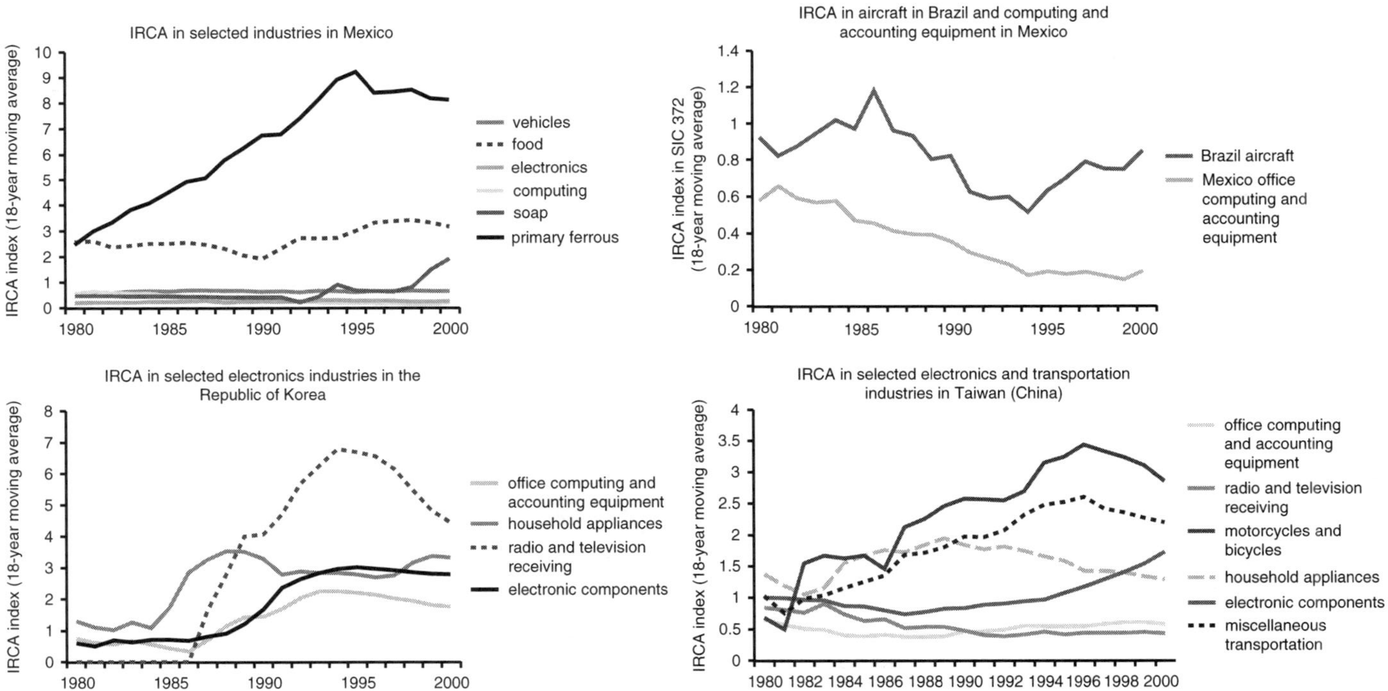

Source: Authors' calculations.

low labor costs to assemble computers rather than developing depth in supporting a knowledge base, these sectors may lose steam in the near future.

The National Innovation System: A Closer Look

The previous sections suggest that both the level and the efficiency of the national R&D effort depend on elements of the national innovation system—the quality of the nonmarket-based institutions developed to resolve market failure and how well they are linked. Mexico's national innovation system appears to be characterized by a lack of coordination among its components.

"Following the money" reveals that all of the government's R&D is self-financed and less than 8 percent of the R&D performed by universities is financed by the productive sector. In turn, neither the government nor universities contribute significant amounts to the productive sector's efforts. The three sectors function more or less independently—a recipe that is unlikely to produce economically meaningful innovation (see Lederman, Maloney, and Servén 2005 for details).

This pattern appears in sharp contrast to some of the Scandinavian countries and newly industrialized countries in Asia, which, while making a massive effort to create an educated scientific community, recognized the importance of a well-coordinated national innovation system to support the private sector in the context of an export-oriented trade policy (de Ferranti and others 2003). Even a cursory review of Finland's technology policy reveals a virtual obsession with coordination among various elements of the national innovation system (between firms and among firms, universities, and think tanks). The emphasis is not only on having the right human capital but on ensuring that it faces the correct incentives to interact and to generate networks that serve as collectors, creators, and disseminators of knowledge that enhances competitiveness.

Firm–Firm Links, Public Research Centers, and State-Owned Enterprises

Firms can internalize many of the externalities of nonappropriability, lumpiness, and risk discussed above by banding together to form industry-level associations and research centers. One of the most famous policy initiatives to promote interfirm linkages in the United States is SEMATECH, a research consortium of semiconductor manufacturers set up in 1987 by 14 U.S. semiconductor firms with the financial assistance of the U.S. government, which has been given credit for reviving the industry in the face of Japanese competition (Irwin and Klenow 1996). In Mexico the Unión Nacional Avícola concerns itself with raising the quality of technological inputs into the production process, particularly by importing foreign technologies

(Mayer 2002). In both cases, the dominant private sector presence ensures the relevance of R&D. But these efforts have developed without the involvement of the public sector, and they have arisen out of the firms' own concerns about their competitiveness.

Theory suggests that market forces alone often do not lead to the establishment of strong knowledge sharing and technology diffusion agreements among private firms, primarily because firms naturally seek to prevent potential competitors from profiting from their own R&D investments and know-how. Mexico, like many other countries, established many public research centers to fill the gap. Rosenberg (2000) argues that such institutions generally have poor track records, because government researchers have relatively little understanding of the needs of the productive sector and face few incentives to address them. This critique appears writ large in Mexico. Many of the 150 public research centers emerged in the import-substituting industrialization period, largely under the logic that fledgling industries lacked in-house capability to undertake the necessary research. Today they enjoy a disproportionate share of national research budgets. However, they are dependent on the secretariat to which they belong and frequently oppose any efforts of the secretariat to contract with firms or outside universities that might be more qualified to investigate a particular question.[30]

Furthermore, the lack of competition has had the usual depressing effects on quality and created disincentives to work with other institutions that might be potential rivals. Proposed reform laws foresee greater autonomy for the centers, and they offer the possibility that research funds will be allocated competitively rather than automatically to a center of investigation. Inducing a greater degree of competition for scarce research funding would move Mexico close to the funding allocation systems of high performers like Finland, which has unified its public research centers under one roof. This ensures the same marginal return to research spending across sectors and permits greater transparency and ease of monitoring.

There are, however, several cases of generally fruitful public research center–led technology transfer to the private sector, all in areas in which Mexico has an IRCA: ferrous metals and food items. In the case of the Center of Research and Technical Assistance (CIATEQ), based in Queretaro, however, successful experiences occurred before the privatization of the industries involved, suggesting the long gestation period needed to establish effective technology transfer relationships. The case of the Center for Research and Advanced Studies, in Irapuato, (CINVESTAV-Irapuato) suggests the importance of establishing international linkages and plugging the domestic national innovation system into the global stock of knowledge. Low levels of human capital in the private sector—related to low tertiary and secondary schooling, especially in rural areas—have also become obstacles to successful technology transfers intermediated by government-funded research centers.

One of the most straightforward ways of ensuring links between productive activities and R&D efforts is through public management of important economic sectors. This has been done in Brazil (EMBRAER before 1994), Chile (CODELCO), Costa Rica (telecommunications, utilities), Mexico (PEMEX), Taiwan (China) (telecommunications), and República Bolivariana de Venezuela (PEDEVESA). The evidence on the efficiency of total R&D expenditures in several of these countries is quite strong, indicating that public management is not necessarily misguided, although that seems not to be the case in Mexico. Furthermore, the initial links of the public research centers referred to above were to public sector firms.

Universities

Universities enrich the innovation network through several channels. First, they produce workers with tertiary education, who are the lifeblood of the national innovation system. The importance of this function cannot be overstated. Interviews with high-tech companies in Costa Rica highlight the issue of the generation of high-quality human capital in a country as an order of magnitude more important than other factors, including R&D incentives and R&D suppliers (de Ferranti and others 2002). Second, universities are well suited for large, long-term, basic research. Third, they are likely to maintain contacts with research centers in industrial countries and hence perform an important role as a link to worldwide scientific and technological know-how. In all cases, the degree to which they remedy the underlying market failures depends on their links to the productive sector.

Higher education plays a dominant role in Mexican and Latin American R&D (table 13.2), which often translates into a bias toward basic as opposed to applied research (Hansen and others 2002).[31] This bias is not necessarily a weakness, provided that the incentives academics face ensure that the work is of high quality and is linked to the productive sector. In Finland 40 percent of firms have collaborative arrangements with universities (Brunner 2001), and these interactions have been vital to the continued dynamism of both the high-tech and more traditional forest industries, as Blomström and Kokko (2001) have documented. Comparable figures are not available for Mexico, but the results of the *Global Competitiveness Report* of the World Economic Forum indicate that businesses perceive university–private sector interactions in Mexico, as in many Latin American countries, as weak (figure 13.8).

These results are consistent with case studies by Mayer (2002) of Avimex, a veterinary pharmaceutical company in Mexico. He notes that although Avimex spends 10–15 percent of sales on R&D and develops world-class innovations in joint projects with U.S. research institutes, the company, like most Mexican firms, lacks research partners in Mexico. This lack has forced the firm to look for partners in the United States.

TABLE 13.2 Structure of R&D Effort in Selected Countries, 1995–2000 (percentage of total R&D expenditures, annual averages)

Country	Financed by productive sector	Financed from abroad	Performed by productive sector	Performed by higher education	Performed by nonproductive public sector
Brazil	39.14	0.00	44.04	44.27	11.69
Canada	44.56	12.69	57.69	28.67	13.64
Chile	19.37	5.75	9.62	47.11	43.27
Costa Rica	—	—	32.58	48.72	18.70
Finland	62.72	4.12	66.95	19.12	13.93
Ireland	66.46	8.79	71.58	20.46	7.96
Rep. of Korea	73.02	0.08	72.53	10.42	17.05
Mexico	19.06	5.22	22.66	39.82	37.52
Sweden	64.37	3.47	75.73	21.57	2.70
Taiwan (China)	60.03	0.10	61.00	12.20	26.80
United States	62.98	0.00	73.88	14.33	11.78
R. B. de Venezuela	38.13	0.00	—	—	—

Source: Authors' calculations based on annual data collected by Lederman and Sáenz (2003).
Note: — = not available.

In her study of 13 high-tech firms in Jalisco, Mexico's Silicon Valley, which employed roughly 249 Mexican engineers involved in research and design, Rivera Vargas (2002) finds that there was limited contact between the firms and their academic counterparts. In particular, no joint research projects in electronics or related areas were being carried out. Hanson (2004) argues that although there are cases in which multinational corporations and universities collaborate on curriculum development, there appear to be no systematic plans to use these relationships to transfer technology, and there appears to be little active learning. More generally, the government, when encouraging manufacturing investment such as the *maquiladoras*, never looked beyond job creation to active programs of technology transfer.

The isolation of the Mexican university arises from both the demand and supply sides. A number of factors—including poor design and faulty incentives—are responsible for the dearth of effective links and collaboration between scientific institutions and the private sector in Latin America and the Caribbean. But there is also a lack of incentives for universities to link with and address private sector knowledge needs. Arguably, the region has more of a liberal arts than a technical culture, with deep historical roots that resonate with critiques of the inadequate U.K. system a few decades ago.[32] But there are also more concrete disincentives.

FIGURE 13.8 Private Sector Perceptions of Quality of Scientific Institutions and University–Private Sector Collaboration, Selected Countries, 1994–2001

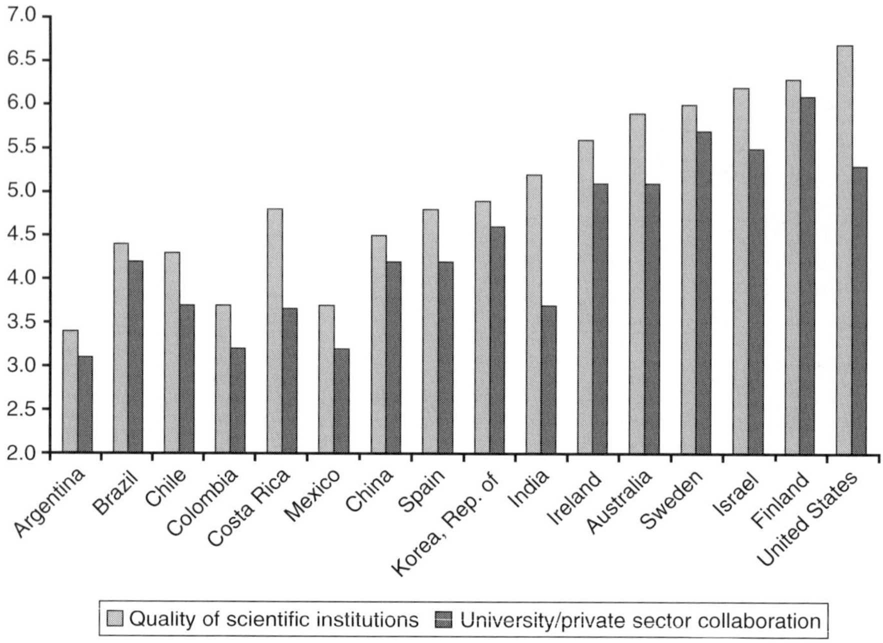

Source: World Economic Forum (2002).

Because researchers cannot appropriate the benefits of innovation, they have little incentive to undertake innovations and link with the private sector. Various industrial countries allow ownership rights to government-funded R&D—often on a case-by-case basis and in some countries, including Japan and the United States, explicitly in the national patent laws. In the United States the Bayh-Dole Act of 1980 allows industry contractors of the government, national laboratories, and academic institutions to automatically retain title to the inventions that come out of their research work, even if the work was or is funded by the government. In return, the government receives from the university or industry a royalty-free license for governmental purposes. There is convincing evidence that these laws have helped speed the rate of patenting by research conducted in U.S. public laboratories (Jaffe and Lerner 2001). The lessons learned regarding incentives for patenting by researchers also apply to Mexican public research centers and universities.

Goldfarb, Henrekson, and Rosenberg (2001) cite differences in academic structures and their influence on researcher involvement with the commercialization of research ideas as an important reason for the much lower spillovers from academia to industry in Sweden than in the United States. Competition for researchers and scientists in the United States has reinforced the need for policies that are attractive to them. Universities have established technology transfer offices, and they have liberal policies on faculty leaves of absence and consulting privileges, which allow faculty to pursue commercial opportunities while keeping their positions on the faculty intact. In contrast, Swedish universities do not gain from commercialization and hence resist faculty involvement with industry. For instance, it is difficult for Swedish professors to take temporary leave to organize firms, as professors in the United States are permitted to do.

If an innovation is to be commercialized, property rights must be allocated to the university or the researcher. How these rights are allocated also has a significant impact. If researchers get the property rights, they are likely to remain at the university; if they do not retain the rights, they are likely to move to the productive sector.

At a more mundane level, Mayer (2002) argues that bureaucratic rigidities in Mexico make it difficult to write contracts with universities or gain access to their laboratories and equipment. Mexico's very centralized and bureaucratic approval process serves as a disincentive to firms attempting to forge ties with universities.

Supporting Capital and Labor Markets and Intellectual Property Regimes

Credit markets and labor markets are critical supporting elements of the national innovation system. Here, too, Mexico may be facing important barriers to raising the innovation effort.

By nature, innovation has long gestation periods and high levels of risk; the absence of long-term credit markets can inhibit long-run innovation-related risk taking. The shallowness of Mexican markets is well documented. Less is known about the specialized capital market institutions required across the innovation cycle, namely, precommercialization financing (seed money) and the subsequent venture capital infusion needed to bring ideas to market. In India in addition to ample fiscal incentives for R&D, a development financial institution (the Industrial Credit and Investment Corporation of India and its subsidiaries) initiated venture capital in 1988. Subsequently, private venture capital firms emerged, albeit on a smaller scale. In Israel innovation policy is essentially credit policy (Trajtenberg 2001). In Mexico there is currently no recognition of the venture capital firm as a legal entity. There are holding companies, or SIMCAS, that hold other firms' assets, but the legal structure does not encourage the association of several entrepreneurs to share risk, and the profits of the SIMCAs enjoy unfavorable fiscal treatment.

That said, other countries in the region, such as Chile, that have established venture capital–type institutions have suffered from a lack of "deal flow"—the ideas are not there, or if they are there, they are unable to escape the institution in which they were generated. A government program to resolve the capital problem may be "pushing on a string" if it does not also address the fundamental dysfunction of the national innovation system.

Parente and Prescott (2000) identify the barriers to the adoption of new technologies embodied in labor codes or other workplace restrictions as explanations for differing levels of development. Whether their largely anecdotal evidence can establish such impediments as primary, it is clear that Mexico's labor code is not friendly to the introduction of new technologies or processes. The dynamic concerns about TFP growth should assume a larger profile in the national debates over labor reform (see OECD 2001).

Although the literature on intellectual property rights is somewhat ambiguous, the strengthening of intellectual property regimes seems important to the innovation effort (see Park 2001, for example). The index of intellectual property regimes is a significant determinant of R&D. In the Mexican case, the rise in the index can account for all of the rebound in R&D, and it is the only determining variable among increased credit intermediation, lower risk, and greater ability to mobilize public resources that improved and thus could have been responsible for the modest recovery in the share of R&D in Mexico's GDP during the second half of the 1990s (see Lederman and Maloney 2003).

Conclusion

The countries to which Mexico often compares itself, particularly the newly industrialized countries of Asia, have followed growth strategies characterized by a very active approach to innovation policy. This has often included an active engagement in world markets, particularly on the export side. NAFTA—and economic integration with Canada and the United States more generally—therefore represents a step in the right direction. Import competition has been associated with improvements in manufacturing TFP, and it may have indirectly increased productivity by improving the efficiency of R&D. Improvements in intellectual property rights and increased credit availability were probably associated with the moderate yet insufficient increase in R&D expenditures in the late 1990s. Nevertheless, the main conclusion of this chapter is that NAFTA is not enough to ensure technological convergence in North America.

On basic indicators of investment in innovation and innovation outcomes, Mexico lags significantly behind comparator countries. The low levels of both innovation effort and efficiency suggest that the country needs to address issues related to the inefficiency of its national innovation system. In particular, it needs to improve the quality of its research institutions by changing both internal promotion/ advancement criteria and incentives to interact with the private sector. In addition to regulatory changes affecting intellectual property regimes, researchers for universities, and public research centers, it is likely that public subsidies will be needed to provide additional incentives for firms to establish such links, especially when there is little previous collaborative experience and thus little mutual trust. The careful design of such mechanisms is essential; simply increasing the financing to existing programs and institutions should emphatically not be an option. Reform of supporting institutions, such as credit and labor markets, should also be considered. Finally, more effort needs to be dedicated to collecting statistics and establishing methods to monitor the functioning of the national innovation system and the success of government interventions. It may be useful to build an information-based monitoring facility that would play a role similar to that of Spain's National Commission for the Evaluation of Research Activity (CNEAI).[33]

Clearly, Mexico's location next to one of the most dynamic sources of innovation in the world offers opportunities for collaboration with northern universities and firms. Within the NAFTA context, Mexico might negotiate the co-financing of joint research or exchange programs. However, such efforts will lead to little if human capital and institutions are not brought up to speed to enable Mexico to participate as a full partner in innovation.

Notes

1. Apologies to Pastor (2002), who first used the title "NAFTA Is Not Enough." We find it especially appropriate to the issues of innovation discussed here.

2. At the firm level, see Cohen and Levinthal (1989); Forbes and Wield (2000); Griffith, Redding, and van Reenen (2003); and Pavitt (2001). At the national level, see Baumol, Nelson, and Wolff (1994).

3. Most of the analyses presented in this chapter are quantitative, relying on internationally comparable indicators of various aspects of innovative activity and technological progress provided by Lederman and Sáenz (2003).

4. The decline in productivity growth in the United States and other high-income countries in the 1970s has been attributed to the oil shock of 1973 and its macroeconomic repercussions (Griliches 1988). It is difficult to blame the decline in productivity in Mexico and other oil exporters on this factor. The East Asia and Pacific region, for example, did not experience such a slowdown, perhaps because some countries in the region, such as Indonesia, are oil exporters. The Republic of Korea, however, did experience the slowdown.

5. In assessing the impact of various factors on TFP growth in the region, Loayza, Fajnzylber, and Calderón (2002) find that for all 20 Latin American countries in their sample, the impact of structural reform policies was positive; for 15 countries in the sample, stabilization policies were positive as well. They note, however, that the estimated combined contribution of the two ranged from 2.5 to 3.0 percent—not insignificant but not likely to transform the region into Asian or Scandinavian growth miracles.

6. Many resource-rich countries—Australia, Canada, Finland, and Sweden—also closed somewhat after the Great Depression, although none did so to the degree of Latin America (Maloney 2002).

7. The USPTO demands that the invention be novel, nontrivial, and have commercial application (Jaffe and Trajtenberg 2002).

8. The patent residuals are derived from a negative binomial regression on GDP, population, and their squares plus a term for trade with the United States, which is known to be correlated with patenting in the United States. Median regression on the same regressors less the U.S. trade variable gives the publications residuals. See Bosch, Lederman, and Maloney (2003) for technical details about the methodologies and data.

9. Interested readers can also consult other literature reviews on these issues, including Navaretti and Tarr (2000) and Saggi (chapter 3 of this volume).

10. The corresponding result from a 30-year cross-section of countries was below this range, falling to 0.005 percent.

11. For a strong critique of cross-country studies that examines the link between trade and economic growth, see Rodriguez and Rodrik (2000).

12. For a more expansive review of the literature, see Keller (2001).

13. High R&D industries are those that have relatively high shares of R&D expenditures over sales in high-income countries.

14. There was no difference between high and low R&D-intensive industries, suggesting that industrial composition is not critical to the benefits of NAFTA.

15. Muendler (2002) argues that this may be explained by the failure among manufacturers to adjust production practices to the increased availability of imported inputs.

16. They do find, however, that the presence of exporters may make it easier for nonexporters to break into foreign markets. In Colombia nonexporters appear to experience cost reductions when export activity increases.

17. Intermediate inputs increased marginal costs and quality among producers of rubber, fertilizer, and pesticide. For pharmaceutical producers, imported intermediates combined with exports or imported capital goods, reduced marginal costs and tended to increase product quality. But these were exceptions to a fairly ambiguous record.

18. The same inconclusiveness appears in the U.S. microeconomic data (Bernard and Jensen 1999). A study of a panel of Spanish firms concludes that there is evidence only in favor of the (Schumpeterian) firm-selection channel; the evidence on the learning-by-exporting hypothesis is very weak (Delgado, Fariñas, and Ruamo 2001). Aw, Chung, and Roberts (2000) report similar results for firms in the Republic of Korea and Taiwan (China). Kraay's study on China (chapter 7 of this volume) is an exception.

19. For less developed countries, see, for example, Calderón, Loayza, and Servén (2002); Carkovic and Levine (2002); and Loayza, Fajnzylber, and Calderón (2002).

20. In a comprehensive review of the literature, Lipsey (2002) argues that the evidence is vast that foreign firms tend to be at least as productive as domestic firms and hence their presence pushes up average productivity. However, the evidence that the presence of foreign firms has positive productivity spillovers is ambiguous. The great majority of studies that find strong effects employ cross-sectional data that cannot control for unobserved country characteristics. Studies using firm-level panels frequently find insignificant or even negative effects (see, for example, Aitken and Harrison 1999 for República Bolivariana de Venezuela). Van Pottelsberghe de la Potterie and Lichtenberg (2001) find that investing in a more technologically advanced country and hence adding foreign production to domestic production increases productivity in the home country but that investing in a technologically less advanced country has insignificant or negative results for the host developing country. Baldwin, Braconier, and Forslid (2000) find mixed results for seven OECD countries. Using panel firm-level data from Sweden, Braconier, Ekholm, and Midelfart Knarvik (2000) discover no spillovers from incoming FDI on productivity; the only variable in their sample affecting TFP is own-country R&D. Using panel data on technological transfer from the United States, Xu (2000) finds a technology transfer effect by U.S. multinational corporations only for advanced countries, although a competition effect does appear to increase productivity. Kinoshita (2000) finds little evidence of positive effects of FDI at the firm level in the Czech Republic

from 1995 to 1998. Javorcik (chapter 10 of this volume) reports no direct impact of FDI in Lithuania on firms in the same industry, although she does find an impact on affiliated upstream suppliers.

21. For these authors, "innovative-effective" human capital is a combination of the level of education and the effort invested by the economy to develop new technologies based on the technological frontier.

22. Similar results were previously suggested by Grossman and Helpman (1991), among others. They propose a model with multiple growth equilibria resulting from intranational R&D externalities. Maloney (2002) argues that this partly explains why Latin America nearly disappeared from some industries it once dominated globally.

23. To derive the relationship between the rate of R&D investment and GDP, we estimated a median regression in which the dependent variable was the log of R&D expenditures over GDP from the early 1960s to 2000 and the arguments were the log of GDP per capita and the log of GDP per capita squared.

24. These are residuals from more general median regressions for both R&D and licensing, which include log GDP, log GDP squared, log labor, log labor squared, and time dummies as arguments.

25. Griliches (1992) estimates social returns to R&D in the United States of 20–60 percent. Jones and Williams (1998) confirm that rates of return in the United States are at least 30 percent and calculate that the optimal resources that should be devoted to R&D could be four times the current level in the United States.

26. See also Lederman and Maloney (2002), who investigate the impact of trade structure on growth and find natural resources to be positively related to growth.

27. For an intuitive discussion of the determinants of the equilibrium private marginal rate of return, see David, Hall, and Toole (2000). The equilibrium return is determined by the marginal costs of and returns to R&D. Some market failures affect the costs (that is, capital markets might be incomplete), others affect the returns (that is, the nonappropriability problem).

28. These estimates were obtained using a pooled regression of 52 countries over a 15-year period (1985–2000) in a negative binomial regression application of Blundell, Griffith, and Windmeijer's (2002) presample mean estimator, which aims to control unobserved country-specific characteristics and the likely endogeneity of some of the explanatory variables.

29. The information and communication technology index used in this analysis is the one provided by Lederman and Xu (2001). It is the result of the first principal component from factor analysis using four indicators of information and communication technology: per capita telephone lines, cellular phones, personal computers, and Internet hosts.

30. This information comes from discussions in 2001 and 2002 with and presentations by Carlos Bazdresch, Centro de Investigacion y Docencia Economicas (CIDE). Mr. Bazdresch is a former head of CONACYT.

31. The data in table 13.1 for Mexico differ from those in table 13.2 because of the different sources of information and different time periods. The data in table 13.2 are internationally comparable.

32. This appears to be the case throughout the region. Agapitova and Holm-Nielsen (2002) argue that overall the university mentality in Chile is not geared toward solving problems on a business time scale. Mullin (2001) argues that overall academic interests tend to be narrow and unapplied. Observers of Costa Rica's two-star technical school stress not so much incentives but the "foundational impulse"—a desire to be patterned more on the Massachusetts Institute of Technology or other technical schools of excellence than on those with a liberal arts bias.

33. This commission has improved the quality and quantity of basic research output in Spain, even during a period of declining public funding of research (Jiménez-Contreras, de Moya Anegón, and López-Cózar 2003). The main monitoring variable used by the CNEAI is the number of publications based on government- and university-funded research, which is made public once or twice a year. Given the reputational rewards sought by researchers, it is likely that the mere publication of the performance index can improve the quality of research. This principle could be used to improve the quality of applied research, by, for example, maintaining an accurate count of patents granted by the Canadian, Mexican, and U.S. governments to researchers residing in Mexico and financed by public funds, through public research centers or universities. Other important variables to monitor are discussed in De Ferranti and others (2003, chapter 8).

References

Acemoglu, Daron, and Fabrizio Zilibotti. 2001. "Productivity Differences." *Quarterly Journal of Economics* 116 (2): 563–606.

Acemoglu, Daron, Philippe Aghion, and Fabrizio Zilibotti. 2002. "Distance to Frontier, Selection, and Economic Growth." Working Paper 9066, National Bureau of Economic Research, Cambridge, MA.

Agapitova, Natalia, and Lauritz Holm-Nielsen. 2002. "Chile: Science, Technology, and Innovation." LCSHD Paper Series No. 79, Latin America and the Caribbean Region, Human Development Department, World Bank, Washington, DC.

Aitken, Brian J., and Ann E. Harrison. 1999. "Do Domestic Firms Benefit from Direct Foreign Investment? Evidence from Venezuela." *American Economic Review* 89 (3): 605–18.

Alvarez, Roberto, and Raymond Robertson. 2001. "Exposure to Foreign Markets and Firm-Level Innovation: Evidence from Chile and Mexico." Departments of Economics, University of Chile, Santiago, and Macalester College, St. Paul, MN.

Aw, Bee Yan, Sukkyun Chung, and Mark J. Roberts. 2000. "Productivity and Turnover in the Export Market: Micro-level Evidence from the Republic of Korea and Taiwan (China)." *World Bank Economic Review* 14 (1): 65–90.

Baldwin, Richard E., and Elena Seghezza. 1996. "Testing for Trade-Induced Investment-Led Growth." Working Paper 5416, National Bureau of Economic Research, Cambridge, MA.

Baldwin, Richard, Henrick Braconier, and Rikard Forslid. 2000. "Multinationals, Endogenous Growth and Technological Spillovers: Theory and Evidence." Discussion Paper 2155, Centre for Economic Policy Research, London.

Barro, Robert J. 1991. "Economic Growth in a Cross-Section of Countries." *Quarterly Journal of Economics* 106 (2): 407–44.

Baumol, William J., Richard R. Nelson, and Edward N. Wolff, eds. 1994. *The Convergence of Productivity, Its Significance and Its Varied Connotations*. Oxford: Oxford University Press.

Bernard, Andrew B., and J. Bradford Jensen. 1999. "Exporting and Productivity." Working Paper 7135, National Bureau of Economic Research, Cambridge, MA.

Blomström, Magnus, and Ari Kokko. 2001. "From Natural Resources to High-Tech Production: The Evolution of Industrial Competitiveness in Sweden and Finland." Stockholm School of Economics.

Blomström, Magnus, and Edward N. Wolff. 1994. "Multinational Corporations and Productivity Convergence in Mexico." In *Convergence of Productivity: Cross-National Studies and Historical Evidence*, ed. William J. Baumol, Richard R. Nelson, and Edward N. Wolff, 263–84. Oxford: Oxford University Press.

Blundell, Richard, Rachel Griffith, and Frank Windmeijer. 2002. "Individual Effects and Dynamics in Count Data Models." *Journal of Econometrics* 108 (1): 113–31.

Bosch, Mariano, Daniel Lederman, and William F. Maloney. 2003. "Patenting and Efficiency: A Global View." Office of the Chief Economist for Latin America and the Caribbean Region, World Bank, Washington, DC.

Braconier, Henrik, Karolina Ekholm, and Karen Helene Midelfart Knarvik. 2000. "Does FDI Work as a Channel for R&D Spillovers? Evidence Based on Swedish Data." Discussion Paper 2469, Centre for Economic Policy Research, London.

Brunner, José Joaquín. 2001. "Chile: Report and Index on Technological Capacity." School of Government, Universidad Adolfo Ibáñez, Santiago de Chile.

Calderón, César, Norman Loayza, and Luis Servén. 2002. "Greenfield FDI vs. Mergers and Acquisitions: Does the Distinction Matter?" Office of the Chief Economist for Latin America and the Caribbean Regino, World Bank, Washington, DC.

Carkovic, Maria, and Ross Levine. 2002. "Does Foreign Direct Investment Accelerate Economic Growth?" Finance Department, University of Minnesota, Minneapolis, MN.

Casas, Rosalba, Rebeca de Gortari, and M. Josefa Santos. 2000. "The Building of Knowledge Spaces in Mexico: A Regional Approach to Networking." *Research Policy* 29 (2): 225–41.

Clerides, Sofronis K., Saul Lach, and James R. Tybout. 1998. "Is Learning by Exporting Important? Micro-Dynamic Evidence from Colombia, Mexico and Morocco." *Quarterly Journal of Economics* 113 (3): 903–47.

Coe, David T., Elhanan Helpman, and Alexander W. Hoffmaister. 1997. "North-South R&D Spillovers." *Economic Journal* 107 (440): 134–49.

Cohen, Wesley M., and Daniel A. Levinthal. 1989. "Innovation and Learning: The Two Faces of R&D." *Economic Journal* 99 (397): 569–96.

David, Paul A., Bronwyn H. Hall, and Andrew A. Toole. 2000. "Is Public R&D a Complement or Substitute for Private R&D? A Review of the Econometric Evidence." *Research Policy* 29 (4–5): 497–529.

de Ferranti, David, Daniel Lederman, William F. Maloney, and Guillermo E. Perry. 2002. "From Natural Resources to the Knowledge Economy: Trade and Job Quality." Latin American and Caribbean Studies Department, World Bank, Washington, DC.

de Ferranti, David, Guillermo E. Perry, Indermit Gill, J. Luis Guasch, William F. Maloney, Carolina Sanchez-Paramo, and Norbert Schady. 2003. "Closing the Gap in Education and Technology." Latin American and Caribbean Studies Department, World Bank, Washington, DC.

Delgado, Miguel, Jose Fariñas, and Sonia Ruamo. 2001. "Firm Productivity and Export Markets: A Non-Parametric Approach." *Journal of International Economics* 57 (2): 397–422.

Dollar, David, and Edward N. Wolf. 1997. "Convergence of Industry Labor Productivity among Advanced Economies, 1963–1982." In *The Economics of Productivity*, Vol. 2, ed. Edward N. Wolff, 39–48. London: Elgar.

Eaton, Jonathan, and Samuel Kortum. 1996. "Trade in Ideas: Patenting and Productivity in the OECD." *Journal of International Economics* 40 (3): 251–78.

Forbes, Naushad, and David Wield. 2000. "Managing R&D in Technology-Followers." *Research Policy* 29 (9): 1095–109.

Furman, Jeffrey L., Michael Porter, and Scott Stern. 2002. "The Determinants of National Innovative Capacity." *Research Policy* 31 (6): 899–933.

Goldfarb, Brent, Magnus Henrekson, and Nathan Rosenberg. 2001. "Demand-vs. Supply-Driven Innovations: U.S. and Swedish Experiences in Academic Entrepreneurship." Working Paper in Economics and Finance 0436, Stockholm School of Economics.

Griffith, Rachel, Stephen Redding, and John van Reenen. 2003. "R&D and Absorptive Capacity: From Theory to Data." Working Paper 01/03, Institute for Fiscal Studies, London.

Griliches, Zvi. 1988. "Productivity Puzzles and R&D: Another Nonexplanation." *Journal of Economic Perspectives* 2 (4): 9–21.

———. 1990. "Patent Statistics as Economic Indicators: A Survey." *Journal of Economic Literature* 28 (4): 1661–707.

———. 1992. "The Search for R&D Spillovers." *Scandinavian Journal of Economics* 94 (S29): 29–47.

———. 1994. "Productivity, R&D and the Data Constraint." *American Economic Review* 84 (1): 1–23.

Griliches, Zvi, and Frank Lichtenberg. 1984a. "Interindustry Technology Flows and Productivity Growth: A Reexamination." *Review of Economics and Statistics* 66 (2): 324–29.

———. 1984b. "R&D and Productivity Growth at the Industry Level: Is There Still a Relationship?" In *R&D, Patents and Productivity*, ed. Zvi Griliches, 465–502. Chicago: University of Chicago Press.

Grossman, Gene M., and Elhanan Helpman.1991. *Innovation and Growth in the Global Economy*. Cambridge, MA: MIT Press.

Hall, Robert, and Charles I. Jones. 1999. "Why Do Some Countries Produce So Much More Output per Worker Than Others?" *Quarterly Journal of Economics* 114 (1): 83–116.

Hansen, Thomas Nokolaj, Natalia Agapitova, Lauritz Holm-Nielsen, and Ognjenka Goga Vukmirovic. 2002. "The Evolution of Science and Technology: Latin America and the Caribbean in Comparative Perspective." World Bank, Department of Human Development, Washington, DC.

Hanson, Mark. 2004. "Transnational Corporations, Knowledge Transfer, and National Development: The Case of Mexico." Graduate School of Education, University of California, Riverside.

Howitt, Peter, and David Mayer-Foulkes. 2002. "R&D, Implementation and Stagnation: A Schumpeterian Theory of Convergence Clubs." Working Paper 9104, National Bureau of Economic Research, Cambridge, MA.

Irwin, Douglas A., and Peter J. Klenow. 1996. "High-Tech R&D Subsidies: Estimating the Effects of Sematech." *Journal of International Economics* 40 (3/4): 323–44.

Jaffe, Adam B., and Josh Lerner. 2001. "Reinventing Public R&D: Patent Policy and the Commercialization of National Laboratory Technologies." *Rand Journal of Economics* 32 (1): 167–98.

Jaffe, Adam B., and Manuel Trajtenberg. 2002. *Patents, Citations and Innovations: A Window on the Knowledge Economy*. Cambridge, MA: MIT Press.

Jiménez-Contreras, Evaristo, Félix de Moya Anegón, and Emilio Delgado López-Cózar. 2003. "The Evolution of Research Activity in Spain: The Impact of the National Commission for the Evaluation of Research Activity (CNEAI)." *Research Policy* 32 (1): 123–42.

Jones, Charles I., and John C. Williams. 1998. Measuring the Social Return to R&D. *Quarterly Journal of Economics* 113 (4): 1119–35.

Keller, Wolfgang. 2001. "International Technology Diffusion." Working Paper 8573, National Bureau of Economic Research, Cambridge, MA.

———. 2002. "Trade and the Transmission of Technology." Working Paper 6113, National Bureau of Economic Research, Cambridge, MA.

Kinoshita, Yuko. 2000. "R&D and Technology Spillovers via FDI: Innovation and Absorptive Capacity." Working Paper 349, William Davidson Institute, Ann Arbor, MI.

Lederman, Daniel, and William F. Maloney. 2003. "Trade Structure and Growth." Office of the Chief Economist for Latin America and the Caribbean Region, World Bank, Washington, DC.

Lederman, Daniel, and Laura Sáenz. 2003. "Innovation around the World: A Cross-Country Data Base of Innovation Indicators." Office of the Chief Economist for Latin America and the Caribbean Region, World Bank, Washington, DC.

Lederman, Daniel, and Lixin Colin Xu. 2001. "Comparative Advantage and Trade Intensity: Are Traditional Endowments Destiny?" Office of the Chief Economist for Latin America and the Caribbean Region, World Bank, Washington, DC.

Lederman, Daniel, William F. Maloney, and Luis Servén. 2005. *Lessons from NAFTA for Latin America and the Caribbean.* Palo Alto, CA: Stanford University Press.

———. 2003. "R&D and Development." World Bank, Office of the Chief Economist for Latin America and the Caribbean Region, Washington, DC.

Levine, Ross, and David Renelt. 1992. "A Sensitivity Analysis of Cross-Country Growth Regressions." *American Economic Review* 82 (4): 942–63.

Lipsey, Robert E. 2002. "Home and Host Country Effects of FDI." Working Paper 9293, National Bureau of Economic Research, Cambridge, MA.

Lloyd-Ellis, Huw, and Joanne Roberts. 2002. "Twin Engines of Growth: Skills and Technology as Equal Partners in Balanced Growth." *Journal of Economic Growth* 7 (2): 87–115.

Loayza, Norman, Pablo Fajnzylber, and César Calderón. 2002. "Economic Growth in Latin America and the Caribbean: Stylized Facts, Explanations, and Forecasts." Office of the Chief Economist for Latin America and the Caribbean Region, Regional Studies Program, World Bank, Washington, DC.

López-Acevedo, Gladys. 2002. "Determinants of Technology Adoption in Mexico." Latin American and the Caribbean Regional Office, Poverty Reduction and Economic Management Sector Unit, World Bank, Washington, DC.

López-Córdova, J. Ernesto. 2002. "NAFTA and Mexico's Manufacturing Productivity: An Empirical Investigation Using Micro-Level Data." Paper presented at the meetings of the Latin American and Caribbean Economic Association, Madrid.

Maloney, William F. 2002. "Missed Opportunities: Innovation, Natural Resources, and Growth in Latin America." *Economia* 3 (1): 111–50.

———. 2005. "Patterns of Innovation." Latin American and the Caribbean Regional Office, World Bank, Washington, DC.

Maloney, William F., and Andrés Rodriguez-Clare. 2005. "Innovation Shortfalls." Latin American and the Caribbean Regional Office, World Bank and Inter-American Development Bank, Washington, DC.

Martin, William, and Devashish Mitra. 2001. "Productivity Growth and Convergence in Agriculture and Manufacturing." *Economic Development and Cultural Change* 49 (2): 403–22.

Maskus, Keith E. 2000. "Intellectual Property Rights in the Global Economy." Institute for International Economics, Washington, DC.

Mayer, David. 2002. "Liberalization, Knowledge, and Technology: Lessons from Veterinary Pharmaceutics and Poultry in Mexico." Centro de Investigacion y Docencia Economicas, Mexico.

Meza González, Liliana, and Ana Belen Mora Yagüe. 2002. "Why Do Mexican Manufacturing Firms Invest in R&D?" Department of Economics, Universidad Iberamericana, Mexico, D.F. Mexico and Department of Economics, Georgetown University, Washington, DC.

Muendler, Marcus. 2002. "Openness and Growth: A Study of Brazilian Manufacturers, 1986–1998." Department of Economics, University of California, Berkeley.

Mullin, James. 2001. *Science, Technology, and Innovation in Chile.* International Development Research Centre, Ottawa.

Navaretti, Giorgio Barba, and David G. Tarr. 2000. "International Knowledge Flows and Economic Performance: A Review of the Evidence." *World Bank Economic Review* 14 (1): 1–15.

Nelson, Richard R. 1993. "A Retrospective." In *National Innovation Systems: A Comparative Analysis*, ed. Richard Nelson, 505–24. New York: Oxford University Press.

Nelson, Richard, and Nathan Rosenberg. 1993. "Technical Innovation and National Systems." In *National Innovation Systems: A Comparative Analysis*, ed. Richard Nelson, 3–22. New York: Oxford University Press.

OECD (Organisation for Economic Co-operation and Development). 1999. *Managing National Innovation Systems.* Paris

———. 2001. *Innovative Clusters: Drivers of National Innovation Systems.* Paris.

———. 2002. "And the Twain Shall Meet: The Cross-Market Effects of Product and Labour Market Policies." In *OECD Employment Outlook* 2002 (9): 257–315.

Parente, Stephen L., and Edward C. Prescott. 2000. *Barriers to Riches.* Cambridge, MA: MIT Press.

Park, Walter G. 2001. "Intellectual Property and Patent Regimes." In *Economic Freedom of the World, 2001 Annual Report*, 101–20. Vancouver: Fraser Institute.

Pastor, Robert A. 2002. "NAFTA Is Not Enough: Steps toward a North American Community." In *The Future of North American Integration beyond NAFTA*, ed. Peter Hakim and Robert E. Litan, 87–118. Washington, DC: Brookings Institution Press.

Patel, Pari, and Keith Pavitt. 1995. "Patterns of Technological Activity: Their Measurement and Interpretation." In *Handbook of the Economics of Innovation and Technological Change*, ed. Paul Stoneman, 14–51. Padstow, United Kingdom: T.J. Press Ltd.

Pavitt, Keith. 2001. "Public Policies to Support Basic Research: What Can the Rest of the World Learn from U.S. Theory and Practice? (And What They Should Not Learn)." *Industrial and Corporate Change* 10 (3): 761–79.

Rivera Vargas, María Isabel. 2002. *Technology Transfer via University-Industry Relationship: The Case of the Foreign High-Technology Electronics Industry in Mexico's Silicon Valley.* New York: Routledge-Falmer.

Rodriguez, Francisco, and Dani Rodrik. 2000. "Trade Policy and Economic Growth: A Skeptic's Guide to the Cross-National Evidence." In *NBER Macroeconomics Annual 2000.* Cambridge, MA: MIT Press.

Romer, Paul. 1990. "Endogenous Technological Change." *Journal of Political Economy* 98 (5): 71–102.

Rosenberg, Nathan. 2000. "Why Do Firms Do Basic Research (with Their Own Money)?" *Economics of Science and Innovation* 2: 197–206.

Scherer, Fredric M. 1982. "Inter-Industry Technology Flows and Productivity Growth." *Review of Economics and Statistics* 64 (4): 627–34.

Schiff, Maurice, and Yanling Wang. 2002. "Regional Integration and Technology Diffusion: The Case of NAFTA." Working Paper 3132, World Bank, Washington, DC.

———. 2004. "Education, Governance and Trade-Related Technology Spillovers in Latin America." IZA Discussion Paper 1028, Institute for the Study of Labor, Bonn, Germany.

Solow, Robert. 1956. "A Contribution to the Theory of Economic Growth. *Quarterly Journal of Economics* 70 (1): 65–94.

———. 1957. "Technical Change and the Aggregate Production Function." *Review of Economics and Statistics* 39: 312–20.

Sveikauskas, Leo. 1981. "Technological Inputs and Multifactor Productivity Growth." *Review of Economics and Statistics* 63 (2): 275–82.

Terleckyj, Nestor E. 1980. "Direct and Indirect Effects of Industrial Research and Development on the Productivity Growth of Industries." In *New Developments in Productivity Measurement and Analysis*, ed. John W. Kendrick and Beatrice N. Vaccara, 359–77. Chicago: University of Chicago Press.

Trajtenberg, Manuel. 2001. "Innovation in Israel 1968–1997: A Comparative Analysis Using Patent Data." *Research Policy* 30 (3): 363–89.

Trefler, Daniel. 1998. "The Long and Short of the Canada-U.S. Free Trade Agreement." Canadian Institute for Advanced Research, University of Toronto and National Bureau of Economic Research, New York.

———. 1999a. "Does Canada Need a Productivity Budget?" *Policy Options* (July–August): 66–71.

———. 1999b. "My Brains and Your Looks: Canada in an Innovative World (A Business Agenda with Heart)." Paper presented at the Canadian Deputy Minister's Retreat, Toronto, January 27.

van Pottelsberghe de la Potterie, Bruno, and Frank Lichtenberg. 2001. "Does Foreign Direct Investment Transfer Technology across Borders?" *Review of Economics and Statistics* 83 (3): 490–97.

Wacziarg, Romain. 2001. "Measuring the Dynamic Gains from Trade." *World Bank Economic Review* 15 (3): 393–429.

World Bank. 2000. "Mexico Export Dynamics and Productivity Analysis of Mexican Manufacturing in the 1990s." Report 19864-ME, Private Sector Development Department, Washington, DC.

World Economic Forum. 2002. "The Global Competitiveness Report 2001–2002." Harvard University Center for International Development. Geneva: World Economic Forum.

Wright, Gavin. 1999. *Can a Nation Learn? American Technology as a Network Phenomenon*. NBER Conference Report Series. Chicago: University of Chicago Press.

Xu, Bin. 2000. "Multinational Enterprises, Technology Diffusion, and Host Country Productivity Growth." *Journal of Development Economics* 62 (2): 477–93.

INDEX